Cooperative Control of Multi-Agent Systems

An Optimal and Robust Perspective

Cooperative Control of Multi-Agent Systems

An Optimal and Robust Perspective

Jianan Wang
Beijing Institute of Technology
School of Aerospace Engineering
Beijing, China

Chunyan Wang
Beijing Institute of Technology
School of Aerospace Engineering
Beijing, China

Ming Xin
University of Missouri
Department of Mechanical and Aerospace Engineering
Lafferre Hall, Columbia, MO, United States

Zhengtao Ding
University of Manchester
Department of Electrical and Electronic Engineering
Manchester, United Kingdom

Jiayuan Shan
Beijing Institute of Technology
School of Aerospace Engineering
Beijing, China

Series Editor

Quan Min Zhu

ELSEVIER

ACADEMIC PRESS
An imprint of Elsevier

Academic Press is an imprint of Elsevier
125 London Wall, London EC2Y 5AS, United Kingdom
525 B Street, Suite 1650, San Diego, CA 92101, United States
50 Hampshire Street, 5th Floor, Cambridge, MA 02139, United States
The Boulevard, Langford Lane, Kidlington, Oxford OX5 1GB, United Kingdom

Library of Congress Cataloging-in-Publication Data
A catalog record for this book is available from the Library of Congress

British Library Cataloguing-in-Publication Data
A catalogue record for this book is available from the British Library

ISBN: 978-0-12-820118-3

For information on all Academic Press publications
visit our website at https://www.elsevier.com/books-and-journals

Publisher: Mara Conner
Acquisitions Editor: Sonnini R. Yura
Editorial Project Manager: Fernanda A. Oliveira
Production Project Manager: Prasanna Kalyanaraman
Designer: Mark Rogers

Typeset by VTeX

Contents

About the authors

Jianan Wang is currently an Associated Professor in the School of Aerospace Engineering at Beijing Institute of Technology, China. He received his B.S. and M.S. in Control Science and Engineering from the Beijing Jiaotong University and Beijing Institute of Technology, Beijing, China, in 2004 and 2007, respectively. He received his Ph.D. in Aerospace Engineering at Mississippi State University, Starkville, MS, USA in 2011. His research interests include cooperative control of multiple dynamic systems, UAV formation control, obstacle/collision avoidance, trustworthy networked system, and estimation of sensor networks. He is a senior member of both IEEE and AIAA.

Chunyan Wang is an Associated Professor in the School of Aerospace Engineering at Beijing Institute of Technology, China. He received the B.Eng. degree in automatic control from Dezhou University, Shandong, China, in 2006, the M.S. degree in control theory and control engineering from Soochow University, Jiangsu, China, in 2009, the M.Sc. degree in electrical and electronic engineering from the University of Greenwich, London, U.K., in 2012, and the Ph.D. degree in control systems from the University of Manchester, Manchester, U.K., in 2016. He was a Research Associate with the School of Electrical and Electronic Engineering, University of Manchester from 2016 to 2018. His current research interests include cooperative control, robust control, and robotics.

Ming Xin is a Professor in the Department of Mechanical and Aerospace Engineering at University of Missouri, US. He received the B.S. and M.S. degrees from the Nanjing University of Aeronautics and Astronautics, Nanjing, China, in 1993 and 1996, respectively, both in automatic control, and the Ph.D. degree in aerospace engineering from the Missouri University of Science and Technology, Rolla, MO, USA, in 2002. He has authored and coauthored more than 120 technical papers in his research areas. His research interests include optimization theory and applications, estimation/filtering and signal processing, and control of networked dynamic systems. Dr. Xin was the recipient of the U.S. National Science Foundation CAREER Award in 2009. He is an Associate Fellow of AIAA and a Senior Member of AAS.

Zhengtao Ding is a Professor in the Department of Electrical and Electronic Engineering at University of Manchester, UK. He received the B.Eng. degree from Tsinghua University, Beijing, China, and the M.Sc. degree in systems and control and the Ph.D. degree in control systems from the University of Manchester Institute of Science and Technology, Manchester, U.K. He was a Lecturer with Ngee Ann Polytechnic, Singapore, for ten years. In 2003, he joined the University of Manchester, Manchester,

U.K., where he is currently the Professor of Control Systems with the Department of Electrical and Electronic Engineering. He has authored the book entitled *Nonlinear and Adaptive Control Systems* (IET, 2013) and a number of journal papers. His current research interests include nonlinear and adaptive control theory, cooperative control, distributed optimization and learning. Prof. Ding has served as an Associate Editor for *IEEE Transactions on Automatic Control, IEEE Control Systems Letters, Transactions of the Institute of Measurement and Control, Control Theory and Technology, Mathematical Problems in Engineering, Unmanned Systems*, and *International Journal of Automation and Computing*.

Jiayuan Shan is a Professor in the School of Aerospace Engineering at Beijing Institute of Technology, China. He received the B.S. degree from Huazhong University of Science and Technology in 1988, and the M.S. and Ph.D. degrees from Beijing Institute of Technology, in 1991 and 1999, respectively. He is currently a Professor at Beijing Institute of Technology. His research interests include guidance, navigation and control of the aircraft and hardware-in-the loop simulation. He has served as the Director of Department of Flight Vehicles Control and the Deputy Director of Flight Dynamics and Control Key Laboratory of Ministry of Education.

Preface

Cooperative control of multi-agent systems has gained an increasing interest in the last decades due to the great potentials in many military and civilian missions. It has been an interdisciplinary research and is widely used in the real world, including wheeled robotic systems, satellites, autonomous underwater vehicles, spacecraft, unmanned aerial vehicles, automated highway systems, sensor network, surveillance and smart grid, etc. Designing, modeling and controlling these agents are quite different from single agent scheme due to the natural demand of information communication therein.

In this book, we present a concise introduction to the latest advances in the cooperative control design for multi-agent systems, especially from an optimal and robust perspective. It covers a wide range of applications, such as Rendezvous, cooperative tracking, formation flying and flocking, etc. Also, it includes worked examples that are helpful for researchers from both academia and industry who want to quickly and efficiently enter the field. The book will build up and shape the understanding of the readers in optimal and robust cooperative control design techniques for multi-agent systems. Readers will also learn new theoretical control challenges and discover unresolved/open problems in multi-agent coordination. It offers a certain amount of analytical mathematics, practical numerical procedures, and actual implementations of some proposed approaches.

The book is organized in three parts as follows. Part I, including Chapters 1 and 2, is about the cooperative control. In Chapter 1, we introduce the background of cooperative control and review some related works. Chapter 2 provides some related preliminaries, including mathematical notations, matrix theory, stability theory, basic algebraic graph theory, and some preliminary results used in this book. Part II, from Chapter 3 to Chapter 6, is about the optimal cooperative control. In Chapter 3, we systematically investigate the optimal consensus control for multiple integrator systems. In Chapter 4, the optimal cooperative tracking and flocking of multi-agent systems are studied. Chapter 5 introduces the optimal formation control of multiple UAVs. In Chapter 6, the optimal coverage control of multi-robot systems is investigated. Part III, from Chapter 7 to Chapter 10, is about the robust cooperative control. In Chapter 7, we systematically investigate the consensus control problem for Lipschitz nonlinear multi-agent systems with input delay. Chapter 8 considers the consensus disturbance rejection problem for multi-agent systems with disturbances. In Chapter 9, the robust consensus problem of nonlinear odd power integrator systems is studied. Chapter 10 investigates the robust cooperative control problem of networked (NI) systems.

The authors are indebted to our colleagues, collaborators, and students for their help through collaboration on the topics of this monograph. In particular, the authors would like to thank Professor Ian R. Petersen in Australian National University, Professor Alexander Lanzon in University of Manchester, Professor Zongli Lin in University of

Virginia, Dr. Zongyu Zuo in Beihang University and Mr. Hilton Tnunay in University of Manchester for their collaboration. The authors would like to thank the former and current students for their assistance in reviewing parts of the manuscript. We would like to extend our thanks to Prasanna Kalyanaraman, and Fernanda Oliveira at Elsevier S&T Books for their professionalism. We wish to thank our families for their support, patience and endless love.

This monograph was typeset by the authors using LaTeX. All simulations and numerical computations were carried out in Matlab®.

Acknowledgments

This work has been supported over years by National Natural Science Foundation of China (NSFC) under Grant Nos. 61503025, 61873031, 61803032, and Science and Technology Facilities Council (STFC) under Newton Fund with grant number ST/N006852/1. In addition, great assistance in numerical simulations and proofreading by graduate students, Mr. Hao Chen, Miss Yunhan Li, Mr. Wei Dong, Mr. Chunyu Li, Mr. Weixiang Shi, and Mr. Xiangjun Ding, is also acknowledged.

Part One

About cooperative control

Introduction

1

1.1 Background

Multi-agent coordination is an emerging engineering field multi-disciplined by many areas as shown in Fig. 1.1. The concept of multi-agent coordination is initially inspired by the observations and descriptions of collective behaviors in nature, such as fish schooling, bird flocking and insect swarming [126]. Fig. 1.2 shows one example of fish schooling. Figs. 1.3 and 1.4 show examples of birds flocking and 'V' formation. These behaviors may have advantages in seeking foods, migrating, or avoiding predators and obstacles, and therefore the study of such behaviors has drawn increased attention from researchers in various fields [87]. In 1987, three simple rules – separation (collision avoidance), alignment (velocity matching) and cohesion (flock centering) – were proposed by Reynolds [151] to summarize the key characteristics of a group of biological agents. After that, a simple model was introduced by Vicsek [179] in 1995 to investigate the emergence of self-ordered motion in systems of particles with biologically motivated interaction. The flocking behaviors were later theoretically studied in [70,127,167,173].

There are many robotic control ideas coming from biological societies. For example, one of them is to use simple local control rules of various biological societies – particularly birds, fishes, bees and ants – to develop similar behaviors in cooperative robot systems. In [66], a number of generations of robotic fishes have been designed for navigation in a 3D environment based on the biologically inspired design. In [7], a number of algorithms have been developed for tracking, recognizing, and learning models of social animal behaviors.

1.1.1 *Motivations*

In recent decades, several researches have been engaged in the multi-agent coordination problem. The motivation of these researches is to discover the benefits compared with single-agent systems. First, it can reduce cost and complexity from hardware platform to software and algorithms, i.e., one large and expensive robot can be replaced with several small and cheap robots on task implementation with lower cost and complexity. Second, multi-agent systems are capable of many tasks which could not be effectively performed by a single-robot system, for example, the surveillance task. Moreover, multi-agent systems with decentralized control have preferred flexibility and robustness and can reduce the signal communication and computational workload by using local neighbor-to-neighbor interaction.

The development of multi-agent systems is also well supported by the technological advancement in sensor, communication, and control. As smaller, more accurate, and more reliable sensor and communication systems are available, the cooperative

Cooperative Control of Multi-Agent Systems. https://doi.org/10.1016/B978-0-12-820118-3.00010-7

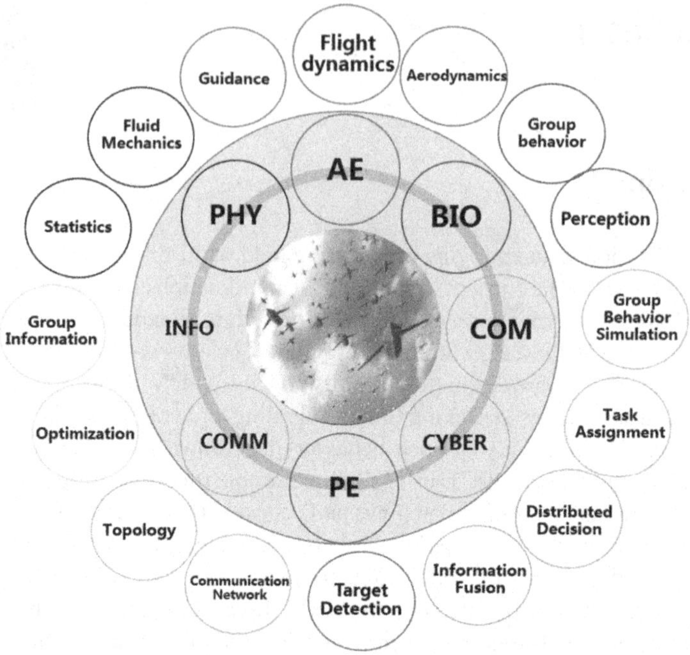

Figure 1.1 Interdisciplinary fields of multi-agent coordination. Acronyms: AE – Aerospace; BIO – Biology; COM – Computer; CYBER – Cybernetics; PE – Photoelectric; COMM – Communication; INFO – Information; PHY – Physics.

Figure 1.2 Fish schooling.

strategies of multi-agents to carry out certain tasks become possible and applicable [18].

Figure 1.3 Birds flocking.

Figure 1.4 Birds flying in 'V' formation.

1.1.2 Control architectures and strategies

For multi-agent systems, various control architectures have been proposed in literature. Most of them can be described as centralized and decentralized schemes. In centralized systems, a central unit that connects all the agents has the global team knowledge, and manages information to guarantee the achievement of the mission. Thus, advanced and expensive equipments are necessary to satisfy all the technological requirements. For decentralized schemes, all the agents are in the same level and have the same equipments. Each agent uses the local sensor to obtain the relative state information of its neighbors, then makes decision for the next step to move and explore the environment. Furthermore, each agent does not need the global information and just communicates with their neighboring agents.

Centralized and decentralized schemes have their own advantages. Regarding the centralized one, the powerful central unit can highly improve the overall performance of the multi-agent systems. Furthermore, the excellent computing capability and high-speed communication ability of the processor can send the command to all the agents quickly and effectively. On the other hand, the whole system highly relies on the central unit. The failure of the central unit will lead to the failure of the whole mission. The robustness of the centralized scheme is insufficient. Moreover, high requirements on the central unit lead to high cost of the whole system. While for decentralized systems, using low-cost sensors and processors to replace the expensive core unit, can reduce the cost effectively. The motion of the agent only relies on the local relative information of neighbors, which reduces the difficulty level of the mission. In addition, decentralized systems are more tolerant to severe environment since failure of partial agents does not affect the performance of the whole system. On the other hand, the decentralized systems rely on more complex control strategies to coordinate and optimize the execution of the mission, which limits the performance of the system. The communication bandwidth and quality limits also affect the overall performance.

1.1.3 Related applications

Cooperative control has broad potential applications in real world including wheeled robotics system [38], satellites formation [144,153], autonomous underwater vehicles (AUVs) [6,131], spacecraft [116], unmanned aerial vehicles (UAVs) [1,165], automated highway systems [95], sensor network [157], surveillance [147] and smart grid [123], and so on. Fig. 1.5 shows formation control of wheeled ground mobile robots in the control system center of University of Manchester. Fig. 1.6 shows a team of AUVs designed by the researchers from the European Union-founded Grex project. Fig. 1.7 shows an UAVs platform designed by the researchers from the cooperative guidance, navigation and control (CGNC) team in Beijing Institute of Technology. All the above examples illustrate the benefits of cooperative control applications.

Cooperative control of multi-agent systems can accomplish many different tasks. Typical tasks include consensus [128,145], flocking [110], swarming [156], formation control [138], and synchronization [157]. Additionally, cooperative control of multiple mobile robots can fulfil some specialized tasks such as distributed manipulation [111], mapping of unknown environments [53,161], rural search and rescue [119], transportation of large objects [198,201], etc.

Figure 1.5 AmigoBots – wheeled ground mobile robots in the control system center of University of Manchester.

Figure 1.6 Autonomous underwater vehicles (AUVs) from The European Union-founded Grex project.

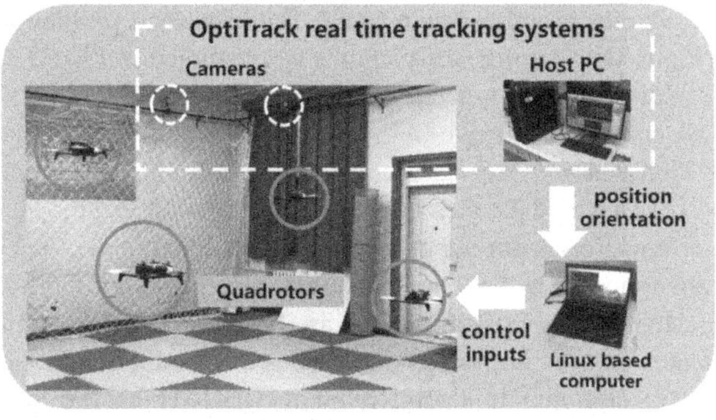

Figure 1.7 UAVs platform designed in the guidance, navigation and control (GNC) team of Beijing Institute of Technology.

1.2 Overview of related works

1.2.1 Consensus control

1.2.1.1 Basic concept

Consensus control is a fundamental problem in cooperative control of multi-agent systems, since many applications are based on the consensus algorithm design. In multi-agent systems, the consensus problem is to design the control strategy for a group of agents to reach a consensus (or agreement) of the states of interest. The basic idea is that each agent updates its information state based on the information states of its local neighbors in such a way that the final information state of each agent converges to a common value [147]. One significant contribution in consensus control is the introduction of graph theory to the conventional control theory [28,48,116]. A general framework of the consensus problem for networks of single integrators is proposed in [128]. Since then, consensus problems have been intensively studied in many directions.

In terms of the dynamics of agents, consensus problems for various systems' dynamics have been massively investigated. The system dynamics has large influence on the final consensus state of the multi-agent systems. For example, the consensus state of multi-agent systems with single-integrator dynamics often converges to a constant value. Meanwhile, consensus for second-order dynamics might converge to a dynamic final value (i.e., a time function) [18]. Many early results on consensus problems are based on simple agent dynamics such as first- or second-order integrator dynamics [16,49,64,71,145,146]. However, in reality a large class of practical physical systems cannot be feedback linearized as the first- or second-order dynamical model. For instance, for a group of UAVs [35], higher-order dynamic models may be needed. More complicated agent dynamics are described by high-order linear multi-agent systems in [68,89,90,122,170,211]. After that, the results were extended to nonlinear multi-agent systems [33,34,92,168,187,204]. Consensus for nonlinear systems is more involved than that for their linear system counterparts. The difficulty of consensus control for nonlinear systems owes to certain restrictions the nonlinearity imposes on using the information of the individual systems. Consensus control for second-order Lipschitz nonlinear multi-agent systems is addressed in [205]. The consensus problems of high-order multi-agent systems with nonlinear dynamics were studied in [33,92,187,204]. The works [34,168] address the consensus output regulation problem of nonlinear multi-agent systems. A common assumption in the previous results is that the dynamics of the agents are identical and precisely known, which might not be practical in many circumstances. Due to the existence of the nonidentical uncertainties, the consensus control of heterogeneous multi-agent systems was studied in [32,76,176,212].

The communication connections between the agents also play an important role in consensus problems. Most of the existing results are based on fixed communication topology, which indicates that the Laplacian matrix $\mathcal{L}$ is a constant matrix (see Chapter 2 for the graph theory notations). It is pointed out in [103,141] that the consensus is reachable if and only if zero is a simple eigenvalue of $\mathcal{L}$. If zero is not a simple eigenvalue of $\mathcal{L}$, the agents cannot reach consensus asymptotically as there exist at least two separate subgroups or at least two agents in the group who do not receive any information [180]. It is also known that zero is a simple eigenvalue of $\mathcal{L}$ if and only if the directed communication topology has a directed spanning tree or the undirected communication topology is connected [128,145]. The results with directed graphs are more involved than those with undirected graphs. The main problem is that the Laplacian matrix associated with a directed graph is generally not positive semi-definite [87]. The decomposition method for the undirected systems cannot be applied to the directed one due to this unfavorable feature. For the consensus control with directed communication graphs, the conditions of being balanced and/or strongly connected are needed in [91,181], which are stronger than the directed spanning tree condition. In practice, the communication between the agents may not be fixed due to technological limitations of sensors or link failures. The consensus control of multi-agent systems with switching topologies has been investigated in [128,187,190]. In [205] and [188], consensus control with communication constraint and Markovian communication failure were studied.

In terms of the number of leaders, the above researches can also be roughly assigned to three groups, that is, leaderless consensus (consensus without a leader) [33, 89] whose agreement value depends on the initial states of the agents, leader-follower consensus (or consensus tracking) [64,187] which has a leader agent to determine the final consensus value, and containment control [16,71] when there is more than one leader in agent networks. Compared to leaderless consensus, consensus tracking and containment control have the advantages in determining the final consensus value in advance [170].

1.2.1.2 Optimal cooperative control

Consensus algorithms from the optimal control perspective have been explored from two aspects: 1) fastest convergence time: the algorithms were designed to achieve the fastest convergence time by finding an optimal weight matrix [193], constructing a proper configuration that maximizes the second smallest eigenvalue [77], and exploring an optimal interaction graph (de Bruijn's graph) for the average consensus [26]; 2) optimal control law design: the consensus problem can also be formulated as an optimal control problem and solved by a number of different optimization methods such as linear matrix inequality (LMI) approach [159], a linear quadratic regulator (LQR) based optimal linear consensus algorithm [15], a distributed subgradient method for multi-agent optimization [120], a locally optimal nonlinear consensus strategy by imposing individual objectives [10], nonlinear optimal control design for multi-agent formation control in [194], an optimal controller in terms of sensor usage in [74], a second-order cone programming techniques based optimal solution to minimize the travel distance [27], a centralized optimal multi-agent coordination problem under tree formation constraints [207], an optimal control of spatially distributed systems [117], a receding horizon control framework [41,44–46,210], [75], a distributed Nash-based game approach [55], a gradient algorithm to find the extreme of a cost function [155], a Pareto optimal coordination to minimize the elapsed time [54], a distance minimizing geodesics based optimal solution for multi-agent coordination [67], a negotiation algorithm that computes an optimal consensus point for agents modeled as linear control systems subject to convex input constraints and linear state constraints [73], a model predictive control strategy [12,45,52,80,85,130,160], an inverse optimality approach to consider the cooperative consensus [25,118,206], a cooperative preview control method for continuous-time multi-agent systems with a time-invariant directed communication topology [96], and a multi-objective optimization approach [31].

1.2.1.3 Robust cooperative control

Robustness of control systems to disturbances and uncertainties has always been a popular topic in feedback control systems. Feedback would not be necessary for most control systems if there were no disturbances and uncertainties [215]. In consensus control of multi-agent systems, modeling errors, external disturbances, time delay caused by communications and some other issues will significantly affect the consensus performance. Thus, robust consensus control problem has been developed as one of the most important topics in this area.

(1) **Uncertainties and disturbances in consensus control**. The practical physical systems often suffer from uncertainties which may be caused by mutations in system parameters, modeling errors or some ignored factors [133]. The robust consensus problem for multi-agent systems with continuous-time and discrete-time dynamics were investigated in [61] and [62], where the weighted adjacency matrix is a polynomial function of uncertain parameters. In particular, H_∞ control has been proved to be effective for disturbance rejection of the multi-agent systems with external disturbances bounded by H_2 norms. The H_∞ robust control problem was investigated in [182] for a group of autonomous agents governed by uncertain general linear dynamics. Most of the existing results on consensus control of uncertain multi-agent systems were often restricted to certain conditions, like single or double integrators [99,175], or undirected network connections [104]. Other than the parameter uncertainties, the agents may also be subject to unknown external disturbances, which might degrade the system performance and even cause the network system to oscillate or diverge. The consensus problems of multi-agent systems with performance requirements have emerged as a challenging topic in recent years. The robust H_∞ consensus problems were investigated for multi-agent systems with the first- and second-order integrator dynamics in [99,100]. The H_∞ consensus problems for general linear dynamics with undirected graphs were studied in [88,105]. The results obtained in [88] were extended to directed graph in [181]. The H_∞ consensus problems for switching directed topologies were investigated in [154,189]. The nonlinear H_∞ consensus problem was studied in [91] with directed graph. Global H_∞ pinning synchronization problem for a class of directed networks with aperiodic sampled-data communications was addressed in [191]. It is worth noting that the directed graphs in [91,181] are restricted to be balanced or strongly connected.

For some systems, disturbances in real engineering problems are often periodic and have inherent characteristics such as harmonics and unknown constant load [20]. For those kinds of disturbances, it is desirable to cancel them directly by utilizing the disturbance information in the design of control input. One common design method is to estimate the disturbance by using the measurements of states or outputs and then use the disturbance estimate to compensate the influence of the disturbance on the system, which is referred to as Disturbance Observer-Based Control (DOBC) [86]. Using DOBC method, consensus of second-order multi-agent dynamical systems with exogenous disturbances was studied in [40,200] for matched disturbances and in [184] for unmatched disturbances. Disturbance observer-based tracking controllers for high-order integrator-type and general multi-agent systems were proposed in [5,13], respectively. A systematic study on consensus disturbance rejection via disturbance observers can be found in [35].

(2) **Delay effects in consensus control**. Time delays widely exist in practical systems due to the time taken for transmission of signals and transportation of materials, etc. The presence of time delays, if not considered in the controller design, may seriously degrade the performance of the controlled system and, in the extreme situations, may even cause the loss of stability. Therefore, the stabilization of time-delay systems has attracted much attention in both academic and industrial communities; see the surveys [59,152], the monographs [58,79,213], and the references therein.

In the formulation of stabilization of time-delay systems, there are two types of feedback methods in the literature: standard (memoryless) feedback and predictive (memory) feedback, respectively. Memoryless controllers are useful for the systems with state delays [22,63,94,196,208]. However, it is known that system with input delay is more difficult to handle in control theory [152]. For predictive feedback, compensation is added in the controller design to offset the adverse effect of the time delay and the stabilization problems are reduced to similar problems for ordinary differential equations. A wide variety of predictor-based methods such as Smith predictor [162], modified Smith predictor [132], finite spectrum assignment, and Artstein–Kwon–Pearson reduction method [3,81] are effective and efficient when the delay is too large to be neglected and a standard (memoryless) feedback would fail. However a drawback of the predictor-based methods is that the controllers involve integral terms of the control input, resulting in difficulty for the control implementation [47]. A halfway solution between these two methods is to ignore the troublesome integral part, and use the prediction based on the exponential of the system matrix, which is known as the truncated prediction feedback (TPF) approach. This idea stemmed from low-gain control of the systems with input saturation [101], then it was developed for linear systems [102,203,214], and nonlinear systems [36,217].

1.2.2 Formation control

Apart from consensus control, which is to drive all agents to the same desired values, another research direction is formation control, in which the agents form a predesigned geometrical configuration through local interactions. Compared to consensus control, the final states of all agents are more diverse under the formation control scenarios. For instance, the agents can reach reference attitude prescribed by a virtual leader in spacecraft swarm through consensus control. In the case of reaching desired formation, the spacecraft positions have to be different.

Formation control is one of the most important applications in cooperative control. Different control strategies have been used for formation control, such as actual leader

[21], virtual leader [42], behavioral [8], etc. A survey on various classifications of formation control is given in [125]. Consensus based formation control strategies have also been investigated [39,141,143,150,169,209]. The role of Laplacian eigenvalues in determining formation stability is investigated in [49]. In [143], it is pointed out that consensus based formation control strategies are more general and contain many existing actual/virtual leader and behavioral approaches as special cases. Most of the early results on formation control focus on simple agent dynamics such as the first- or second-order integrator dynamics. In reality, some practical physical systems cannot be feedback linearized as the first- or second-order dynamical model. Formation control for a class of high-order linear multi-agent systems with time delays is studied in [39]. The applications of formation control in various areas could be found in [141, 169].

By different types of sensing and controlled variables, the formation control problem can be classified into position-, displacement-, and distance-based control [124]. When agents receive specified positions with respect to the global coordinate system, they only sense their own positions, which is called position-based control. When agents receive orientation in the global coordinate system, they sense relative positions of their neighboring agents, which is called displacement-based control. When agents receive desired inter-agent distances, they sense relative positions of their neighboring agents with respect to their own local coordinate systems, which is called distance-based control. Since specifying desired position for each agent is unrealistic, and providing global information for each agent is resource-consuming, it is more practical to adopt the distance-based control algorithm.

To implement distance-based control, agents are required to sense relative positions of their neighbors, which implies interactions among agents. Since the desired formation is specified by the desired distances between any pair of agents, the formation can be treated as a given rigid body.

1.2.3 Other related research

The cooperative control promises several advantages in performing cooperative group tasks, such as strong adaptivity, high robustness, flexibility, scalability, and low operational costs. Besides consensus control and formation control, there are also some other subjects that have been well studied. In what follows, we introduce some other related directions in cooperative control.

- **Flocking control**. As aforementioned about the motivation of cooperative control, animals' behaviors in groups attracted scientists' attention. Through relatively simple interactions among individuals, they are able to demonstrate collective behaviors. The study of biological environment later developed into the flocking problem in engineering applications. It is concerned with the process of coordinating multi-agent velocities to a common velocity. It was first investigated by Reynolds in 1987 who introduced an agent model following three heuristic rules in [151] to create flocking: cohesion, separation, and alignment. Similar to the behavior-based scheme, the cohesion and separation rules have been usually implemented by designing artificial potential functions to control inter-agent distances.

As for the alignment rule, it is achieved by velocity synchronization among the agents. The majority of related literatures can be viewed as implementations of the Reynolds rules. Flocking has been studied on double-integrator models in the plane [171] and n-dimensional space [127], systems with fixed and switching topologies [172,174], and later extended to the obstacle avoidance scheme [129].

- **Rendezvous**. Rendezvous is concerned with enabling multiple agents in a network to reach destination at the same time. Cooperative strike or cooperative jamming are two application scenarios for the rendezvous problem [177]. From the control point of view, the rendezvous problem is to design local control strategies individually for each agent, which enables all agents in a group to eventually rendezvous at a single unspecified location without any active communication among agents [97]. An early formulation and algorithmic solution of the rendezvous problem is introduced in [2], where agents have limited range sensing capabilities. Stop-and-go strategies extending the algorithm are proposed in [97] and [98], which cover various synchronous and asynchronous formulations. An n-dimensional rendezvous problem was approached via proximity graphs in [23].
- **Synchronization**. Concurrent synchronization is a regime where diverse groups of fully synchronized dynamic systems stably coexist. In the scientific approach, the synchronization of coupled nonlinear oscillators is a typical problem to solve. Synchronization involves, at least, two elements in interaction, and the behavior of a few interacting oscillators has been intensively studied in recent literatures. In the pioneering work [134], the synchronization phenomenon of two master-slave chaotic systems was observed and applied to secure communications. References [107] and [135] addressed the synchronization stability of a network of oscillators by using the master stability function method. Recently, the synchronization of complex dynamical networks, such as small world and scale-free networks, has been widely studied (see [43,78,106,108,136,186,192], and the references therein).
- **Containment control**. In the leader-follower scheme, the followers need to stay around the leader as a formation. A simpler way to achieve this is containment control. By driving the followers into a convex hull spanned by multiple leaders, there is no need of relative velocity information with respect to neighbor followers. In the collision avoidance scenario, the leaders can form a (moving) safety area by detecting the position of obstacles. Then the group can arrive at the destination safely given that the followers always stay within the safety area. Containment control has been studied for single-integrators with fixed [51,71], switching [14], and state-dependent topology [19], double-integrators [17], unicycles [29], Euler–Lagrangian systems [30,112,113], and general linear systems [93].

1.2.4 Future research topics

This monograph reports recent advances in optimal and robust cooperative control of multi-agent systems. There still exist several issues that need to be addressed in the near future. Some future research topics are highlighted as follows.

- **Game theory and reinforcement learning based optimal cooperative control**. For a large group of agents, each agent is required to perform a special task with consideration of an optimal performance. Game theory is a powerful tool to achieve optimal decision making with the efficient mathematical representation. For example, in a multi-agent pursuit-evasion case, the evader could be smart to make intelligent move such that the pursuer needs to move accordingly to achieve the Nash equilibrium. In addition, reinforcement learning based on Q-learning technique is also one important research topic in optimal cooperative control. With the wide availability of data, reinforcement learning has demonstrated the capability to deal with incomplete information in an uncertain environment and realize full automomy.

- **Robust cooperative control based on composite disturbance rejection technique**. In practice, multi-agent systems generally have the characteristics of nonlinear coupling or model uncertainties. The performance can be affected by network communication delay as well as external disturbances, etc. Moreover, faults induced by actuators, sensors and other components are inevitable. In many cases, the coupling of multiple disturbances and faults will make the situation worse and even lead to instability of multi-agent systems. These disturbances and uncertainties require the research on advanced disturbance rejection techniques.
- **Extended applications based on optimal and robust cooperative control**. More application scenarios, such as cooperative pursuit-evasion problem, cooperative surveillance of forest fire, cooperative assembling of multiple robotic arms, panoramic monitoring inter-disciplined with image processing, salvo attack guidance of multiple missiles, and consensus filtering of multiple sensors, are highly expected in both optimal and robust perspective.

1.3 Objectives of this book

This book presents a concise introduction to the latest advances in cooperative control design for multi-agent systems, especially from perspectives of optimality and robustness. It covers a wide range of applications, such as cooperative tracking of multiple robots, formation flying of fixed-wing UAVs, etc. Also, it includes ready examples suitable for researchers from both academia and industry who want to quickly and efficiently enter the field. It will offer an opportunity for researchers to present an extended exposition of their recent work in all aspects of multi-agent technology for a wide and rapid dissemination. The main benefits to audience include the following:

- The book will build up and shape the understanding of the reader in optimal and robust cooperative control design techniques for multi-agent systems.
- Readers will learn new theoretical control challenges; discover unresolved/open problems and future research trends in multi-agent systems.
- It offers a certain amount of analytical mathematics, practical numerical procedures, and actual implementations of some proposed approaches.
- Readers will find extended contents on formation control, cooperative tracking, flocking, and coverage, etc.

1.4 Book outline

In this book, we focus on optimal and robust cooperative control design techniques for multi-agent systems. The book is organized as follows in three parts: Part I is about the cooperative control including Chapter 1 and 2. In Chapter 1, we review the background of cooperative control of multi-agent systems and some related work. Chapter 2 provides some related preliminaries, including mathematical notations, matrix theory, stability theory, basic algebraic graph theory, and preliminary results used in this book. Part II is about optimal cooperative control from Chapter 3 to Chapter 6. In Chapter 3, we systematically investigate the optimal consensus control for multiple integrator systems. In Chapter 4, optimal cooperative tracking and flocking of multi-agent systems are studied. Chapter 5 introduces optimal formation control of multiple UAVs. In Chapter 6, optimal coverage control of multi-robot systems is investigated. Part III is about the robust cooperative control from Chapter 7 to Chapter 10. In Chapter 7, we systematically investigate the consensus control problem for Lipschitz nonlinear multi-agent systems with input delay. Chapter 8 considers the consensus disturbance rejection problem for multi-agent systems with disturbances and directed graph. In Chapter 9, robust consensus problem of nonlinear p-order integrator systems is studied. Chapter 10 investigates the robust cooperative control problem of networked (NI) systems.

Preliminaries

2

Useful preliminaries are generally introduced in this chapter for later use, such as matrix theory, stability theory and graph theory.

2.1 Matrix theory

In this section, some mathematical notations and basic definitions that will be used in the remainder of this book are provided.

Definition 2.1. The Kronecker product of matrices $A \in \mathbb{R}^{m\times n}$ and $B \in \mathbb{R}^{p\times q}$ is defined as

$$A \otimes B = \begin{bmatrix} a_{11}B & \cdots & a_{1n}B \\ \vdots & \ddots & \vdots \\ a_{m1}B & \cdots & a_{mn}B \end{bmatrix},$$

and there are the following properties:

1. $(A \otimes B)(C \otimes D) = (AC) \otimes (BD)$,
2. $(A \otimes B) + (A \otimes C) = A \otimes (B + C)$,
3. $(A \otimes B)^{-1} = A^{-1} \otimes B^{-1}$,
4. $(A + B) \otimes C = (A \otimes C) + (B \otimes C)$,
5. $(A \otimes B)^T = A^T \otimes B^T$,
6. If $A \in \mathbb{R}^{m\times n}$ and $B \in \mathbb{R}^{p\times q}$ are both positive definite (positive semi-definite), so is $A \otimes B$,

where the matrices are assumed to be compatible for multiplication.

Lemma 2.1 (Gershgorin's Disk Theorem [65]). *Let $A = [a_{ij}] \in \mathbb{R}^{n\times n}$, let*

$$R_i'(A) \equiv \sum_{j=1, j\neq i}^{n} \left|a_{ij}\right|, i = 1, 2, \cdots, n$$

denote the deleted absolute row sums of A, and consider the n Gershgorin disks

$$\left\{z \in \mathbb{C} : |z - a_{ii}| \leq R_i'(A)\right\}, i = 1, 2, \cdots, n.$$

The eigenvalues of A in the union of Gershgorin disks are given by

$$G(A) = \bigcup_{i=1}^{n} \left\{z \in \mathbb{C} : |z - a_{ii}| \leq R_i'(A)\right\}.$$

Cooperative Control of Multi-Agent Systems. https://doi.org/10.1016/B978-0-12-820118-3.00011-9

Furthermore, if the union of k of the n disks that comprise $G(A)$ forms a set $G_k(A)$ that is disjoint from the remaining $n-k$ disks, then $G_k(A)$ contains exactly k eigenvalues of A, counted according to their algebraic multiplicities.

Definition 2.2 ([65]). A matrix $A=[a_{ij}]\in\mathbb{R}^{n\times n}$ is diagonally dominant if

$$|a_{ii}|\geq\sum_{j=1,j\neq i}^{n}|a_{ij}|=R_i'(A),\ \forall i=1,2,\cdots,n.$$

It is strictly diagonally dominant if

$$|a_{ii}|>\sum_{j=1,j\neq i}^{n}|a_{ij}|=R_i'(A),\ \forall i=1,2,\cdots,n.$$

Definition 2.3 (M-matrix, Definition 6 in [87]). A square matrix $A\in\mathbb{R}^{n\times n}$ is called a singular (nonsingular) M-matrix, if all its off-diagonal elements are nonpositive and all eigenvalues of A have nonnegative (positive) real parts.

Lemma 2.2 (Schur Complement Lemma). *For any constant symmetric matrix*

$$S=\begin{bmatrix} S_{11} & S_{12}\\ S_{12} & S_{22}\end{bmatrix},$$

the following statements are equivalent:

(1) $S<0$,
(2) $S_{11}<0,\ S_{22}-S_{12}^TS_{11}^{-1}S_{12}<0$,
(3) $S_{22}<0,\ S_{11}-S_{12}S_{22}^{-1}S_{12}^T<0$.

2.2 Stability theory

In this section, some basic concepts of stability based on Lyapunov functions are provided. Consider a nonlinear system

$$\dot{x}=f(x), \tag{2.1}$$

where $x\in\mathcal{D}\subset\mathbb{R}^n$ is the state of the system, and $f:\mathcal{D}\subset\mathbb{R}^n\longrightarrow\mathbb{R}^n$ is a continuous function with $x=0$ as an equilibrium point, that is $f(0)=0$, and with $x=0$ as an interior point of $\mathcal{D}$. $\mathcal{D}$ denotes a domain around the equilibrium $x=0$.

Definition 2.4 (Lyapunov stability). For the system (2.1), the equilibrium point $x=0$ is said to be Lyapunov stable if for any given positive real number R, there exists a positive real number r such that $\|x(t)\|<R$ for all $t>0$ if $\|x(0)\|<r$. Otherwise, the equilibrium point is unstable

Definition 2.5 (Asymptotic stability). For the system (2.1), the equilibrium point $x = 0$ is asymptotically stable if it is (Lyapunov) stable and furthermore $\lim_{t\to\infty} x(t) = 0$.

Definition 2.6 (Exponential stability). For the system (2.1), the equilibrium point $x = 0$ is exponentially stable if there exist two positive real numbers α and λ such that the inequality

$$\|x(t)\| < \alpha \|x(0)\| e^{-\lambda t}$$

holds for $t > 0$ in some neighborhood $\mathcal{D} \subset \mathbb{R}^n$ containing the equilibrium point.

Definition 2.7 (Global asymptotic stability). If the asymptotic stability defined in Definition 2.5 holds for any initial state in $\mathbb{R}^n$, the equilibrium point is said to be globally asymptotically stable.

Definition 2.8 (Global exponential stability). If the exponential stability defined in Definition 2.6 holds for any initial state in $\mathbb{R}^n$, the equilibrium point is said to be globally exponentially stable.

Definition 2.9 (Positive definite function). A function $V(x) \in \mathcal{D} \subset \mathbb{R}^n$ is said to be locally positive definite if $V(x) > 0$ for $x \in \mathcal{D}$ except at $x = 0$ where $V(x) = 0$. If $\mathcal{D} = \mathbb{R}^n$, i.e., the above property holds for the entire state space, $V(x)$ is said to be globally positive definite.

Definition 2.10 (Lyapunov function). If in $\mathcal{D} \in \mathbb{R}^n$ containing the equilibrium point $x = 0$, the function $V(x)$ is positive definite and has continuous partial derivatives, and if its time derivative along any state trajectory of system (2.1) is nonpositive, i.e.,

$$\dot{V}(x) \leq 0,$$

then $V(x)$ is a Lyapunov function.

Definition 2.11 (Radially unbounded function). A positive definite function $V(x)$: $\mathbb{R}^n \longrightarrow \mathbb{R}$ is said to be radially unbounded if $V(x) \longrightarrow \infty$ as $\|x\| \longrightarrow \infty$.

Theorem 2.1 (Lyapunov theorem for global stability, Theorem 4.3 in [37]). *For the system (2.1) with $\mathcal{D} \in \mathbb{R}^n$, if there exists a function $V(x) : \mathbb{R}^n \longrightarrow \mathbb{R}$ with continuous first order derivatives such that*

- *$V(x)$ is positive definite*
- *$\dot{V}(x)$ is negative definite*
- *$V(x)$ is radially unbounded*

then the equilibrium point $x = 0$ is globally asymptotically stable.

The optimal consensus control in this book is designed via the inverse optimal control theory as follows.

Lemma 2.3 ([11]). *Consider the nonlinear controlled dynamical system*

$$\dot{\hat{X}}(t) = f(\hat{X}(t), U(t)), \quad \hat{X}(0) = \hat{X}_0, \quad t \geq 0 \tag{2.2}$$

with $f(0,0)=0$ and a cost functional given by

$$J\left(\hat{X}_0, U(\cdot)\right) \triangleq \int_0^{\infty} T\left(\hat{X}(t), U(t)\right) dt \tag{2.3}$$

where $U(\cdot)$ is an admissible control. Let $D \subseteq \mathbb{R}^n$ be an open set and $\Omega \subseteq \mathbb{R}^m$. Assume that there exists a continuously differentiable function $V : D \rightarrow \mathbb{R}$ and a control law $\phi : D \rightarrow \Omega$ such that

$$V(\mathbf{0}) = 0 \tag{2.4}$$

$$V(\hat{X}) > 0, \quad \hat{X} \in D, \quad \hat{X} \neq 0 \tag{2.5}$$

$$\phi(\mathbf{0}) = 0 \tag{2.6}$$

$$V'(\hat{X}) f(\hat{X}, \phi(\hat{X})) < 0, \quad \hat{X} \in D, \quad \hat{X} \neq 0 \tag{2.7}$$

$$H(\hat{X}, \phi(\hat{X})) = 0, \quad \hat{X} \in D \tag{2.8}$$

$$H(\hat{X}, U) \geq 0, \quad \hat{X} \in D, \quad U \in \Omega \tag{2.9}$$

where $H(\hat{X}, U) \triangleq T(\hat{X}, U) + V'(\hat{X}) f(\hat{X}, U)$ is the Hamiltonian function. The superscript $'$ denotes partial differentiation with respect to $\hat{X}$.

Then, with the feedback control

$$U(\cdot) = \phi(\hat{X}(\cdot)) \tag{2.10}$$

the solution $\hat{X}(t) \equiv 0$ of the closed-loop system is locally asymptotically stable and there exists a neighborhood of the origin $D_0 \subseteq D$ such that

$$J\left(\hat{X}_0, \phi(\hat{X}(\cdot))\right) = V(\hat{X}_0), \quad \hat{X}_0 \in D_0 \tag{2.11}$$

In addition, if $\hat{X}_0 \in D_0$ then the feedback controller (2.10) minimizes $J\left(\hat{X}_0, U(\cdot)\right)$ in the sense that

$$J\left(\hat{X}_0, \phi(\hat{X}(\cdot))\right) = \min_{U(\cdot) \in S\left(\hat{X}_0\right)} J\left(\hat{X}_0, U(\cdot)\right) \tag{2.12}$$

where $S\left(\hat{X}_0\right)$ denotes the set of asymptotically stabilizing controllers for each initial condition $\hat{X}_0 \in D$. Finally, if $D = \mathbb{R}^n$, $\Omega = \mathbb{R}^m$, and

$$V(\hat{X}) \rightarrow \infty \text{ as } \|\hat{X}\| \rightarrow \infty \tag{2.13}$$

the solution $\hat{X}(t) \equiv 0$ of the closed-loop system is globally asymptotically stable.

Proof. Omitted. Refer to [11]. □

Remark 2.1. This Lemma underlines the fact that the steady-state solution of the Hamilton–Jacobi–Bellman equation is a Lyapunov function for the nonlinear system and thus guarantees both stability and optimality.

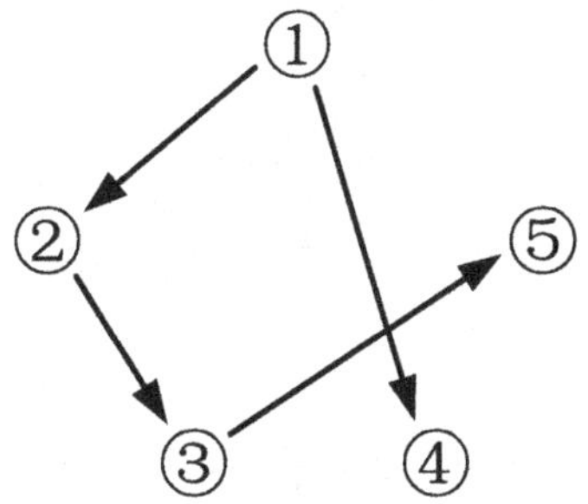

Figure 2.1 An example of directed graph.

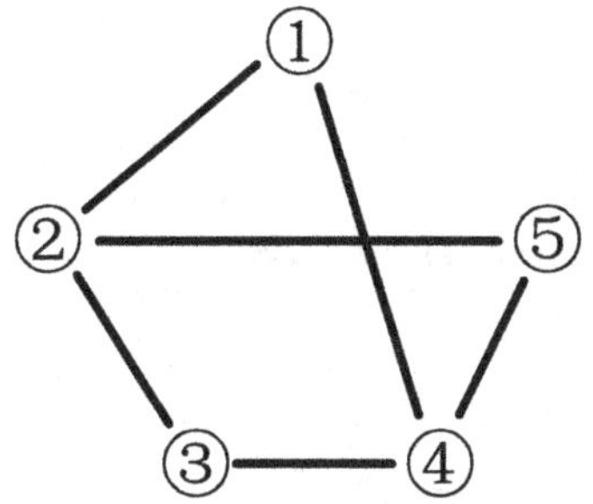

Figure 2.2 An example of undirected graph.

2.3 Basic algebraic graph theory

This section introduces some knowledge related to graph theory, which is vital in the consensus control design.

2.3.1 Basic definitions

The graph theory has been introduced by Leonard Euler in year 1736. Generally, it is convenient to model the information exchanges among agents by directed or undirected graphs. A directed graph $\mathcal{G} \triangleq (\mathcal{V}, \mathcal{E})$, in which $\mathcal{V} \triangleq \{v_1, v_2, \cdots, v_N\}$ is the set of nodes, and $\mathcal{E} \subseteq \mathcal{V} \times \mathcal{V}$ is the set of edges with ordered pairs of nodes. A vertex represents an agent, and each edge represents a connection. A weighted graph associates a weight with every edge in the graph. Self loops in the form of (v_i, v_i) are excluded unless otherwise indicated. The edge (v_i, v_j) in the edge set $\mathcal{E}$ denotes that agent v_j can obtain information from agent v_i, but not necessarily vice versa (see Fig. 2.1). A graph with the property that $(v_i, v_j) \in \mathcal{E}$ implies $(v_j, v_i) \in \mathcal{E}$ for any $v_i, v_j \in \mathcal{V}$ is said to be undirected, where the edge (v_i, v_j) denotes that agents v_i and v_j can obtain information from each other (see Fig. 2.2).

For an edge (v_i, v_j), node v_i is called the parent node, v_j is the child node, and v_i is a neighbor of v_j. The set of neighbors of node v_i is denoted as $\mathcal{N}_i$, whose cardinality

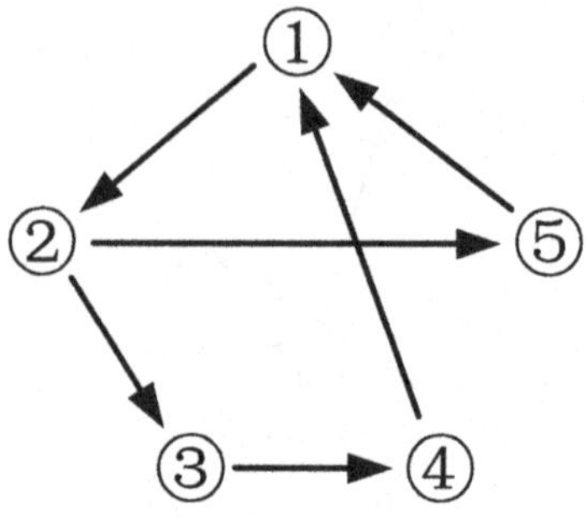

Figure 2.3 A case of strongly connected graph.

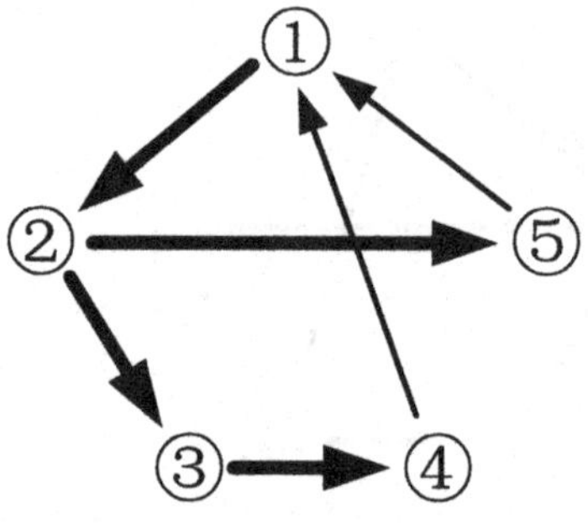

Figure 2.4 A spanning tree for the graph in the strongly connected graph with root node 1.

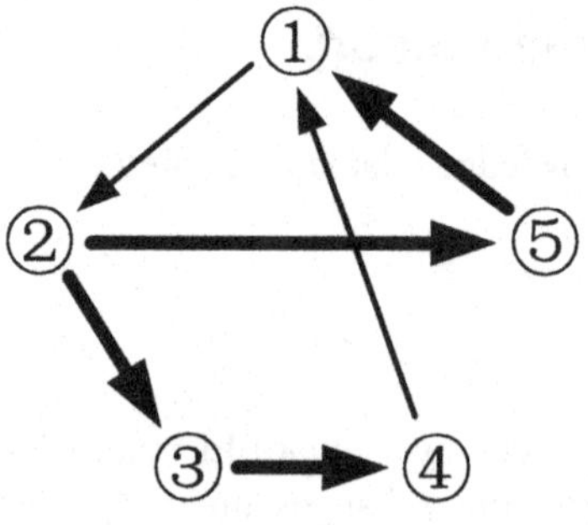

Figure 2.5 A spanning tree for the graph in the strongly connected graph with root node 2.

is called the in-degree of node v_i. A directed graph is strongly connected if there is a directed path from every node to every other node (see Figs. 2.3–2.6). Note that for an undirected graph, strong connectedness is simply termed connectedness. A graph is defined as being balanced when it has the same number of ingoing and outgoing edges for all the nodes (see Fig. 2.7). Clearly, an undirected graph is a special balanced graph.

A directed path from node v_{i_1} to node v_{i_l} is a sequence of ordered edges of the form $(v_{i_k}, v_{i_{k+1}})$, $k = 1, 2, \cdots, l-1$. An undirected path in an undirected graph is defined

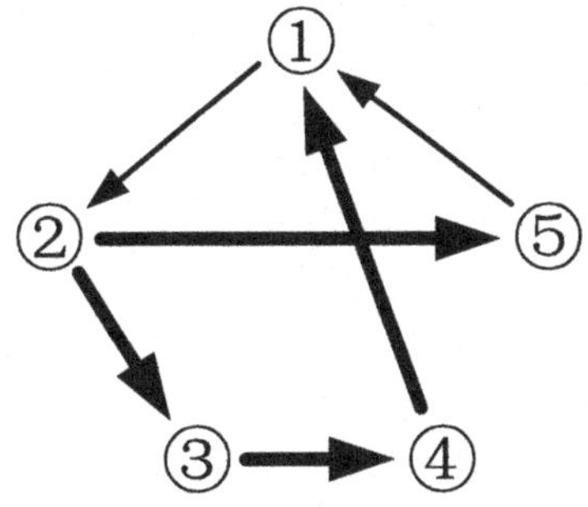

Figure 2.6 A spanning tree for the graph in the strongly connected graph with root node 2 (another situation).

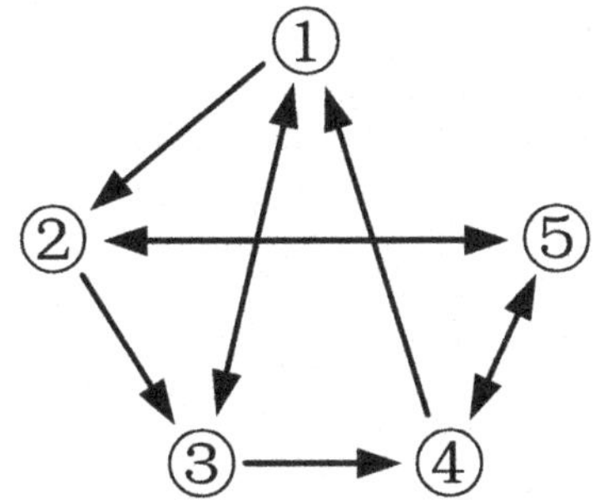

Figure 2.7 A case of balanced and strongly connected graph.

analogously. A cycle is a directed path that starts and ends at the same node. A directed graph is complete if there is an edge from every node to every other node. A undirected tree is an undirected graph where all the nodes can be connected by a single undirected path. A directed graph is said to contain a spanning tree if there exists a node called the root, and this root has a directed path to every other node of the graph. (Fig. 2.1, for example). For undirected graphs, the existence of a directed spanning tree is equivalent to being connected. However, in directed graphs, the existence of a directed spanning tree is a weaker condition than being strongly connected. A strongly connected graph contains more than one directed spanning tree. For example, Figs. 2.4–2.6 show three different spanning trees.

2.3.2 Graph matrices

The adjacent matrix associated with the communication graph is defined as $\mathcal{A} = [a_{ij}] \in \mathbb{R}^{N \times N}$, where the element a_{ij} denotes the connection between the agent i and agent j. $a_{ij} = 1$ if $(j, i) \in \mathcal{E}$, otherwise it is zero. $a_{ii} = 0$ for all nodes with the assumption that there exists no self loop. The Laplacian matrix $\mathcal{L} = [l_{ij}] \in \mathbb{R}^{N \times N}$ is defined by $l_{ii} = \sum_{j=1}^{N} a_{ij}$ and $l_{ij} = -a_{ij}$ when $i \neq j$. For the undirected graph in Fig. 2.2, the adjacency matrix is given by

$$\mathcal{A} = \begin{bmatrix} 0 & 1 & 0 & 1 & 0 \\ 1 & 0 & 1 & 0 & 1 \\ 0 & 1 & 0 & 1 & 0 \\ 1 & 0 & 1 & 0 & 1 \\ 0 & 1 & 0 & 1 & 0 \end{bmatrix},$$

and the resultant Laplacian matrix is obtained as

$$\mathcal{L} = \begin{bmatrix} 2 & -1 & 0 & -1 & 0 \\ -1 & 3 & -1 & 0 & -1 \\ 0 & -1 & 2 & -1 & 0 \\ -1 & 0 & -1 & 3 & -1 \\ 0 & -1 & 0 & -1 & 2 \end{bmatrix}.$$

The eigenvalues of $\mathcal{L}$ are $\{0, 2, 2, 3, 5\}$. For the directed graph in Fig. 2.3, the adjacency matrix is given by

$$\mathcal{A} = \begin{bmatrix} 0 & 0 & 0 & 1 & 1 \\ 1 & 0 & 0 & 0 & 0 \\ 0 & 1 & 0 & 0 & 0 \\ 0 & 0 & 1 & 0 & 0 \\ 0 & 1 & 0 & 0 & 0 \end{bmatrix},$$

and the resultant Laplacian matrix is obtained as

$$\mathcal{L} = \begin{bmatrix} 2 & 0 & 0 & -1 & -1 \\ -1 & 1 & 0 & 0 & 0 \\ 0 & -1 & 1 & 0 & 0 \\ 0 & 0 & -1 & 1 & 0 \\ 0 & -1 & 0 & 0 & 1 \end{bmatrix}.$$

The eigenvalues of $\mathcal{L}$ are $\{0, 1, 1.5 \pm j0.866, 2\}$.

From the definition of the Laplacian matrix and also the above examples, it can be seen that $\mathcal{L}$ is diagonally dominant and has nonnegative diagonal entries. Since $\mathcal{L}$ has zero row sums, 0 is an eigenvalue of $\mathcal{L}$ with an associated right eigenvector $\mathbf{1}$. According to Gershgorin's disk theorem, all nonzero eigenvalues of $\mathcal{L}$ are located within a disk in the complex plane centered at $d_{\max}$ and having radius of $d_{\max}$, where $d_{\max}$ denotes the maximum in-degree of all nodes. According to the definition of M-matrix in the last subsection, we know that the Laplacian matrix $\mathcal{L}$ is a singular M-matrix.

2.3.3 Properties

Lemma 2.4. *For an undirected graph $\mathcal{G}$ with N agents, if it is connected, the Laplacian matrix $\mathcal{L}$ is symmetric and has a single zero eigenvalue with $\mathcal{L}\mathbf{1}_N = 0$, and all* *the other eigenvalues of $\mathcal{L}$ are real and positive. Furthermore, we have the following properties for the Laplacian matrix $\mathcal{L}$ from [145]:*

1. $x^T(\mathcal{L}\otimes I_n)x = \frac{1}{2}\sum_{i=1}^{N}\sum_{j=1}^{N} a_{ij}(x_j - x_i)^2$;

2. $\lambda_2(\mathcal{L}) = \min_{x\neq 0, 1_N^T x=0} \frac{x^T\mathcal{L}x}{x^T x}$;

3. $x^T(\mathcal{L}\otimes I_n)x \geq \lambda_2(\mathcal{L})x^T x$, *where* $\lambda_2(\mathcal{L})$ *is the smallest positive eigenvalue of* $\mathcal{L}$.

Proof. For the sake of simplicity, we only give the proof of 3. Let $x = [x_1^T, \ldots, x_N^T]^T$, we can write

$$\Delta := \sum_{i=1}^{N}\sum_{j=1}^{N} l_{ij}x_i^T x_j = x^T(\mathcal{L}\otimes I_n)x$$

Since $\mathcal{L}$ is symmetric, we have $\mathcal{L} = \mathcal{U}^T\mathcal{D}\mathcal{U}$ with $\mathcal{U}$ being a unitary matrix and $\mathcal{D}$ being a diagonal matrix. Since $\mathcal{L}$ has only one zero eigenvalue and the connectivity is $\alpha = \lambda_2(\mathcal{L})$, we can arrange $D = diag(d_1, d_2, \ldots, d_N)$ with $d_1 = 0$ and $d_i \geq \alpha$ for $i > 1$. Hence, we have the first row of $\mathcal{U}$ as $\mathcal{U}_{(1)} = [1, \ldots, 1]/\sqrt{N}$. Clearly, we have $\left(\mathcal{U}_{(1)}\otimes I_n\right)x = 0$, and we can partition $(\mathcal{U}\otimes I_n)x = \begin{bmatrix} 0 \\ y \end{bmatrix}$ with $y \in R^{n(N-1)}$. This leads to

$$\begin{aligned}\Delta &= x^T(\mathcal{U}^T\otimes I_n)(D\otimes I_n)(\mathcal{U}\otimes I_n)x \\ &= [0, y^T](D\otimes I_n)[0, y^T]^T \geq \alpha\|y\|^2 = \alpha\|x\|^2.\end{aligned}$$

The last equality in the above follows from the fact that $\mathcal{U}$ is a unitary matrix. This completes the proof. □

Proposition 2.1. *$\mathcal{L}^2$ is positive semi-definite and $\mathcal{L}^2\mathbf{1}_{N\times 1} = \mathbf{0}_{N\times 1}$ if the graph is undirected and connected.*

Proof. Based on the definition, $\mathcal{L}$ is a symmetric Laplacian matrix and is positive semi-definite. For a given Laplacian matrix $\mathcal{L}$, there exists a matrix Q composed of the eigenvectors of L such that

$$\mathcal{L} = Q\Lambda Q^{-1}, \tag{2.14}$$

where Λ is the diagonal matrix with the diagonal elements being the eigenvalues of $\mathcal{L}$. Since $\mathcal{L}$ is a symmetric Laplacian matrix, all the eigenvalues are real numbers. Additionally, one of the eigenvalues is zero and all others are positive. Then, it is straightforward to see

$$\mathcal{L}^2 = Q\Lambda Q^{-1}Q\Lambda Q^{-1} = Q\Lambda^2 Q^{-1}. \tag{2.15}$$

It can be seen that $\mathcal{L}^2$ has the eigenvalues that are squares of the eigenvalues of $\mathcal{L}$ and has the same corresponding eigenvectors. Therefore, $\mathcal{L}^2$ is positive semi-definite. Furthermore, it can be seen that

$$\mathcal{L}^2\mathbf{1}_{N\times 1} = \mathcal{L}\cdot\mathcal{L}\mathbf{1}_{N\times 1} = \mathcal{L}\cdot\mathbf{0}_{N\times 1} = \mathbf{0}_{N\times 1}. \tag{2.16}$$

This completes the proof. □

Lemma 2.5. *For an undirected graph with a leader indexed by 0, $\mathcal{L}$ is the Laplacian matrix between followers and $\mathcal{B}$ is a diagonal matrix which represents the connections between the leader and the followers. The matrix $\mathcal{H} = \mathcal{L} + \mathcal{B} = \mathcal{H}^T \in \mathbb{R}^{N \times N}$ is positive definite if the graph is connected and at least one follower can receive the leader's information.*

Lemma 2.6 ([145,87]). *The Laplacian matrix $\mathcal{L}$ of a directed graph $\mathcal{G}$ has at least one zero eigenvalue with a corresponding right eigenvector $\mathbf{1} = [1, 1, \ldots, 1]^T$ and all nonzero eigenvalues have positive real parts. Furthermore, zero is a simple eigenvalue of $\mathcal{L}$ if and only if $\mathcal{G}$ has a directed spanning tree. In addition, there exists a nonnegative left eigenvector r of $\mathcal{L}$ associated with the zero eigenvalue, satisfying $r^T \mathcal{L} = 0$ and $r^T 1 = 1$. Moreover, r is unique if $\mathcal{G}$ has a directed spanning tree.*

Lemma 2.7 ([33]). *For a Laplacian matrix $\mathcal{L}$ having zero as its simple eigenvalue, there exists a similarity transformation T, with its first column being $T_1 = \mathbf{1}$, such that*

$$T^{-1} \mathcal{L} T = J, \tag{2.17}$$

with J being a block diagonal matrix in the real Jordan form

$$J = \begin{bmatrix} 0 & & & & & & \\ & J_1 & & & & & \\ & & \ddots & & & & \\ & & & J_p & & & \\ & & & & J_{p+1} & & \\ & & & & & \ddots & \\ & & & & & & J_q \end{bmatrix}, \tag{2.18}$$

where $J_k \in \mathbb{R}^{n_k}$, $k = 1, 2, \ldots, p$, are the Jordan blocks for real eigenvalues $\lambda_k > 0$ with the multiplicity n_k in the form

$$J_k = \begin{bmatrix} \lambda_k & 1 & & & \\ & \lambda_k & 1 & & \\ & & \ddots & \ddots & \\ & & & \lambda_k & 1 \\ & & & & \lambda_k \end{bmatrix},$$

and $J_k \in \mathbb{R}^{2n_k}$, $k = p+1, p+2, \ldots, q$, are the Jordan blocks for conjugate eigenvalues $\alpha_k \pm j\beta_k$, $\alpha_k > 0$ and $\beta_k > 0$, with the multiplicity n_k in the form

$$J_k = \begin{bmatrix} \nu(\alpha_k, \beta_k) & I_2 & & & \\ & \nu(\alpha_k, \beta_k) & I_2 & & \\ & & \ddots & \ddots & \\ & & & \nu(\alpha_k, \beta_k) & I_2 \\ & & & & \nu(\alpha_k, \beta_k) \end{bmatrix},$$

with I_2 *being the identity matrix in* $\mathbb{R}^{2\times 2}$ *and*

$$\nu(\alpha_k, \beta_k) = \begin{bmatrix} \alpha_i & \beta_i \\ -\beta_i & \alpha_i \end{bmatrix} \in \mathbb{R}^{2\times 2}.$$

Lemma 2.8 ([35]). *In the communication topology* $\mathcal{G}$*, suppose that N followers contain a directed spanning tree with the leader indexed by 0 as the root. Since the leader has no neighbors, the Laplacian matrix* $\mathcal{L}$ *has the following structure*

$$\mathcal{L} = \begin{bmatrix} 0 & 0_{1\times N} \\ \mathcal{L}_2 & \mathcal{L}_1 \end{bmatrix},$$

where $\mathcal{L}_1 \in \mathbb{R}^{N\times N}$ *and* $\mathcal{L}_2 \in \mathbb{R}^{N\times 1}$*. It can be seen that* $\mathcal{L}_1$ *is a nonsingular M-matrix and there exists a positive diagonal matrix Q such that*

$$Q\mathcal{L}_1 + \mathcal{L}_1^T Q \geq \rho_0 I, \tag{2.19}$$

for some positive constant ρ_0*. Q can be constructed as* $Q = \mathrm{diag}\{q_1, q_2, \cdots, q_N\}$*, where* $q = [q_1, q_2, \cdots, q_N]^T = \left(\mathcal{L}_1^T\right)^{-1}[1, 1, \cdots, 1]^T$.

2.4 Useful lemmas on inequalities

The following lemmas on the inequalities are needed.

Lemma 2.9 (Young's Inequality). *For any given* $a, b \in \mathbb{R}^n$*, we have*

$$2a^T SQb \leq a^T SPS^T a + b^T Q^T P^{-1} Qb,$$

where $P > 0$*, S and Q have appropriate dimensions.*

Lemma 2.10 (Hölder's Inequality). *For* $x \in \mathbb{R}^n$ *and* $y \in \mathbb{R}^n$*, if* $p > 1$ *and* $q > 1$ *are real numbers such that* $1/p + 1/q = 1$*, then*

$$\sum_{i=1}^{n} |x_i y_i| \leq \left\{\sum_{i=1}^{n} |x_i|^p\right\}^{1/p} \left\{\sum_{i=1}^{n} |y_i|^q\right\}^{1/q}.$$

Lemma 2.11 ([199]). *Given positive real numbers* x, y, m, n, a, b*, the following inequality holds:*

$$ax^m y^n \leq bx^{m+n} + \frac{n}{m+n}(\frac{m+n}{m})^{-\frac{m}{n}} a^{\frac{m+n}{n}} b^{-\frac{m}{n}} y^{m+n}. \tag{2.20}$$

Lemma 2.12 ([140]). *Given* $x, y \in \mathbb{R}$*,* $p \geq 1$ *being an odd integer, the following inequality holds:*

$$|x - y|^p \leq 2^{p-1}|x^p - y^p|.$$

Lemma 2.13 ([140]). *Given $x, y \in \mathbb{R}$, $p \geq 1$ being an odd integer, there exists a positive constant c such that*

$$|(x+y)^p - x^p| \leq c|x^{p-1} + y^{p-1}||y|.$$

Lemmas 2.11 and 2.12 are straightforward due to Young's Inequality. The proof of Lemma 2.13 can be obtained by applying Lemma 2.11 to each element of the polynomial $g(x, y)$, where $|(x+y)^p - x^p| = |(x+y) - x||g(x, y)|$.

Lemma 2.14 (Jensen's Inequality in [58]). *For a positive definite matrix P, and a function $x : [a, b] \to \mathbb{R}^n$, with $a, b \in \mathbb{R}$ and $b > a$, the following inequality holds:*

$$\left(\int_a^b x^T(\tau)\mathrm{d}\tau\right) P \left(\int_a^b x(\tau)\mathrm{d}\tau\right) \leq (b-a)\int_a^b x^T(\tau)Px(\tau)\mathrm{d}\tau. \tag{2.21}$$

Lemma 2.15 ([36]). *For a positive definite matrix P, the following identity holds*

$$e^{A^T t}Pe^{At} - e^{\omega_1 t}P = -e^{\omega_1 t}\int_0^t e^{-\omega_1 \tau}e^{A^T \tau}Re^{A\tau}\mathrm{d}\tau,$$

where

$$R = -A^T P - PA + \omega_1 P.$$

Furthermore, if R is positive definite, $\forall t > 0$,

$$e^{A^T t}Pe^{At} < e^{\omega_1 t}P. \tag{2.22}$$

Part Two

Optimal cooperative control

Jianan Wang, Chunyan Wang, Ming Xin, Zhengtao Ding

Optimal consensus control of multiple integrator systems

3

3.1 Problem formulation

Generally speaking, agents in systems can be described by various types of dynamics, such as single-integrator dynamics when the position loop is considered, or double-integrator dynamics when both position and velocity loops are considered, etc.

For the single-integrator dynamics of n agents, the system equations are given as follows:

$$\dot{\boldsymbol{p}}_i = \boldsymbol{u}_i, \quad i = 1, \ldots, n \tag{3.1}$$

where $\boldsymbol{p}_i(t) \in \mathbb{R}^m$ and $\boldsymbol{u}_i(t) \in \mathbb{R}^m$ are, respectively, the position and control input of the ith agent.

Alternatively, for the double-integrator dynamics of n agents, the system equations are given as follows:

$$\begin{cases} \dot{\boldsymbol{p}}_i = \boldsymbol{v}_i \\ \dot{\boldsymbol{v}}_i = \boldsymbol{u}_i \end{cases}, \quad i = 1, \ldots, n \tag{3.2}$$

where $\boldsymbol{p}_i(t) \in R^m$, $\boldsymbol{v}_i(t) \in R^m$ and $\boldsymbol{u}_i(t) \in R^m$ are, respectively, the position, velocity, and control input of the ith agent.

Fig. 3.1 shows an example of four agents' consensus problem. Four agents start their motions from different initial positions and velocities. R_{dj} denotes the radius of the jth obstacle detection region and r_j denotes the radius of the jth obstacle, where $\boldsymbol{O}_{bj}$ is the location vector of the jth obstacle. The dashed line in the figure denotes the original consensus trajectory without concern of the obstacle. The proposed consensus law will be able to not only drive all the agents along the solid lines to reach consensus but also avoid the obstacle with an optimal control effort.

For the convenience of the problem formulation, the following regions are defined.

Collision region for the jth obstacle:

$$\Lambda_j \triangleq \left\{ \boldsymbol{x} \,\middle|\, \boldsymbol{x} \in \mathbb{R}^m, \left\| \boldsymbol{x} - \boldsymbol{O}_{bj} \right\| \le r_j \right\} \tag{3.3}$$

Detection region for the jth obstacle:

$$\Psi_j \triangleq \left\{ \boldsymbol{x} \,\middle|\, \boldsymbol{x} \in \mathbb{R}^m, \left\| \boldsymbol{x} - \boldsymbol{O}_{bj} \right\| \le R_{d_j} \right\} \tag{3.4}$$

Reaction region for the jth obstacle:

$$\Gamma_j \triangleq \left\{ \boldsymbol{x} \,\middle|\, \boldsymbol{x} \in \mathbb{R}^m, r_j < \left\| \boldsymbol{x} - \boldsymbol{O}_{bj} \right\| \le R_{d_j} \right\} \tag{3.5}$$

Cooperative Control of Multi-Agent Systems. https://doi.org/10.1016/B978-0-12-820118-3.00013-2

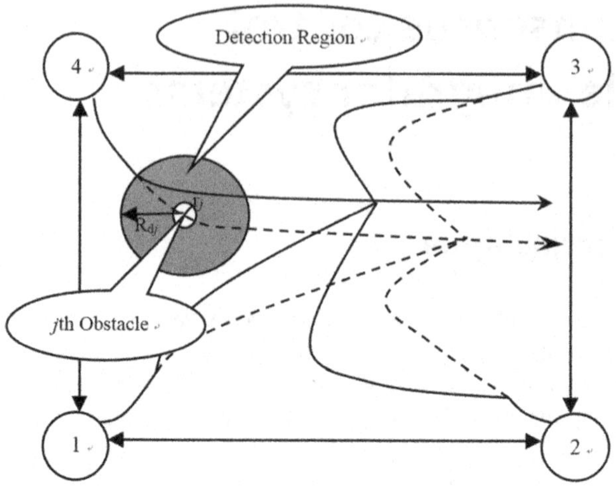

Figure 3.1 Multi-agent consensus scenario with an obstacle.

Accordingly, the entire safety region can be denoted as $\Theta = \left(\bigcup_j \Lambda_j\right)^c$, and the entire outside detection region can be denoted as $\Pi = \left(\bigcup_j \Psi_j\right)^c$. The symbol $\bigcup$ denotes the union of sets and the superscript c stands for the complement of a set.

In the scope of this chapter, we impose the following assumptions.

Assumption 3.1. All the obstacles can be modeled as circle-shaped objects.

Assumption 3.2. $\Psi_j \bigcap \Psi_k = \varnothing,\ j \neq k.$

Assumption 3.3. The information exchange topology among agents is assumed to be undirected and connected.

Assumption 3.2 means that the detection regions of multiple obstacles have no intersection. This assumption is to exclude the scenario in which an agent steps into the intersection region and cannot autonomously determine which obstacle to avoid. It implies that each agent will face only one obstacle at any instant. Future research will be conducted to handle the special scenario with a dense pack of obstacles appearing on the trajectory at the same time.

The consensus control problem in this chapter is to design a distributed optimal control law $\boldsymbol{u}_i(t)$ based on the information topology such that all the agents' states converge to the same value, i.e. $\left\|\boldsymbol{p}_i(t) - \boldsymbol{p}_j(t)\right\| \to 0$ in the single-integrator case, or at the same time $\left\|\boldsymbol{v}_i(t) - \boldsymbol{v}_j(t)\right\| \to 0$ for the double-integrator case. More importantly, each agent can avoid the obstacle along its trajectory.

3.2 Optimal consensus control with obstacle avoidance for single-integrator case

3.2.1 Optimal consensus algorithm: single-integrator case

In order to further study this case, multiple single-integrator systems given in (3.1) can be written in a matrix form:

$$\dot{\boldsymbol{P}} = \boldsymbol{U} \tag{3.6}$$

with

$$\boldsymbol{P} = \left[\boldsymbol{p}_1^T, \ldots, \boldsymbol{p}_n^T\right]^T, \boldsymbol{U} = \left[\boldsymbol{u}_1^T, \ldots, \boldsymbol{u}_n^T\right]^T$$

$\boldsymbol{P} \in \mathbb{R}^{nm}$ and $\boldsymbol{U} \in \mathbb{R}^{nm}$ are, respectively, the aggregate state and control input of all agents.

For the convenience of formulation, the error state is defined as follows:

$$\hat{\boldsymbol{P}} = [\hat{\boldsymbol{p}}_1^T \quad \hat{\boldsymbol{p}}_2^T \quad \cdots \quad \hat{\boldsymbol{p}}_n^T]^T \triangleq \boldsymbol{P} - \boldsymbol{P}_{cs} \tag{3.7}$$

where $\boldsymbol{P}_{cs} = \left[\mathbf{1}_{1\times n} \otimes \boldsymbol{p}_{cs}^T\right]^T$ is the final consensus state. For instance, in a planar motion, $\boldsymbol{p}_{cs} = \left[\alpha_x \quad \alpha_y\right]^T$, where α_x, α_y are the final consensus position along x-axis and y-axis, respectively. According to the property of the Laplacian matrix $\mathcal{L}$

$$\mathcal{L}\mathbf{1} = \mathbf{0} \tag{3.8}$$

when the agents reach consensus, we have

$$(\mathcal{L} \otimes I_m)\, \boldsymbol{P}_{cs} = \mathbf{0}_{nm\times 1} \tag{3.9}$$

Note that the final consensus state $\boldsymbol{P}_{cs}$ will be a constant and the consensus law $\boldsymbol{U}_s$ becomes zero when the agents reach consensus. Then, the error dynamics becomes

$$\dot{\hat{\boldsymbol{P}}} = \boldsymbol{U} \tag{3.10}$$

The consensus is achieved when the system (3.10) is asymptotically stable.

The optimal obstacle avoidance consensus formulation consists of three cost function components:

$$\begin{aligned} &Min: \mathcal{J} = \mathcal{J}_1 + \mathcal{J}_2 + \mathcal{J}_3 \\ &S.t. \quad \dot{\hat{\boldsymbol{P}}} = \boldsymbol{U} \end{aligned} \tag{3.11}$$

where $\mathcal{J}_1, \mathcal{J}_2, \mathcal{J}_3$ represent consensus cost, obstacle avoidance cost, and control effort cost, respectively.

The consensus cost has the form of

$$\mathcal{J}_1 = \int_0^{\infty} \hat{\boldsymbol{P}}^T R_1 \hat{\boldsymbol{P}} dt = \int_0^{\infty} \left[\hat{\boldsymbol{P}}^T \left(w_p^2 \mathcal{L}^2 \otimes I_m \right) \hat{\boldsymbol{P}} \right] dt \tag{3.12}$$

where $R_1 = w_p^2 \mathcal{L}^2 \otimes I_m$, $\mathcal{L}$ is the symmetric Laplacian matrix established by the undirected and connected graph. w_p represents the weight on the consensus error.

Remark 3.1. Proposition 2.1 shows that R_1 is positive semi-definite and $\mathcal{J}_1$ is a meaningful consensus cost. The formulation of R_1 in Eq. (3.12) is to guarantee that the proposed optimal control law is a linear function of the Laplacian matrix $\mathcal{L}$, and is thus completely dependent on the information topology, which will be shown in the proof of Theorem 3.1. A form of the consensus cost similar to Eq. (3.12) has also been analyzed in [15].

The obstacle avoidance cost has the form of

$$\mathcal{J}_2 = \int_0^{\infty} h(\hat{\boldsymbol{P}}) dt \tag{3.13}$$

where $h(\hat{\boldsymbol{P}})$ will be constructed from an inverse optimal control approach in Theorem 3.1.

The control effort cost has the regular quadratic form of

$$\mathcal{J}_3 = \int_0^{\infty} \boldsymbol{U}^T R_2 \boldsymbol{U} dt \tag{3.14}$$

where $R_2 = w_c^2 I_n \otimes I_m$ is positive definite and w_c is a scalar weighting parameter.

Lemma 2.3 in Chapter 2 is adopted here to derive the first main result and is used to prove both asymptotic stability and optimality of the proposed obstacle avoidance consensus algorithm as shown below.

Theorem 3.1. *For multi-agent systems (3.1) under Assumptions 3.1–3.3, there always exist parameters w_p and w_c such that the feedback control law*

$$\phi(\boldsymbol{P}) = -\frac{w_p}{w_c} (\mathcal{L} \otimes I_m) \boldsymbol{P} - \frac{1}{2w_c^2} g'(\boldsymbol{P}) \tag{3.15}$$

is an optimal control for the consensus problem (3.11) with

$$h\left(\hat{\boldsymbol{P}}\right) = \frac{w_p}{w_c} g'^T \left(\hat{\boldsymbol{P}}\right) (\mathcal{L} \otimes I_m) \hat{\boldsymbol{P}} + \frac{1}{4w_c^2} g'^T \left(\hat{\boldsymbol{P}}\right) g' \left(\hat{\boldsymbol{P}}\right) \tag{3.16}$$

in (3.13) where $g\left(\hat{\boldsymbol{P}}\right)$ is the obstacle avoidance potential function defined by

$$g\left(\hat{\boldsymbol{P}}\right) = \sum_{i=1}^{n} m\left(\boldsymbol{p}_i\right) = g\left(\boldsymbol{P}\right) \tag{3.17}$$

with

$$m(\boldsymbol{p}_i) = \begin{cases} 0 & \boldsymbol{p}_i \in \Pi \\ \left(\dfrac{R_{d_j}^2 - \|\boldsymbol{p}_i - \boldsymbol{O}_{bj}\|^2}{\|\boldsymbol{p}_i - \boldsymbol{O}_{bj}\|^2 - r_j^2} \right)^2 & \boldsymbol{p}_i \in \Gamma_j \\ \textit{not defined} & \boldsymbol{p}_i \in \Lambda_j \end{cases} \tag{3.18}$$

and

$$\begin{aligned} g'(\hat{\boldsymbol{P}}) &= \left[\left(\frac{dm(\boldsymbol{p}_1)}{d\hat{\boldsymbol{p}}_1} \right)^T \; \left(\frac{dm(\boldsymbol{p}_2)}{d\hat{\boldsymbol{p}}_2} \right)^T \; \cdots \; \left(\frac{dm(\boldsymbol{p}_n)}{d\hat{\boldsymbol{p}}_n} \right)^T \right]^T \\ &= \left[\left(\frac{dm(\boldsymbol{p}_1)}{d\boldsymbol{p}_1} \right)^T \; \left(\frac{dm(\boldsymbol{p}_2)}{d\boldsymbol{p}_2} \right)^T \; \cdots \; \left(\frac{dm(\boldsymbol{p}_n)}{d\boldsymbol{p}_n} \right)^T \right]^T = g'(\boldsymbol{P}). \end{aligned} \tag{3.19}$$

$g'\left(\hat{\boldsymbol{P}}\right)$ *is the derivative of* $g\left(\hat{\boldsymbol{P}}\right)$ *with respect to* $\hat{\boldsymbol{P}}$.

In addition, the closed-loop system is globally asymptotically stable or the consensus is guaranteed with $\boldsymbol{P}(t) \to \boldsymbol{P}_{cs}$.

Proof. Using Lemma 2.3 in Chapter 2 for this optimal consensus problem, we have the following equations:

$$T\left(\hat{\boldsymbol{P}}, \boldsymbol{U}\right) = \hat{\boldsymbol{P}}^T R_1 \hat{\boldsymbol{P}} + h\left(\hat{\boldsymbol{P}}\right) + \boldsymbol{U}^T R_2 \boldsymbol{U} \tag{3.20}$$

$$\boldsymbol{f}\left(\hat{\boldsymbol{P}}, \boldsymbol{U}\right) = \boldsymbol{U} \tag{3.21}$$

and $\boldsymbol{f}$ satisfies $\boldsymbol{f}(\boldsymbol{0}_{nm\times 1}, \boldsymbol{0}_{nm\times 1}) = \boldsymbol{0}_{nm\times 1}$.

A candidate Lyapunov function $V\left(\hat{\boldsymbol{P}}\right)$ is chosen to be

$$V\left(\hat{\boldsymbol{P}}\right) = \hat{\boldsymbol{P}}^T \mathcal{P} \hat{\boldsymbol{P}} + g\left(\hat{\boldsymbol{P}}\right) \tag{3.22}$$

where $\mathcal{P}$ is the solution of a Riccati equation, which will be shown afterwards.

In order for the function $V\left(\hat{\boldsymbol{P}}\right)$ in Eq. (3.22) to be a valid Lyapunov function, it must be continuously differentiable with respect to $\hat{\boldsymbol{P}}$ or equivalently $g\left(\hat{\boldsymbol{P}}\right)$ must be continuously differentiable with respect to $\hat{\boldsymbol{P}}$. From Eqs. (3.17) and (3.18), it suffices to show that $m\left(\boldsymbol{p}_i\right)$ is continuously differentiable in the safety region Θ. In fact, this is true if $m\left(\boldsymbol{p}_i\right)$ and $\frac{dm(\boldsymbol{p}_i)}{d\boldsymbol{p}_i}$ are continuous at the boundary of Ψ_j. Since Eq. (3.18) implies that $\lim_{\|\boldsymbol{p}_i - \boldsymbol{O}_{bj}\| \to R_{d_j}^-} m(\boldsymbol{p}_i) = 0 = \lim_{\|\boldsymbol{p}_i - \boldsymbol{O}_{bj}\| \to R_{d_j}^+} m(\boldsymbol{p}_i)$, $m\left(\boldsymbol{p}_i\right)$ is continuous at the boundary of Ψ_j and thus continuous over Θ. In addition, since

$$\frac{dm(\boldsymbol{p}_i)}{d\boldsymbol{p}_i} = \begin{cases} \boldsymbol{0} & \boldsymbol{p}_i \in \Pi \\ \frac{-4(R_{d_j}^2 - r_j^2)(R_{d_j}^2 - \|\boldsymbol{p}_i - \boldsymbol{O}_{bj}\|^2)}{\left(\|\boldsymbol{p}_i - \boldsymbol{O}_{bj}\|^2 - r_j^2\right)^3}(\boldsymbol{p}_i - \boldsymbol{O}_{bj}) & \boldsymbol{p}_i \in \Gamma_j \\ \text{not defined} & \boldsymbol{p}_i \in \Lambda_j \end{cases} \tag{3.23}$$

it is easy to see that $\lim_{\|\boldsymbol{p}_i - \boldsymbol{O}_{bj}\| \to R_{d_j}^-} \frac{dm(\boldsymbol{p}_i)}{d\boldsymbol{p}_i} = \boldsymbol{0}_{m\times 1} = \lim_{\|\boldsymbol{p}_i - \boldsymbol{O}_{bj}\| \to R_{d_j}^+} \frac{dm(\boldsymbol{p}_i)}{d\boldsymbol{p}_i}$, which implies that $\frac{dm(\boldsymbol{p}_i)}{d\boldsymbol{p}_i}$ is continuous at the boundary of Ψ_j, and thus continuous over Θ. Therefore, $g\left(\hat{\boldsymbol{P}}\right)$ and the Lyapunov function $V\left(\hat{\boldsymbol{P}}\right)$ are continuously differentiable with respect to $\hat{\boldsymbol{P}}$ in the safety region Θ.

The Hamiltonian for the consensus problem becomes

$$\begin{aligned} H\left(\hat{\boldsymbol{P}}, \boldsymbol{U}\right) &= T\left(\hat{\boldsymbol{P}}, \boldsymbol{U}\right) + V^T\left(\hat{\boldsymbol{P}}\right) f\left(\hat{\boldsymbol{P}}, \boldsymbol{U}\right) \\ &= \hat{\boldsymbol{P}}^T R_1 \hat{\boldsymbol{P}} + h\left(\hat{\boldsymbol{P}}\right) + \boldsymbol{U}^T R_2 \boldsymbol{U} + \left[2\hat{\boldsymbol{P}}^T \mathcal{P} + g'^T\left(\hat{\boldsymbol{P}}\right)\right]\boldsymbol{U} \end{aligned} \tag{3.24}$$

Setting $(\partial/\partial \boldsymbol{U}) H\left(\hat{\boldsymbol{P}}, \boldsymbol{U}\right) = \boldsymbol{0}$ yields the optimal control law:

$$\boldsymbol{U}^* = \phi\left(\hat{\boldsymbol{P}}\right) = -\frac{1}{2}R_2^{-1}V'\left(\hat{\boldsymbol{P}}\right) = -R_2^{-1}\mathcal{P}\hat{\boldsymbol{P}} - \frac{1}{2}R_2^{-1}g'\left(\hat{\boldsymbol{P}}\right) \tag{3.25}$$

With Eq. (3.25), it follows that

$$\begin{aligned} V'^T\left(\hat{\boldsymbol{P}}\right) f\left(\hat{\boldsymbol{P}}, \phi\left(\hat{\boldsymbol{P}}\right)\right) &= -2\hat{\boldsymbol{P}}^T \mathcal{P} R_2^{-1} \mathcal{P} \hat{\boldsymbol{P}} - \hat{\boldsymbol{P}}^T \mathcal{P} R_2^{-1} g'\left(\hat{\boldsymbol{P}}\right) \\ &\quad - g'^T\left(\hat{\boldsymbol{P}}\right) R_2^{-1} \mathcal{P} \hat{\boldsymbol{X}} - \frac{1}{2} g'^T\left(\hat{\boldsymbol{P}}\right) R_2^{-1} g'\left(\hat{\boldsymbol{P}}\right) \end{aligned} \tag{3.26}$$

Substituting Eqs. (3.25) and (3.26) into Eq. (3.24) yields

$$\begin{aligned} H\left(\hat{\boldsymbol{P}}, \phi\left(\hat{\boldsymbol{P}}\right)\right) &= \hat{\boldsymbol{P}}^T\left(R_1 - \mathcal{P} R_2^{-1} \mathcal{P}\right)\hat{\boldsymbol{P}} - g'^T\left(\hat{\boldsymbol{P}}\right) R_2^{-1} \mathcal{P} \hat{\boldsymbol{P}} + h\left(\hat{\boldsymbol{P}}\right) \\ &\quad - \frac{1}{4} g'^T\left(\hat{\boldsymbol{P}}\right) R_2^{-1} g'\left(\hat{\boldsymbol{P}}\right) \end{aligned} \tag{3.27}$$

In order to prove that the control law (3.25) is an optimal solution to the consensus problem (3.11) using the Lemma 2.3 in Chapter 2, the conditions expressed in Eqs. (2.4)–(2.9) need to be verified. In order to satisfy the condition (2.8) or let Eq. (3.27) be zero, we can let

$$R_1 - \mathcal{P} R_2^{-1} \mathcal{P} = 0 \tag{3.28}$$

and require that

$$-g'^T\left(\hat{\boldsymbol{P}}\right)R_2^{-1}\mathcal{P}\hat{\boldsymbol{P}}+h\left(\hat{\boldsymbol{P}}\right)-\frac{1}{4}g'^T\left(\hat{\boldsymbol{P}}\right)R_2^{-1}g'\left(\hat{\boldsymbol{P}}\right)=0 \tag{3.29}$$

With Eqs. (3.25), (3.28), and (3.29), it can be shown that

$$\begin{aligned}
&H\left(\hat{\boldsymbol{P}},\boldsymbol{U},V'^T\left(\hat{\boldsymbol{P}}\right)\right)\\
&=\boldsymbol{U}^TR_2\boldsymbol{U}+h\left(\hat{\boldsymbol{P}}\right)+\hat{\boldsymbol{P}}^TR_1\hat{\boldsymbol{P}}+\left(2\hat{\boldsymbol{P}}^T\mathcal{P}+g'^T\left(\hat{\boldsymbol{P}}\right)\right)\boldsymbol{U}\\
&=\boldsymbol{U}^TR_2\boldsymbol{U}+h\left(\hat{\boldsymbol{P}}\right)+\hat{\boldsymbol{P}}^TR_1\hat{\boldsymbol{P}}+\left(2\hat{\boldsymbol{P}}^T\mathcal{P}+g'^T\left(\hat{\boldsymbol{P}}\right)\right)\boldsymbol{U}\\
&-\hat{\boldsymbol{P}}^T\left(R_1-\mathcal{P}R_2^{-1}\mathcal{P}\right)\hat{\boldsymbol{P}}\\
&=\boldsymbol{U}^TR_2\boldsymbol{U}+h\left(\hat{\boldsymbol{P}}\right)+g'^T\left(\hat{\boldsymbol{P}}\right)\boldsymbol{U}+2\hat{\boldsymbol{P}}^T\mathcal{P}\boldsymbol{U}+\hat{\boldsymbol{P}}^T\mathcal{P}R_2^{-1}\mathcal{P}\hat{\boldsymbol{P}}\\
&=\boldsymbol{U}^TR_2\boldsymbol{U}+\frac{1}{4}g'^T\left(\hat{\boldsymbol{P}}\right)R_2^{-1}g'\left(\hat{\boldsymbol{P}}\right)+g'^T\left(\hat{\boldsymbol{P}}\right)R_2^{-1}\mathcal{P}\hat{\boldsymbol{P}}+\hat{\boldsymbol{P}}^T\mathcal{P}R_2^{-1}\mathcal{P}\hat{\boldsymbol{P}}\\
&+\left(2\hat{\boldsymbol{P}}^T\mathcal{P}+g'^T\left(\hat{\boldsymbol{P}}\right)\right)\boldsymbol{U}\\
&=\boldsymbol{U}^TR_2\boldsymbol{U}+\frac{1}{4}\left(2\hat{\boldsymbol{P}}^T\mathcal{P}+g'^T\left(\hat{\boldsymbol{P}}\right)\right)R_2^{-1}\left(2\hat{\boldsymbol{P}}^T\mathcal{P}+g'^T\left(\hat{\boldsymbol{P}}\right)\right)^T\\
&+\left(2\hat{\boldsymbol{P}}^T\mathcal{P}+g'^T\left(\hat{\boldsymbol{P}}\right)\right)\boldsymbol{U}\\
&=\boldsymbol{U}^TR_2\boldsymbol{U}+\frac{1}{4}V'^T\left(\hat{\boldsymbol{P}}\right)R_2^{-1}V'\left(\hat{\boldsymbol{P}}\right)+\boldsymbol{U}^TV'\left(\hat{\boldsymbol{P}}\right)\\
&=\boldsymbol{U}^TR_2\boldsymbol{U}+\phi\left(\hat{\boldsymbol{P}}\right)^TR_2\phi\left(\hat{\boldsymbol{P}}\right)-2\boldsymbol{U}^TR_2\phi\left(\hat{\boldsymbol{P}}\right)\\
&=\left[\boldsymbol{U}-\phi\left(\hat{\boldsymbol{P}}\right)\right]^TR_2\left[\boldsymbol{U}-\phi\left(\hat{\boldsymbol{P}}\right)\right]\geq 0
\end{aligned} \tag{3.30}$$

which validates the condition (2.9).

Substituting the expressions of R_1, R_2 in Eq. (3.28), one can solve for $\mathcal{P}$:

$$\mathcal{P}=w_pw_c\mathcal{L}\otimes I_m \tag{3.31}$$

Then the Lyapunov function (3.22) becomes

$$V(\hat{\boldsymbol{P}})=\hat{\boldsymbol{P}}^T\mathcal{P}\hat{\boldsymbol{P}}+g(\hat{\boldsymbol{P}})=\begin{cases} w_pw_c\boldsymbol{P}^T(\mathcal{L}\otimes I_m)\boldsymbol{P} & \boldsymbol{p}_i\in\Pi\\ w_pw_c\boldsymbol{P}^T(\mathcal{L}\otimes I_m)\boldsymbol{P}+g(\boldsymbol{P}) & \boldsymbol{p}_i\in\Gamma_j\\ \text{not defined} & \boldsymbol{p}_i\in\Lambda_j\,. \end{cases} \tag{3.32}$$

Note that we use the property of $\mathcal{L}$, i.e. Eq. (3.9), to convert $V(\hat{\boldsymbol{P}})$ to $V(\boldsymbol{P})$. If $\hat{\boldsymbol{P}}\neq\boldsymbol{0}$ or $\boldsymbol{P}\neq\boldsymbol{P}_{cs}$, $\boldsymbol{P}^T(\mathcal{L}\otimes I_m)\boldsymbol{P}$ will not be zero but positive according to the property of the Laplacian matrix. Note that $\boldsymbol{P}=\boldsymbol{0}$ that leads to $\boldsymbol{P}^T(\mathcal{L}\otimes I_m)\boldsymbol{P}=0$ is a special case of $\boldsymbol{P}=\boldsymbol{P}_{cs}$ when $\boldsymbol{P}_{cs}=\boldsymbol{0}$, which implies $\hat{\boldsymbol{P}}=\boldsymbol{0}$ as well. Hence,

$\boldsymbol{P}^T(\mathcal{L}\otimes I_m)\boldsymbol{P} > 0$ if $\hat{\boldsymbol{P}} \neq \boldsymbol{0}$. In addition, from the definition of $g(\boldsymbol{P})$ in Eqs. (3.17) and (3.18), it is concluded that $g(\boldsymbol{P}) \geq 0$. Therefore, the condition (2.5), i.e. $V(\hat{\boldsymbol{P}}) > 0$ for $\hat{\boldsymbol{P}} \neq \boldsymbol{0}$ can be satisfied since $w_p w_c \boldsymbol{P}^T(\mathcal{L}\otimes I_m)\boldsymbol{P} + g(\boldsymbol{P}) > 0$ for $\hat{\boldsymbol{P}} \neq \boldsymbol{0}$.

Next, $h(\hat{\boldsymbol{P}})$ in $\mathcal{J}_2$ is constructed by solving Eq. (3.29):

$$h\left(\hat{\boldsymbol{P}}\right) = \frac{w_p}{w_c} g'^T\left(\hat{\boldsymbol{P}}\right)(\mathcal{L}\otimes I_m)\,\hat{\boldsymbol{P}} + \frac{1}{4w_c^2} g'^T\left(\hat{\boldsymbol{P}}\right) g'\left(\hat{\boldsymbol{P}}\right) \tag{3.33}$$

which turns out to be Eq. (3.16). Note that $h(\hat{\boldsymbol{P}}) \geq 0$ can be guaranteed by choosing proper values of the weighting parameters. Specifically, one can always find a small enough w_p for a given w_c to let the positive-definite term $\frac{1}{4{w_c}^2} g'^T(\hat{\boldsymbol{P}})g'(\hat{\boldsymbol{P}})$ dominate the sign-indefinite term $\frac{w_p}{w_c} g'^T(\hat{\boldsymbol{P}})(\mathcal{L}\otimes I_m)\hat{\boldsymbol{P}}$.

Using Eqs. (3.28) and (3.29), Eq. (3.26) becomes

$$\begin{aligned} &V'^T\left(\hat{\boldsymbol{P}}\right) f\left(\hat{\boldsymbol{P}}, \phi\left(\hat{\boldsymbol{P}}\right)\right) \\ &= -\left[\hat{\boldsymbol{P}}^T R_1 \hat{\boldsymbol{P}} + h\left(\hat{\boldsymbol{P}}\right) + \left(\hat{\boldsymbol{P}}^T \mathcal{P} + \tfrac{1}{2} g'^T\left(\hat{\boldsymbol{P}}\right)\right) R_2^{-1} \left(\mathcal{P}\hat{\boldsymbol{P}} + \tfrac{1}{2} g'\left(\hat{\boldsymbol{P}}\right)\right)\right] \end{aligned} \tag{3.34}$$

Hence, the condition (2.7) can be satisfied since $\hat{\boldsymbol{P}}^T R_1 \hat{\boldsymbol{P}} \geq 0$, $h(\hat{\boldsymbol{P}}) \geq 0$, and

$$\left(\hat{\boldsymbol{P}}^T \mathcal{P} + \frac{1}{2} g'^T\left(\hat{\boldsymbol{P}}\right)\right) R_2^{-1} \left(\mathcal{P}\hat{\boldsymbol{P}} + \frac{1}{2} g'\left(\hat{\boldsymbol{P}}\right)\right) > 0$$

when $\hat{\boldsymbol{P}} \neq \boldsymbol{0}$.

It remains to verify the conditions (2.4) and (2.6). It can be seen from Eqs. (3.22) and (3.25) that the conditions (2.4) and (2.6) are met if $g(\hat{\boldsymbol{P}}) = 0$ and $g'(\hat{\boldsymbol{P}}) = \boldsymbol{0}$ when $\hat{\boldsymbol{P}} = \boldsymbol{0}$. Eqs. (3.15), (3.19) and (3.23) imply that if all agents converge inside the reaction region, the avoidance force is not zero and the agents will be driven outside the reaction region until they reach a new consensus point. When the consensus point ($\hat{\boldsymbol{P}} = \boldsymbol{0}$) is outside the reaction region of the obstacle, we have $g(\hat{\boldsymbol{P}}) = 0$ and $g'(\hat{\boldsymbol{P}}) = \boldsymbol{0}$, which validates the conditions (2.4) and (2.6).

Substituting Eq. (3.31) and $\hat{\boldsymbol{P}} = \boldsymbol{P} - \boldsymbol{P}_{cs}$ into Eq. (3.25) leads to the optimal control law (3.15). Note that the terms containing the final consensus state $\boldsymbol{P}_{cs}$ become zero due to Eq. (3.9). Thus, the control law (3.15) is only dependent on $\boldsymbol{P}$ without $\boldsymbol{P}_{cs}$. This is desired because $\boldsymbol{P}_{cs}$ is not known a priori.

Now, all the conditions (2.4)–(2.9) can be satisfied. Therefore, the control law (3.15) is an optimal control law for the problem (3.11) in the sense of Eqs. (2.11) and (2.12) according to Lemma 2.3 in Chapter 2, and the closed-loop system is asymptotically stable. It implies $\boldsymbol{P} = \boldsymbol{P}_{cs}$ and the consensus is achieved.

In addition, it can be easily seen from Eq. (3.22) that $V(\hat{\boldsymbol{P}}) \to \infty$ as $\left\|\hat{\boldsymbol{P}}\right\| \to \infty$. Thus, the closed-loop system is globally asymptotically stable. Note that the region of global asymptotic stability excludes the collision region Λ_j since no agent can start from inside the obstacle. □

Remark 3.2. From the proof of Theorem 3.1, one can note that the optimal consensus algorithm is developed from an inverse optimal control approach since the cost function $h(\hat{\boldsymbol{P}})$ in $\mathcal{J}_2$ is not specified a priori but constructed from the optimality condition (3.29). The obstacle avoidance through $h(\hat{\boldsymbol{P}})$ can be seen from Eqs. (3.16), (3.18), and (3.19): if the agent is outside the detection region, it implies $h(\hat{\boldsymbol{P}}) = 0$ and it becomes a regular optimal consensus problem without obstacle, i.e. $\mathcal{J}_2 = 0$; if the agent is inside the reaction region and approaches the obstacle, the denominator $\left\|\boldsymbol{p}_i - \boldsymbol{O}_{bj}\right\|^2 - r_j^2$ in $h(\hat{\boldsymbol{P}})$ (see $m(\boldsymbol{p}_i)$ in (Eq. (3.18))) will approach zero, so $h(\hat{\boldsymbol{P}})$ and $\mathcal{J}_2$ will increase. Since the optimal control always minimizes the cost function including $\mathcal{J}_2 = \int_0^\infty h(\hat{\boldsymbol{P}})dt$, and minimizing $\mathcal{J}_2 = \int_0^\infty h(\hat{\boldsymbol{P}})dt$ is equivalent to increasing the denominator $\left\|\boldsymbol{p}_i - \boldsymbol{O}_{bj}\right\|^2 - r_j^2$ in $h(\hat{\boldsymbol{P}})$, it implies that the agent is driven away from the obstacle. Therefore, if the optimality and asymptotic stability can be proven in Theorem 3.1, the obstacle avoidance is guaranteed.

Remark 3.3. To recapitulate the design of the optimal consensus algorithm, w_p and w_c are adjustable weighting parameters to penalize the consensus error and the control effort, respectively. This is also used to ensure the condition of $h(\hat{\boldsymbol{P}}) \geq 0$. Tuning these parameters is analogous to tuning the weighting matrices Q and R in the conventional LQR problem:

$$\int_0^\infty \left[\hat{\boldsymbol{X}}^T Q\hat{\boldsymbol{X}} + \boldsymbol{U}^T R\boldsymbol{U}\right]dt \tag{3.35}$$

It is not difficult to tune w_p and w_c since we are dealing with a linear single-integrator system. On the other hand, this is not a conventional linear optimal control problem like LQR because the obstacle avoidance cost function is a nonquadratic nonlinear function. However, the only constraint for w_p and w_c is to satisfy the condition $h(\hat{\boldsymbol{P}}) \geq 0$ in order the obstacle avoidance cost function to be meaningful. Since there are only two terms in $h(\hat{\boldsymbol{P}})$, i.e. $h(\hat{\boldsymbol{P}}) = \frac{w_p}{w_c} g'^T(\hat{\boldsymbol{P}})(\mathcal{L} \otimes I_m)\hat{\boldsymbol{P}} + \frac{1}{4{w_c}^2} g'^T(\hat{\boldsymbol{P}})g'(\hat{\boldsymbol{P}})$, the basic selection rule for these two weighting parameters other than balancing the consensus error and the control effort is to choose a small enough w_p for a given w_c such that the positive term $\frac{1}{4{w_c}^2} g'^T(\hat{\boldsymbol{P}})g'(\hat{\boldsymbol{P}})$ be always greater than the sign-indefinite term $\frac{w_p}{w_c} g'^T(\hat{\boldsymbol{P}})(\mathcal{L} \otimes I_m)\hat{\boldsymbol{P}}$ to meet the condition $h(\hat{\boldsymbol{P}}) \geq 0$.

Remark 3.4. It can be seen from $\phi(\boldsymbol{P})$ in Eq. (3.15) that the optimal control law contains two components, the consensus law and the obstacle avoidance law. The consensus law of each agent is a linear function of $(\mathcal{L} \otimes I_m)\boldsymbol{P}$ and thus only requires the local information from its neighbors based on the communication topology, rather than all agents' information. The obstacle avoidance law $g'(\boldsymbol{P})$ only needs the relative position information between the agent and the obstacle. Thus, to implement the proposed optimal control law only local information is required.

Remark 3.5. Note that Eq. (3.18) is the same as the potential function used in the reference [166]. Nevertheless, the actual avoidance function in this section is the cost function $h(\hat{\boldsymbol{P}})$ in Eq. (3.16), which is different from the one in [166]. This section addresses the consensus problem with obstacle avoidance and each agent does not know their final consensus point a priori, whereas the reference [166] addresses the problem when multiple agents approach the predefined points with collision avoidance and each agent is assumed to know its desired position a priori. The more important difference is that the communication topology is considered in the current formulation and consensus law, which is not the case in the reference [166]. Moreover, a consistent optimality can be achieved by this inverse optimal control approach, whereas the optimality is not guaranteed in the reference [166] when the agent steps into the detection region.

Remark 3.6. The reference [15] also addressed the consensus problem from the optimal control perspective using the LQR approach. However, the proposed algorithm in this section is significantly different from the one in the reference in the following aspects:

1) From the perspective of the problem, this section integrates the obstacle avoidance with the consensus algorithm in one optimal control framework, whereas the reference [15] does not consider obstacle avoidance. The integrated optimal control problem is more challenging because obstacle avoidance cannot be formulated by a quadratic type of cost function and thus the regular LQR method used in the reference cannot solve this problem.

2) From the perspective of the control theory, the inverse optimal control theory is adopted to tackle the nonquadratic obstacle avoidance cost function such that an analytical solution can be obtained. This is an important extension of the conventional LQR approach to solve the optimal consensus problem.

3) From the perspective of design, a weighted consensus cost is used in this section with adjustable parameters to bring more flexibility for tuning system performance and feasibility of the stability conditions.

3.2.2 Numerical examples

In this section, several simulation scenarios are provided to validate the proposed optimal obstacle avoidance consensus law. Without loss of generality, consider the scenario in Fig. 3.1 with $n = 4$ agents (single-integrator dynamics) in planar motion ($m = 2$). By the definition of the Laplacian matrix, $\mathcal{L}$ is given by

$$\mathcal{L} = \begin{bmatrix} 2 & -1 & 0 & -1 \\ -1 & 2 & -1 & 0 \\ 0 & -1 & 2 & -1 \\ -1 & 0 & -1 & 2 \end{bmatrix} \tag{3.36}$$

The initial positions of four agents are arbitrarily given by $(-7.5, -5)$ m, $(2.5, -7.5)$ m, $(10, 10)$ m and $(-3, 4)$ m, respectively. The weights in the optimal control formulation are set to $w_p = 0.8$ and $w_c = 4$.

3.2.2.1 *Consensus without obstacles on the trajectories of agents*

In this scenario, an obstacle is assumed to appear on (10, 0) m, which is not on the trajectory of any agent. The radii of the obstacle and detection region are assumed to be $r_1 = 0.5$ m and $R_{d_1} = 2.5$ m, respectively. The simulation results of the four agents' motion under the proposed optimal consensus law are shown in Figs. 3.2–3.4. Fig. 3.2

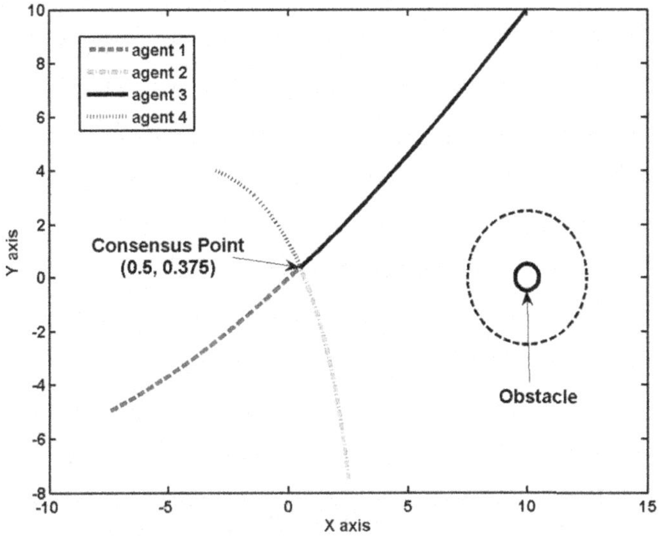

Figure 3.2 Consensus trajectories of four agents without obstacles on the trajectories of agents.

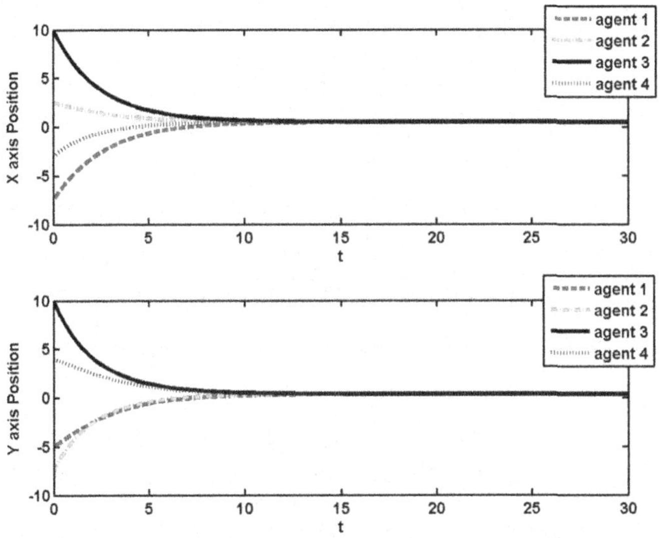

Figure 3.3 Time histories of four agents' positions.

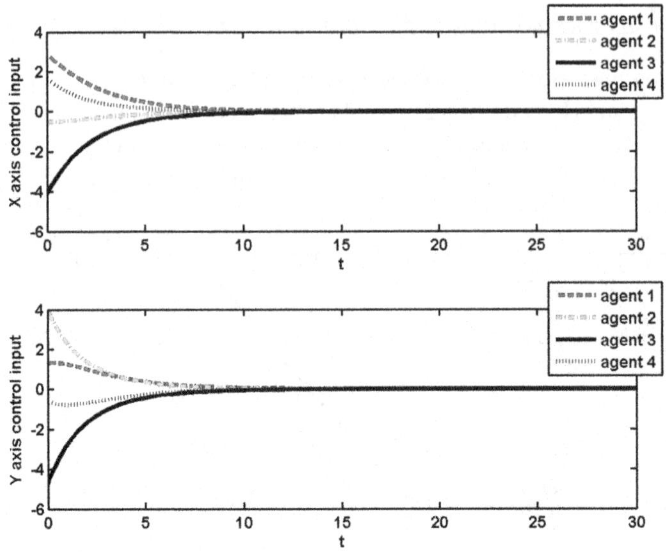

Figure 3.4 Time histories of four agents' control inputs.

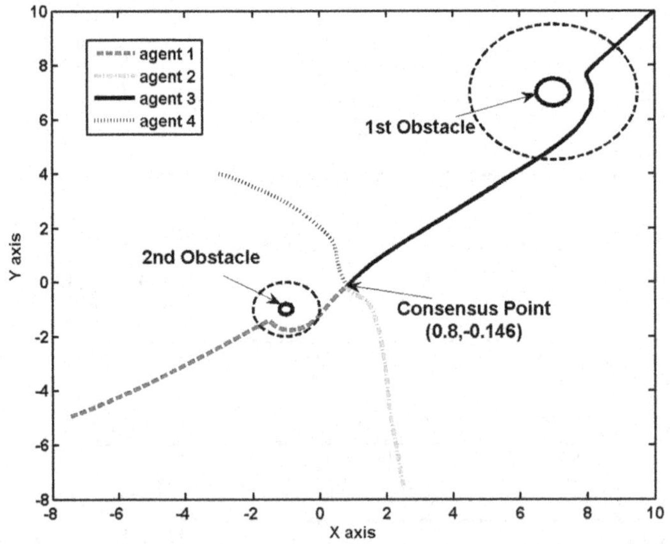

Figure 3.5 Consensus trajectories of four agents with multiple obstacles on the trajectories of agent 1 and 3.

gives the trajectories; Figs. 3.3 and 3.4 show the time histories of positions and optimal control inputs, respectively. The final consensus point is located at (0.5, 0.375) m.

Note that $h(\hat{\boldsymbol{P}})$ in the obstacle avoidance cost function (3.13) is always equal to zero since all the agents are outside the detection region, which implies that the

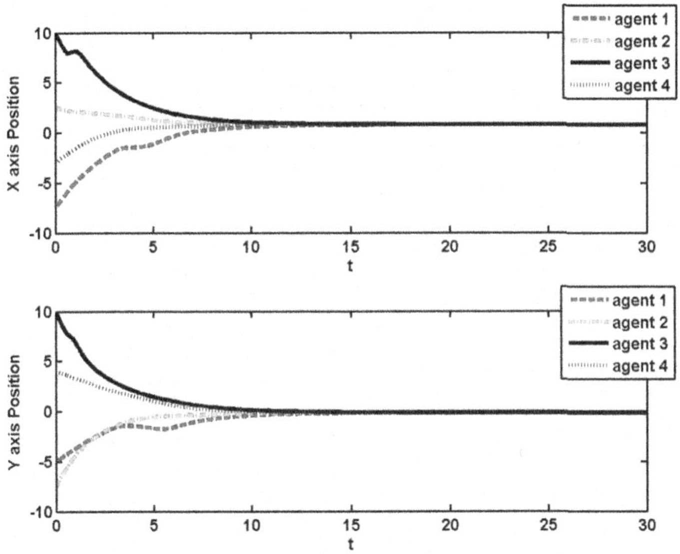

Figure 3.6 Time histories of four agents' positions.

problem is just a normal consensus problem. The optimal control law for the networked single-integrator system becomes $\boldsymbol{U}^* = -\frac{w_p}{w_c}(\mathcal{L} \otimes I_m)\boldsymbol{P}$, which has the same form as the conventional consensus algorithm for networked single-integrator systems [147] except the weighting parameters. Therefore, it shows that the conventional consensus algorithm in [147] is optimal in the sense of minimizing the cost functional $\mathcal{J}_1 + \mathcal{J}_3$.

3.2.2.2 Consensus with multiple obstacles on the trajectories of agents

In this scenario, two obstacles appearing at different locations and time are considered. The first obstacle is assumed to appear on the trajectory of agent 3 at $(7, 7)$ m with the radii of the obstacle and the detection region being $r_1 = 0.5$ m and $R_{d_1} = 2.5$ m. The other one is assumed to appear on the trajectory of agent 1 at $(-1, -1)$ m with the radii of the obstacle and the detection region being $r_2 = 0.2$ m and $R_{d_2} = 1$ m. The simulation results can be seen in Figs. 3.5–3.7.

Fig. 3.5 demonstrates that all agents are able to avoid the multiple obstacles on their trajectories and reach the final consensus point at (0.8, −0.144) m, which is different from the previous consensus point (0.5, 0.375) m. Figs. 3.6 and 3.7 present the time histories of the agents' positions and control inputs, respectively. It can be seen that the obstacle avoidance control occurs in the time interval of [0.2, 2] s and [2.5, 6] s. The proposed optimal control law is capable of achieving consensus as well as multiple obstacles avoidance.

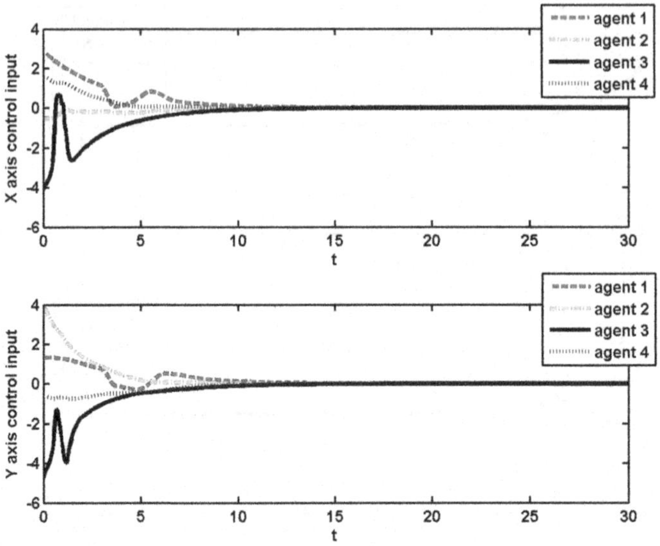

Figure 3.7 Time histories of four agents' control inputs.

3.3 Optimal consensus control with obstacle avoidance for double-integrator case

3.3.1 Optimal consensus algorithm: double-integrator case

In order to further study this case, multiple double-integrator systems given in Eq. (3.2) can be written in a matrix form:

$$\dot{X} = AX + BU \tag{3.37}$$

with

$$A = \begin{bmatrix} 0_{n\times n} & I_n \\ 0_{n\times n} & 0_{n\times n} \end{bmatrix} \otimes I_m,\ B = \begin{bmatrix} 0_{n\times n} \\ I_n \end{bmatrix} \otimes I_m$$

$$X = [\underbrace{\boldsymbol{p}_1{}^T, \ldots, \boldsymbol{p}_n{}^T}_{\boldsymbol{p}^T}, \underbrace{\boldsymbol{v}_1{}^T, \ldots, \boldsymbol{v}_n{}^T}_{\boldsymbol{v}^T}]^T, U = [\boldsymbol{u}_1{}^T, \ldots, \boldsymbol{u}_n{}^T]^T$$

$\boldsymbol{X} \in R^{2nm}$ and $\boldsymbol{U} \in R^{nm}$ are, respectively, the aggregate state and control input of all agents. $\otimes$ denotes the Kronecker product, which is used to extend the dimension. I_n and I_m denote the identity matrices with dimensions of n and m respectively, and $0_{n\times n}$ denotes the zero matrix.

For the convenience of formulation, we need to define the error state

$$\hat{\boldsymbol{X}} = [\ \hat{\boldsymbol{p}}^T \quad \hat{\boldsymbol{v}}^T\ ]^T \triangleq \boldsymbol{X} - \boldsymbol{X}_{cs} \tag{3.38}$$

where

$$\boldsymbol{X}_{cs} = \begin{bmatrix} \boldsymbol{p}_{cs}^T & \boldsymbol{v}_{cs}^T \end{bmatrix}^T \tag{3.39}$$

is the final consensus state. For instance, in a planar motion,

$$\boldsymbol{X}_{cs} = \begin{bmatrix} \boldsymbol{p}_{cs}^T & \boldsymbol{v}_{cs}^T \end{bmatrix}^T = \begin{bmatrix} \mathbf{1}_{1\times n} \otimes \begin{bmatrix} \alpha_x & \alpha_y \end{bmatrix} & \mathbf{1}_{1\times n} \otimes \begin{bmatrix} \beta_x & \beta_y \end{bmatrix} \end{bmatrix}^T \tag{3.40}$$

where α_x, α_y are the final consensus position along x axis and y axis, respectively; β_x, β_y are the final consensus velocity along x axis and y axis, respectively. Note that the consensus state $\boldsymbol{X}_{cs}$ is not known a priori.

According to the property of the Laplacian $\mathcal{L}$ in Eq. (3.8), if the undirected graph is connected, when the agents reach consensus,

$$\begin{aligned} (\mathcal{L} \otimes I_m)\, \boldsymbol{p}_{cs} &= \mathbf{0}_{nm\times 1} \\ (\mathcal{L} \otimes I_m)\, \boldsymbol{v}_{cs} &= \mathbf{0}_{nm\times 1} \end{aligned} \tag{3.41}$$

The final consensus state satisfies the dynamic equation

$$\dot{\boldsymbol{X}}_{cs} = A\boldsymbol{X}_{cs} + B\boldsymbol{U}_{cs} = A\boldsymbol{X}_{cs} \tag{3.42}$$

since $\boldsymbol{U}_{cs} = \mathbf{0}_{nm\times 1}$ when the agents reach consensus.

Then, from Eqs. (3.37) and (3.42) the error dynamics becomes

$$\dot{\hat{\boldsymbol{X}}} = A\hat{\boldsymbol{X}} + B\boldsymbol{U} \tag{3.43}$$

The consensus is achieved when the system (3.43) is asymptotically stable.

In this section, the consensus problem is formulated as an optimal control problem including three cost function components:

$$\begin{aligned} &Min: \mathcal{J} = \mathcal{J}_1 + \mathcal{J}_2 + \mathcal{J}_3 \\ &S.t. \quad \dot{\hat{\boldsymbol{X}}} = A\hat{\boldsymbol{X}} + B\boldsymbol{U} \end{aligned} \tag{3.44}$$

where $\mathcal{J}_1, \mathcal{J}_2, \mathcal{J}_3$ represent consensus cost, obstacle avoidance cost, and control effort cost, respectively.

The consensus cost has the form of

$$\mathcal{J}_1 = \int_0^\infty \hat{\boldsymbol{X}}^T R_1 \hat{\boldsymbol{X}} dt = \int_0^\infty \left\{ \hat{\boldsymbol{X}}^T \left(\begin{bmatrix} w_p^2 \mathcal{L}^2 & 0_{n\times n} \\ 0_{n\times n} & w_v^2 \mathcal{L}^2 - 2 w_p w_c \mathcal{L} \end{bmatrix} \otimes I_m \right) \hat{\boldsymbol{X}} \right\} dt \tag{3.45}$$

where $\mathcal{L}$ is the symmetric Laplacian matrix established by the undirected and connected graph. w_p, w_v, and w_c represent the weights on position consensus, velocity consensus, and control effort, respectively. It is necessary to show that R_1 is positive semi-definite. Proposition 2.1 and the following proposition are needed.

Proposition 3.1. *${w_v}^2\mathcal{L}^2 - 2w_pw_c\mathcal{L}$ is positive semi-definite and $\left({w_v}^2\mathcal{L}^2 - 2w_pw_c\mathcal{L}\right)\cdot \mathbf{1}_{n\times 1} = \mathbf{0}_{n\times 1}$ if the graph is undirected and connected and*

$$ {w_v}^2 e_i^2 - 2w_pw_c e_i \geq 0 \tag{3.46}$$

where e_i is the eigenvalue of $\mathcal{L}$.

Proof. From Proposition 2.1,

$$\begin{aligned}
&{w_v}^2\mathcal{L}^2 - 2w_pw_c\mathcal{L} \\
&= {w_v}^2 Q\Lambda^2 Q^{-1} - 2w_pw_c Q\Lambda Q^{-1} = Q\left({w_v}^2\Lambda^2 - 2w_pw_c\Lambda\right)Q^{-1} \\
&= Q\begin{bmatrix} {w_v}^2e_1^2 - 2w_pw_ce_1 & 0 & \cdots & 0 \\ 0 & {w_v}^2e_2^2 - 2w_pw_ce_2 & \cdots & 0 \\ \vdots & \vdots & \ddots & \vdots \\ 0 & 0 & \cdots & {w_v}^2e_n^2 - 2w_pw_ce_n \end{bmatrix} Q^{-1}
\end{aligned} \tag{3.47}$$

where e_i is the eigenvalue of $\mathcal{L}$, $i = 1, ..., n$. Since $\mathcal{L}$ is a symmetric Laplacian matrix by Assumption 3.3, $e_i \geq 0$. Therefore, ${w_v}^2\mathcal{L}^2 - 2w_pw_c\mathcal{L}$ is positive semi-definite if ${w_v}^2e_i^2 - 2w_pw_ce_i \geq 0$.

It can also be seen from Eqs. (2.16) and (3.8) that

$$\left(w_v^2\mathcal{L}^2 - 2w_pw_c\mathcal{L}\right)\cdot \mathbf{1}_{n\times 1} = w_v^2\mathcal{L}^2 \cdot \mathbf{1}_{n\times 1} - 2w_pw_c\mathcal{L}\cdot \mathbf{1}_{n\times 1} = \mathbf{0}_{n\times 1} \tag{3.48}$$

□

Remark 3.7. One can always find proper weights to satisfy Eq. (3.47). For instance, a large w_v and small enough w_p and w_c are applicable due to $e_i \geq 0$. It is obvious that R_1 in the consensus cost (3.45) is positive semi-definite because the diagonal elements of R_1 are both positive semi-definite.

Remark 3.8. The term $-2w_pw_c\mathcal{L}$ in the consensus cost (3.45) is used to guarantee that the analytical solution of the Riccati equation for the optimal control law is a linear function of the Laplacian matrix $\mathcal{L}$, and is thus completely dependent on the information topology, which will be shown in the proof of Theorem 3.2.

Remark 3.9. It is worth noting that the position consensus cost in Eq. (3.45) corresponding to the weight of ${w_p}^2\mathcal{L}^2$ is defined in the same way as that in [15]. However, the LQR based consensus algorithm in [15] only applies to the single-integrator kinematics. Our method not only extends it to the double-integrator dynamics, but also addresses the obstacle avoidance.

The obstacle avoidance cost has the form of

$$\mathcal{J}_2 = \int_0^\infty h\left(\hat{X}\right)dt \tag{3.49}$$

where $h(\hat{X})$ will be constructed from the inverse optimal control approach in Theorem 3.2.

The control effort cost has the regular quadratic form of

$$\mathcal{J}_3 = \int_0^{\infty} U^T R_2 U dt \tag{3.50}$$

where $R_2 = {w_c}^2 I_n \otimes I_m$ is positive definite and w_c is the weighting parameter.

Lemma 2.3 is adopted again to derive the second result and is used to prove both asymptotic stability and optimality of the proposed obstacle avoidance consensus algorithm.

Theorem 3.2. *For a multi-agent system (3.2) with Assumptions 3.1–3.3, there always exist a large enough w_v, small enough w_p and w_c, such that the feedback control law*

$$U = \phi(X) = -\frac{w_p}{w_c}(\mathcal{L} \otimes I_m)\,p - \frac{w_v}{w_c}(\mathcal{L} \otimes I_m)\,v - \frac{1}{2w_c^2} g_v'(X) \tag{3.51}$$

is an optimal control law for the consensus problem (3.44) with

$$\begin{aligned} h(\hat{X}) &= \tfrac{w_p}{w_c}\hat{v}^T(\mathcal{L} \otimes I_m)(G_p \otimes I_m)(\mathcal{L} \otimes I_m)\hat{p} \\ &+ \tfrac{w_v}{w_c}\hat{v}^T(\mathcal{L} \otimes I_m)(G_p \otimes I_m)(\mathcal{L} \otimes I_m)\hat{v} \\ &- {g_p}'^T(\hat{X})\hat{v} + \tfrac{1}{4{w_c}^2}\hat{v}^T(\mathcal{L} \otimes I_m)(G_p^2 \otimes I_m)(\mathcal{L} \otimes I_m)\hat{v} \end{aligned} \tag{3.52}$$

in Eq. (3.49), where ${g_v}'(X)$ and ${g_p}'^T(\hat{X})$ in Eqs. (3.51) and (3.52) are derived from the obstacle avoidance potential function defined by

$$g\left(\hat{X}\right) = \frac{1}{2}\hat{v}^T\left(G_p \otimes I_m\right)(\mathcal{L} \otimes I_m)\,\hat{v} \tag{3.53}$$

with $G_p = \operatorname{diag}\left(m\left(p_1\right) \quad m\left(p_2\right) \cdots m\left(p_n\right)\right)$

$$m(p_i) = \begin{cases} 0 & p_i \in \Pi \\ \left(\dfrac{R_{d_j}^2 - \|p_i - O_{bj}\|^2}{\|p_i - O_{bj}\|^2 - r_j^2}\right)^2 & p_i \in \Gamma_j \qquad i = 1, \dots, n \\ \text{not defined} & p_i \in \Lambda_j \end{cases} \tag{3.54}$$

and

$$\begin{aligned} g'(\hat{X}) &= \left[{g_p}'^T(\hat{X}) \quad {g_v}'^T(\hat{X})\right]^T \\ {g_p}'(\hat{X}) &= \left[\left(\tfrac{\partial g(\hat{X})}{\partial \hat{p}_1}\right)^T \quad \left(\tfrac{\partial g(\hat{X})}{\partial \hat{p}_2}\right)^T \quad \cdots \quad \left(\tfrac{\partial g(\hat{X})}{\partial \hat{p}_n}\right)^T\right]^T \\ {g_v}'(\hat{X}) &= (G_p \otimes I_m)(\mathcal{L} \otimes I_m)\hat{v} = (G_p \otimes I_m)(\mathcal{L} \otimes I_m)v = {g_v}'(X) \end{aligned} \tag{3.55}$$

where $g_p'(\hat{X})$ and $g_v'(X)$ represent the partial differentiation of $g(\hat{X})$ with respect to the position error $\hat{\mathbf{p}}$ and the velocity error $\hat{\mathbf{v}}$ respectively. In addition, the closed-loop system (3.44) is globally asymptotically stable.

Proof. Specific to this optimal consensus problem, we will have the following equations corresponding to Lemma 2.3 in Chapter 2:

$$T\left(\hat{X}, U\right) = \hat{X}^T R \hat{X} + h\left(\hat{X}\right) + U^T R_2 U \tag{3.56}$$

$$f\left(\hat{X}, U\right) = A\hat{X} + BU \tag{3.57}$$

A candidate Lyapunov function $V\left(\hat{X}\right)$ is chosen to be

$$V\left(\hat{X}\right) = \hat{X}^T \mathcal{P} \hat{X} + g\left(\hat{X}\right) \tag{3.58}$$

where $\mathcal{P}$ is the solution of a Riccati equation, which will be shown afterwards.

In order for the function $V(\hat{X})$ in Eq. (3.58) to be a valid Lyapunov function, it must be continuously differentiable with respect to $\hat{X}$ or equivalently $g(\hat{X})$ must be continuously differentiable with respect to $\hat{X}$. From Eqs. (3.53) and (3.54), it suffices to show that $m(\mathbf{p}_i)$ is continuously differentiable in the safety region Θ. In fact, this is true if $m(\mathbf{p}_i)$ and $\frac{dm(\mathbf{p}_i)}{d\mathbf{p}_i}$ are continuous at the boundary of Ψ_j. Since Eq. (3.54) implies that $\lim\limits_{\|\mathbf{p}_i - \mathbf{O}_{bj}\| \to R_{d_j}^-} m(\mathbf{p}_i) = 0 = \lim\limits_{\|\mathbf{p}_i - \mathbf{O}_{bj}\| \to R_{d_j}^+} m(\mathbf{p}_i)$, $m(\mathbf{p}_i)$ is continuous at the boundary of Ψ_j by the definition of continuity and thus continuous over the safety region Θ. In addition,

$$\frac{dm(\mathbf{p}_i)}{d\mathbf{p}_i} = \begin{cases} \mathbf{0} & \mathbf{p}_i \in \Pi \\ \dfrac{-4(R_{d_j}^2 - r_j^2)(R_{d_j}^2 - \|\mathbf{p}_i - \mathbf{O}_{bj}\|^2)}{\left(\|\mathbf{p}_i - \mathbf{O}_{bj}\|^2 - r_j^2\right)^3}(\mathbf{p}_i - \mathbf{O}_{bj})^T & \mathbf{p}_i \in \Gamma_j \\ \text{not defined} & \mathbf{p}_i \in \Lambda_j \end{cases} \tag{3.59}$$

It is easy to see that $\lim\limits_{\|\mathbf{p}_i - \mathbf{O}_{bj}\| \to R_{d_j}^-} \frac{dm(\mathbf{p}_i)}{d\mathbf{p}_i} = 0_{m\times 1} = \lim\limits_{\|\mathbf{p}_i - \mathbf{O}_{bj}\| \to R_{d_j}^+} \frac{dm(\mathbf{p}_i)}{d\mathbf{p}_i}$. Hence, $\frac{dm(\mathbf{p}_i)}{d\mathbf{p}_i}$ is continuous at the boundary of Ψ_j and thus continuous over the safety region Θ. Therefore, $g(\hat{X})$ and the Lyapunov function $V(\hat{X})$ are continuously differentiable with respect to $\hat{X}$ in the safety region Θ.

After showing $V(\hat{X})$ is continuously differentiable, the Hamiltonian for this optimal consensus problem can be written as

$$\begin{aligned} H(\hat{X}, U, V'^T(\hat{X})) &= T(\hat{X}, U) + V'^T(\hat{X}) f(\hat{X}, U) \\ &= \hat{X}^T R_1 \hat{X} + h(\hat{X}) + U^T R_2 U + [2\hat{X}^T \mathcal{P} + g'^T(\hat{X})][A\hat{X} + BU] \end{aligned} \tag{3.60}$$

Setting $(\partial/\partial U)\,H\left(\hat{X}, U, V'^T\left(\hat{X}\right)\right) = 0$ yields the optimal control law

$$U^* = \phi\left(\hat{X}\right) = -\frac{1}{2}R_2^{-1}B^T V'\left(\hat{X}\right) = -R_2^{-1}B^T\mathcal{P}\hat{X} - \frac{1}{2}R_2^{-1}B^T g'\left(\hat{X}\right) \quad (3.61)$$

With Eq. (3.61), it follows that

$$\begin{aligned} &V'^T(\hat{X})f(\hat{X},\phi(\hat{X})) \\ &= \hat{X}^T(A^T\mathcal{P} + \mathcal{P}A - 2\mathcal{P}S\mathcal{P})\hat{X} - \hat{X}^T\mathcal{P}Sg'(\hat{X}) \\ &\quad + g'^T(\hat{X})(A - S\mathcal{P})\hat{X} - \frac{1}{2}g'^T(\hat{X})Sg'(\hat{X}) \end{aligned} \quad (3.62)$$

where $S \triangleq B(R_2)^{-1}B^T$. Substituting Eqs. (3.61) and (3.62) into Eq. (3.60) yields

$$\begin{aligned} &H(\hat{X},\phi(\hat{X}), V'^T(\hat{X})) \\ &= \hat{X}^T(A^T\mathcal{P} + \mathcal{P}A + R_1 - \mathcal{P}S\mathcal{P})\hat{X} + g'^T(\hat{X})(A - S\mathcal{P})\hat{X} + h(\hat{X}) \\ &\quad - \frac{1}{4}g'^T(\hat{X})Sg'(\hat{X}) \end{aligned} \quad (3.63)$$

In order to prove that the control law (3.61) is an optimal solution for the consensus problem (3.44) using the Lemma 2.3 in Chapter 2, the conditions (2.4)–(2.9) need to be verified.

Since $B = \begin{bmatrix} 0_{n\times n} \\ I_n \end{bmatrix} \otimes I_m$, it can be seen that $\phi(\hat{X}) = -R_2^{-1}B^T\mathcal{P}\hat{X} - \frac{1}{2}R_2^{-1}g_v{}'(\hat{X})$. Therefore, from the form of $g_v{}'(\hat{X})$ in Eq. (3.55), the condition (2.6), i.e. $\phi(\mathbf{0}) = \mathbf{0}$, is satisfied.

In order to satisfy the condition (2.8) in Lemma 2.3 in Chapter 2 or let Eq. (3.63) be zero, we can let

$$A^T\mathcal{P} + \mathcal{P}A + R_1 - \mathcal{P}S\mathcal{P} = 0 \quad (3.64)$$

and require that

$$g'^T\left(\hat{X}\right)(A - S\mathcal{P})\hat{X} + h\left(\hat{X}\right) - \frac{1}{4}g'^T\left(\hat{X}\right)Sg'\left(\hat{X}\right) = 0 \quad (3.65)$$

With Eqs. (3.61), (3.64), and (3.65), it can be shown that

$$
\begin{aligned}
& H\left(\hat{X}, U, V'^T\left(\hat{X}\right)\right) \\
&= U^T R_2 U + h\left(\hat{X}\right) + \hat{X}^T R_1 \hat{X} + \left(2\hat{X}^T \mathcal{P} + g'^T\left(\hat{X}\right)\right)\left(A\hat{X} + BU\right) \\
&= U^T R_2 U + h\left(\hat{X}\right) + \hat{X}^T R_1 \hat{X} + \left(2\hat{X}^T \mathcal{P} + g'^T\left(\hat{X}\right)\right)\left(A\hat{X} + BU\right) \\
&\quad - \hat{X}^T\left(A^T \mathcal{P} + \mathcal{P}A + R_1 - \mathcal{P}S\mathcal{P}\right)\hat{X} \\
&= U^T R_2 U + h\left(\hat{X}\right) + g'^T\left(\hat{X}\right)\left(A\hat{X} + BU\right) + 2\hat{X}^T \mathcal{P}BU + \hat{X}^T \mathcal{P}S\mathcal{P}\hat{X} \\
&= U^T R_2 U + \frac{1}{4} g'^T\left(\hat{X}\right) S g'\left(\hat{X}\right) + g'^T\left(\hat{X}\right) S\mathcal{P}\hat{X} + \hat{X}^T \mathcal{P}S\mathcal{P}\hat{X} \\
&+ \left(2\hat{X}^T \mathcal{P} + g'^T\left(\hat{X}\right)\right) BU \\
&= U^T R_2 U + \frac{1}{4}\left(2\hat{X}^T \mathcal{P} + g'^T\left(\hat{X}\right)\right) S \left(2\hat{X}^T \mathcal{P} + g'^T\left(\hat{X}\right)\right)^T \\
&+ \left(2\hat{X}^T \mathcal{P} + g'^T\left(\hat{X}\right)\right) BU \\
&= U^T R_2 U + \frac{1}{4} V'^T\left(\hat{X}\right) S V'\left(\hat{X}\right) + U^T B^T V'\left(\hat{X}\right) \\
&= U^T R_2 U + \phi\left(\hat{X}\right)^T R_2 \phi\left(\hat{X}\right) - 2U^T R_2 \phi\left(\hat{X}\right) \\
&= \left[U - \phi\left(\hat{X}\right)\right]^T R_2 \left[U - \phi\left(\hat{X}\right)\right] \geq 0
\end{aligned}
\tag{3.66}
$$

Therefore, the condition (2.9) is satisfied.

Next, substituting A, B, R_1, R_2 in (3.64) and assuming $\mathcal{P} = \begin{bmatrix} P_1 & P_2 \\ P_2 & P_3 \end{bmatrix} \otimes I_m$ yields

$$
\begin{bmatrix} -\frac{1}{w_c^2} P_2^2 & P_1 - \frac{1}{w_c^2} P_2 P_3 \\ P_1 - \frac{1}{w_c^2} P_3 P_2 & 2P_2 - \frac{1}{w_c^2} P_3^2 \end{bmatrix} + \begin{bmatrix} w_p{}^2 \mathcal{L}^2 & 0_{n\times n} \\ 0_{n\times n} & w_v{}^2 \mathcal{L}^2 - 2w_p w_c \mathcal{L} \end{bmatrix} = 0 \tag{3.67}
$$

Then, $\mathcal{P}$ can be solved in the analytical form

$$
\mathcal{P} = \begin{bmatrix} w_p w_v \mathcal{L}^2 & w_p w_c \mathcal{L} \\ w_p w_c \mathcal{L} & w_c w_v \mathcal{L} \end{bmatrix} \otimes I_m \tag{3.68}
$$

Note that the motivation of introducing the term $-2w_p w_c \mathcal{L}$ in R_1 (see Eq. (3.45)) is to let it cancel $2P_2$ when solving P_3 from Eq. (3.67) in order to make P_3 a linear function of the Laplacian matrix $\mathcal{L}$ as mentioned in Remark 3.8.

Next, the cost function term $h(\hat{X})$ in $\mathcal{J}_2$ is constructed from solving Eq. (3.65) and using Eq. (3.68):

$$\begin{aligned}
h(\hat{X}) = & \frac{w_p}{w_c}\hat{v}^T(\mathcal{L}\otimes I_m)(G_p\otimes I_m)(\mathcal{L}\otimes I_m)\hat{p} \\
& + \frac{w_v}{w_c}\hat{v}^T(\mathcal{L}\otimes I_m)(G_p\otimes I_m)(\mathcal{L}\otimes I_m)\hat{v} - {g_p}'^T(\hat{X})\hat{v} \\
& + \frac{1}{4{w_c}^2}\hat{v}^T(\mathcal{L}\otimes I_m)(G_p^2\otimes I_m)(\mathcal{L}\otimes I_m)\hat{v}
\end{aligned} \tag{3.69}$$

which turns out to be Eq. (3.52).

Using Eqs. (3.64) and (3.65), Eq. (3.62) becomes

$$V'^T(\hat{X})f(\hat{X},\phi(\hat{X})) = -[\hat{X}^T R_1\hat{X} + h(\hat{X}) + (\hat{X}^T\mathcal{P} + \frac{1}{2}g'^T(\hat{X}))S(\mathcal{P}\hat{X} + \frac{1}{2}g'(\hat{X}))] \tag{3.70}$$

It can be seen from Eq. (3.70) that the condition (2.7) is guaranteed when $h(\hat{X}) \geq 0$ since $\hat{X}^T R_1\hat{X}$ is positive semi-definite and $(\hat{X}^T\mathcal{P} + \frac{1}{2}g'^T(\hat{X}))S(\mathcal{P}\hat{X} + \frac{1}{2}g'(\hat{X}))$ is positive definite. By selecting proper values of the weights w_p, w_v, and w_c, one can always make $h(\hat{X}) \geq 0$. Specifically, if all the agents are outside the detection region, $h(\hat{X}) = 0$ by the definition of G_p in (3.54). $h(\hat{X}) > 0$ can be guaranteed if one can choose a large enough w_v, small enough w_p and w_c such that the positive terms $\hat{v}^T(\mathcal{L}\otimes I_m)(G_p\otimes I_m)(\mathcal{L}\otimes I_m)\hat{v}$ and $\hat{v}^T(\mathcal{L}\otimes I_m)(G_p^2\otimes I_m)(\mathcal{L}\otimes I_m)\hat{v}$ in Eq. (3.69) are always greater than the other sign-indefinite terms. Note that G_p and ${g_p}'^T(\hat{X})$ are both finite in the safety region.

Next we will verify the conditions (2.4) and (2.5). Note that $\hat{X}^T\mathcal{P}\hat{X}$ can be written as

$$\begin{aligned}
\hat{X}^T\mathcal{P}\hat{X} &= \hat{X}^T\left(\begin{bmatrix} w_p w_v\mathcal{L}^2 & w_p w_c\mathcal{L} \\ w_p w_c\mathcal{L} & w_c w_v\mathcal{L}\end{bmatrix}\otimes I_m\right)\hat{X} \\
&= w_p w_v\hat{p}^T(\mathcal{L}^2\otimes I_m)\hat{p} + w_c w_v\hat{v}^T(\mathcal{L}\otimes I_m)\hat{v} + 2w_p w_c\hat{p}^T(\mathcal{L}\otimes I_m)\hat{v} \\
&= w_p w_v p^T(\mathcal{L}^2\otimes I_m)p + w_c w_v v^T(\mathcal{L}\otimes I_m)v + 2w_p w_c p^T(\mathcal{L}\otimes I_m)v
\end{aligned} \tag{3.71}$$

The Lyapunov function finally turns out to be

$$\begin{aligned}
V(\hat{X}) &= \hat{X}^T\mathcal{P}\hat{X} + g(\hat{X}) \\
&= \begin{cases}
w_p w_v p^T(\mathcal{L}^2\otimes I_m)p + w_c w_v v^T(\mathcal{L}\otimes I_m)v + 2w_p w_c p^T(\mathcal{L}\otimes I_m)v & p_i \in \Pi \\
w_p w_v p^T(\mathcal{L}^2\otimes I_m)p + w_c w_v v^T(\mathcal{L}\otimes I_m)v + 2w_p w_c p^T(\mathcal{L}\otimes I_m)v & \\
\qquad + \frac{1}{2}\hat{v}^T(G_p\otimes I_m)(\mathcal{L}\otimes I_m)\hat{v} & p_i \in \Gamma_j \\
\text{not defined} & p_i \in \Lambda_j
\end{cases}
\end{aligned} \tag{3.72}$$

It can be seen from Eqs. (3.71) and (3.72) that $V(\hat{X}) = 0$ when $\hat{X} = \mathbf{0}$ such that the condition (2.4) is satisfied. Moreover, if $\hat{X} \neq \mathbf{0}$, which indicates $X \neq X_{cs}$ or $p \neq p_{cs}$

and $\boldsymbol{v} \neq \boldsymbol{v}_{cs}$, $\boldsymbol{p}^T(\mathcal{L}^2 \otimes I_m)\boldsymbol{p}$ and $\boldsymbol{v}^T(\mathcal{L} \otimes I_m)\boldsymbol{v}$ will not be equal to zero but positive according to the property of the Laplacian matrix, i.e., Eq. (3.41). Note that $\boldsymbol{p} = \mathbf{0}$ and $\boldsymbol{v} = \mathbf{0}$ that leads to $\boldsymbol{p}^T(\mathcal{L}^2 \otimes I_m)\boldsymbol{p} = 0$ and $\boldsymbol{v}^T(\mathcal{L} \otimes I_m)\boldsymbol{v} = 0$ is a special case of $\boldsymbol{p} = \boldsymbol{p}_{cs}$ and $\boldsymbol{v} = \boldsymbol{v}_{cs}$ when $\boldsymbol{p}_{cs} = \mathbf{0}$ and $\boldsymbol{v}_{cs} = \mathbf{0}$, which implies $\hat{\boldsymbol{X}} = \mathbf{0}$ as well. Hence, $\boldsymbol{p}^T(\mathcal{L}^2 \otimes I_m)\boldsymbol{p} > 0$ and $\boldsymbol{v}^T(\mathcal{L} \otimes I_m)\boldsymbol{v} > 0$ if $\hat{\boldsymbol{X}} \neq \mathbf{0}$. Therefore, $V(\hat{\boldsymbol{X}}) > 0$ when $\hat{\boldsymbol{X}} \neq \mathbf{0}$ can be guaranteed by selecting a large enough w_v for given w_p and w_c such that the positive terms $w_p w_v \boldsymbol{p}^T(\mathcal{L}^2 \otimes I_m)\boldsymbol{p}$ and $w_c w_v \boldsymbol{v}^T(\mathcal{L} \otimes I_m)\boldsymbol{v}$ are always greater than the sign-indefinite terms and thus the condition (2.5) is satisfied.

Substituting $\mathcal{P}$ and $g'(\hat{\boldsymbol{X}})$ into Eq. (3.61) leads to the optimal control law

$$\phi\left(\hat{\boldsymbol{X}}\right) = -\frac{w_p}{w_c}(\mathcal{L} \otimes I_m)\,\hat{\boldsymbol{p}} - \frac{w_v}{w_c}(\mathcal{L} \otimes I_m)\,\hat{\boldsymbol{v}} - \frac{1}{2w_c^2} g_{\boldsymbol{v}}'\left(\hat{\boldsymbol{X}}\right) \tag{3.73}$$

which turns out to be Eq. (3.51)

$$\boldsymbol{U} = \phi(\boldsymbol{X}) = -\frac{w_p}{w_c}(\mathcal{L} \otimes I_m)\,\boldsymbol{p} - \frac{w_v}{w_c}(\mathcal{L} \otimes I_m)\,\boldsymbol{v} - \frac{1}{2w_c^2} g_v'(\boldsymbol{X}) \tag{3.74}$$

by substituting $\hat{\boldsymbol{p}} = \boldsymbol{p} - \boldsymbol{p}_{cs}$ and $\hat{\boldsymbol{v}} = \boldsymbol{v} - \boldsymbol{v}_{cs}$ into Eq. (3.73) and noting that $g_v{}'(\hat{\boldsymbol{X}}) = g_v{}'(\boldsymbol{X})$ in Eq. (3.55).

Now, all the conditions (2.4)–(2.9) in Lemma 2.3 in Chapter 2 can be satisfied by selecting a large enough w_v and small enough w_p and w_c. Furthermore, this rule of weight selection also applies to satisfy the condition (3.46) in order for the cost functional (3.42) to be meaningful as discussed in Remark 3.7. Therefore, according to Lemma 2.3 in Chapter 2, the control law (3.51) is an optimal control law for the problem (3.44) in the sense of Eqs. (2.11) and (2.12), and the closed-loop system is asymptotically stable. It implies $\boldsymbol{X} = \boldsymbol{X}_{cs}$ and the consensus is achieved.

In addition, it can be easily seen from Eq. (3.58) that $V(\hat{\boldsymbol{X}}) \to \infty$ as $\left\|\hat{\boldsymbol{X}}\right\| \to \infty$. Therefore, the closed-loop system is globally asymptotically stable. Note that the region of the global asymptotic stability excludes the undefined area, which is physically meaningful because no agent can start from inside the obstacle. □

Remark 3.10. Note that the terms in Eq. (3.73) that contain the final consensus states $\boldsymbol{p}_{cs}$ and $\boldsymbol{v}_{cs}$ become zeros according to the property of the Laplacian matrix, i.e., Eq. (3.41). Then, the optimal control law (3.51) is only a function of $\boldsymbol{X}$ without $\boldsymbol{X}_{cs}$. This is desired because $\boldsymbol{X}_{cs}$ is not known a priori.

Remark 3.11. Similar to Remark 3.2, the optimal consensus control is developed from the inverse optimal control approach since the cost function $h(\hat{\boldsymbol{X}})$ is constructed from the optimality condition (3.65), which results in a nonquadratic cost function due to G_p in $h(\hat{\boldsymbol{X}})$. Note that $h(\hat{\boldsymbol{X}})$ is physically meaningful in the sense of obstacle avoidance. From Eqs. (3.52), (3.54), and (3.55) one can see the following: if the agents are outside the detection region, it implies $h(\hat{\boldsymbol{X}}) = 0$. Thus, the agents will keep approaching each other without the concern of obstacles, i.e. $\mathcal{J}_2 = 0$; if the agent is inside the detection region and approaches the obstacle, the denominator $(\left\|\boldsymbol{p}_i - \boldsymbol{O}_{bj}\right\|^2 - r_j^2)^2$ in $h(\hat{\boldsymbol{X}})$ (see G_p and $g(\boldsymbol{p}_i)$ in Eqs. (3.52) and (3.54)) will become smaller such that the penalty of $h(\hat{X})$ and $\mathcal{J}_2$ increase. Then the obstacle avoidance takes effect since the optimal control will drive the agent away from the obstacle to make $\mathcal{J}_2$ smaller.

Remark 3.12. From Eq. (3.51) and $g_v'(X)$ in Eq. (3.55), it can be also seen that the optimal control law of each agent only requires the information from its neighbors since it is a linear function of the Laplacian matrix $\mathcal{L}$. This feature offers a great implementation advantage.

3.3.2 Numerical examples

In this section, two simulation scenarios are used to validate the proposed optimal consensus algorithm given by Theorem 3.2. Consider the same scenario in Fig. 3.1 with $n = 4$ agents (double-integrator dynamics). Planar motion is considered in this section and thus $m = 2$. $\mathcal{L}$ is given the same as Eq. (3.36). The initial positions of four agents are given by $(-2, -2)$ m, $(2, -2)$ m, $(2, 2)$ m and $(-2, 2)$ m, respectively. The initial velocities of the four agents are assumed to be $(0.2, 0.4)$ m/s, $(-0.4, 0.2)$ m/s, $(-0.2, -0.4)$ m/s, and $(0.2, -0.2)$ m/s. The associated weights in the consensus algorithm are set to $w_p = 0.04$, $w_v = 1.2$, and $w_c = 0.8$ to meet the conditions on weights in Theorem 3.2 and Proposition 3.1.

3.3.2.1 *Consensus without obstacles on the trajectories of the agents*

In this scenario, an obstacle is assumed to appear on $(2, 0)$ m, which is not on the trajectory of any agent. The radii of the obstacle and the detection region are set to 0.1 m and 0.5 m, respectively. The simulation results of the four agents' motion under the proposed optimal consensus law are shown in Figs. 3.8–3.10. Fig. 3.8 shows that the four agents achieve the consensus. Fig. 3.9 presents the time histories of the four agents' positions and velocities. The optimal control inputs are shown in Fig. 3.10 and the two bottom subfigures of Fig. 3.10 show the zoom-in transient response for better illustration of control efforts. As can be seen, the obstacle avoidance does not take effect since no agent steps into the detection region.

Note that $h(\hat{X})$ in the obstacle avoidance cost function (3.49) is always equal to zero since all the agents are outside the detection region, which implies that the problem is just a normal optimal consensus problem. The optimal control law for the multiple double-integrator systems becomes $U^* = -\frac{w_p}{w_c}(\mathcal{L} \otimes I_m)\boldsymbol{p} - \frac{w_v}{w_c}(\mathcal{L} \otimes I_m)\boldsymbol{v}$, which has the same form as the conventional consensus algorithm for networked double-integrator systems [147] except the weighting parameters. Therefore, the conventional consensus algorithm is optimal in the sense of minimizing the cost functional $\mathcal{J}_1 + \mathcal{J}_3$.

3.3.2.2 *Consensus with multiple obstacles on the trajectories of the agents*

In this scenario, two obstacles are considered. The first obstacle appears on $(1, 1.3)$ m, which is on the trajectory of agent 3. The radius of this obstacle and the radius of the detection region are set as 0.1 m and 0.5 m, respectively. The other obstacle is

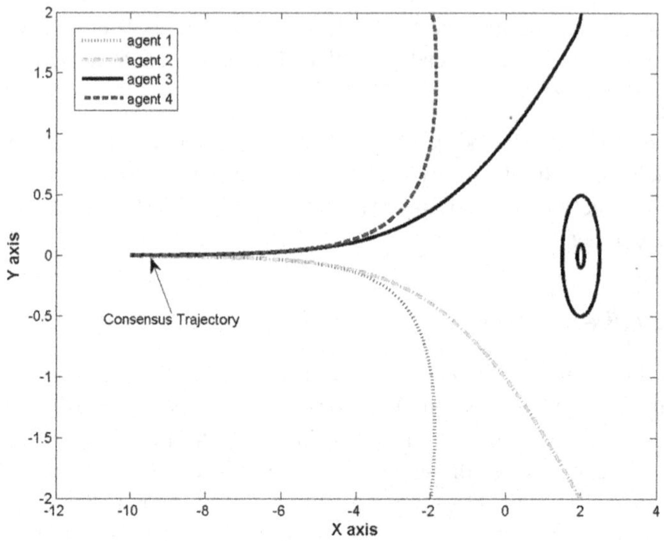

Figure 3.8 Trajectories of four agents without obstacles on the trajectory of any agent.

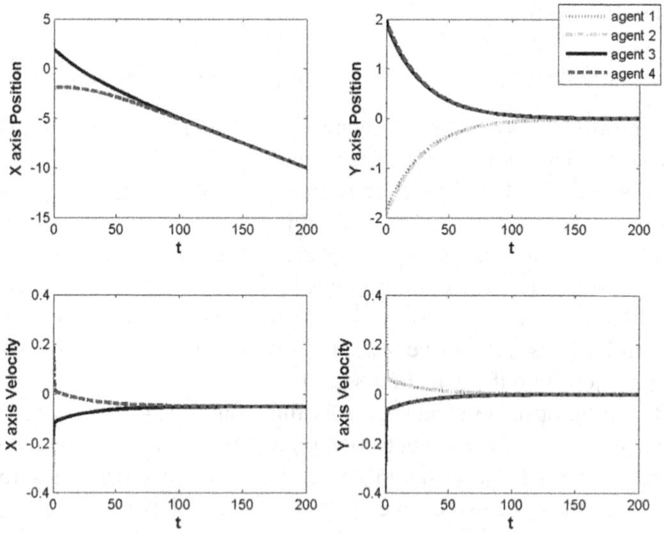

Figure 3.9 Time histories of the four agents' positions and velocities.

assumed to appear on (0.5, 3.2) m, which is on the trajectories of agent 4 and agent 1. The radius of this obstacle and the radius of the detection region are set as 0.2 m and 0.8 m, respectively.

The simulation results are shown in Figs. 3.11–3.13. Fig. 3.11 demonstrates that all agents are able to avoid the multiple obstacles on their trajectories and reach the final consensus. Fig. 3.12 presents the time histories of the agents' positions and velocities,

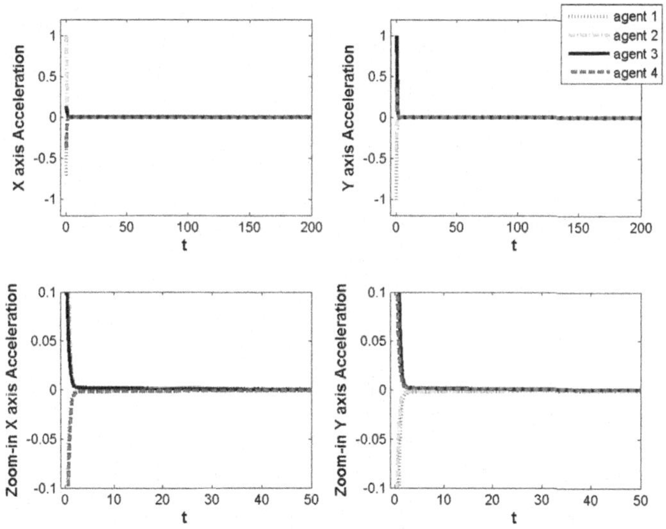

Figure 3.10 Time histories of the four agents' optimal control inputs.

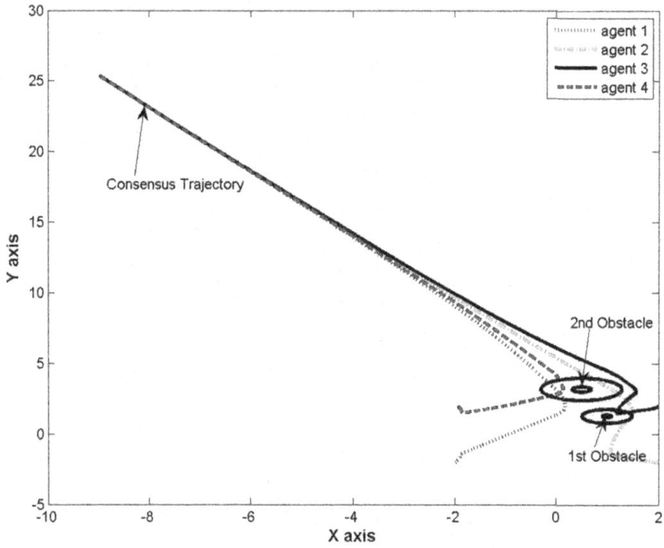

Figure 3.11 Trajectories of the four agents with two obstacles on the trajectory of three agents.

respectively. The optimal control inputs are shown in Fig. 3.13. In the two bottom subfigures of Fig. 3.13, the time histories in the first 50 seconds are shown for better illustration of the transient responses. From Fig. 3.13, it can be seen that the obstacle avoidance control occurs in the intervals of $[5, 13]$ s and $[27, 44]$ s. Furthermore, the velocity response and the control response show that the optimal obstacle avoidance control law does not require large control effort.

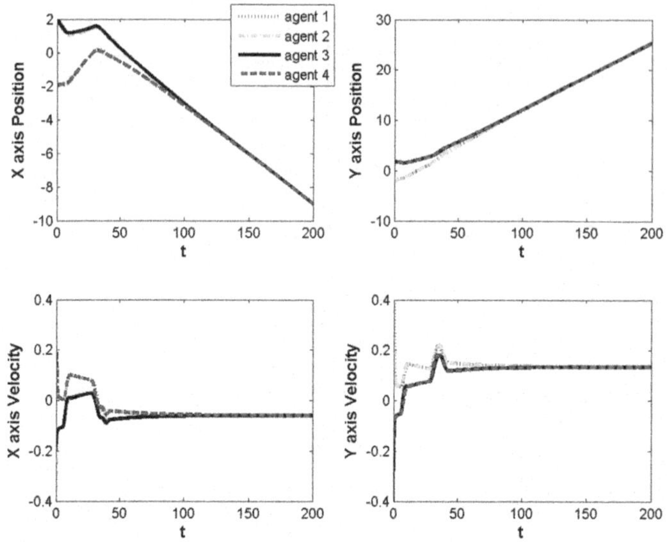

Figure 3.12 Time histories of the four agents' positions and velocities.

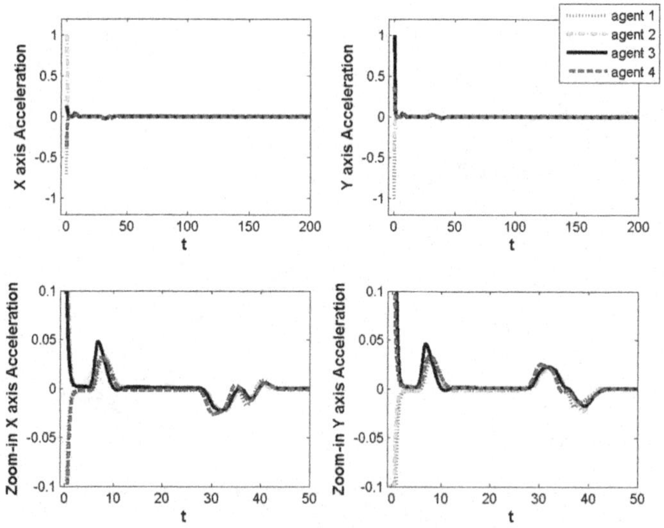

Figure 3.13 Time histories of the four agents' optimal control inputs.

3.4 Conclusion remarks

In this chapter, we have investigated the multi-agent optimal consensus control with obstacle avoidance for both multiple single-integrator systems and double-integrator systems. Two consensus cost functions are innovatively proposed to guarantee the convergence to the consensus point and furthermore the analytical form of solutions.

The inverse optimal control framework enables the possibility of dealing nonquadratic yet meaningful cost functions (such as obstacle avoidance in this chapter) in the future applications. The stability and optimality of the closed-loop system have been provided as well. In the end, simulation results validate the capability of the proposed optimal consensus algorithms in handling both consensus and obstacle avoidance.

Optimal cooperative tracking and flocking of multi-agent systems

4

4.1 Optimal rendezvous and cooperative tracking control with obstacle avoidance

Rendezvous problem is a class of cooperative control problems that a group of networked agents are required to reach a prespecified point simultaneously. Obviously, rendezvous can be formulated as a direct application of consensus. The variables of interest for a rendezvous problem are normally the positions of the multi-agents. The positions are required to be coordinated to the desired position, and the velocities are generally required to be zero when the mission is accomplished. The cooperative tracking problem considers the case that all the agents are cooperatively controlled to track a dynamic reference simultaneously.

4.1.1 Problem formulation

The rendezvous problem in this chapter is to design a distributed optimal control law $\boldsymbol{u}_i$ in Eq. (3.2) based on local information such that all agents' positions converge to a prespecified rendezvous point simultaneously with the final velocities being $\mathbf{0}$. In addition, each agent is able to avoid the obstacle along its trajectory.

Fig. 4.1 shows an example scenario of four agents' rendezvous problem. R_{dj} denotes the radius of the detection region; r_j denotes the radius of the obstacle; $\boldsymbol{O}_{bj}$ is the coordinate of the obstacle center, and $\boldsymbol{p}_r$ is the coordinate of the prespecified rendezvous point. Four agents start their motions from different initial conditions. The dashed line in the figure denotes the original rendezvous trajectory without concern of the obstacle. The proposed control law will be able to not only drive all the agents along the solid lines to reach the same rendezvous point but also avoid the obstacle with an optimal control effort.

In the scope of this chapter, we also make Assumptions 3.1–3.3 of Chapter 3. Another assumption is also needed as follows:

Assumption 4.1. $\boldsymbol{p}_r(t) \in \Pi$, $t \to +\infty$ (the rendezvous point is not in the detection region).

4.1.2 Optimal rendezvous algorithm with obstacle avoidance

For the convenience of formulation, we define the error state

$$\hat{\boldsymbol{X}} = [\ \hat{\boldsymbol{p}}^T \quad \hat{\boldsymbol{v}}^T\ ]^T = \boldsymbol{X} - \boldsymbol{X}_r \tag{4.1}$$

where $\boldsymbol{X}_r = \left[\mathbf{1}_{1\times n} \otimes \boldsymbol{p}_r^T \quad \mathbf{1}_{1\times n} \otimes \boldsymbol{v}_r^T\right]^T = \left[\mathbf{1}_{1\times n} \otimes \boldsymbol{p}_r^T \quad \mathbf{0}_{1\times nm}\right]^T$.

Cooperative Control of Multi-Agent Systems. https://doi.org/10.1016/B978-0-12-820118-3.00014-4

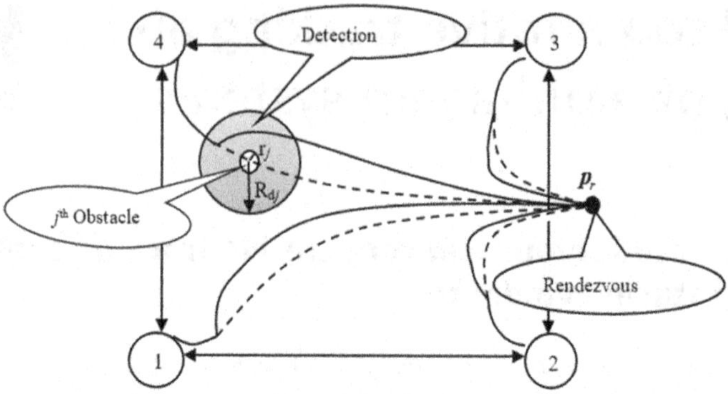

Figure 4.1 Multi-agent rendezvous scenario with an obstacle.

By the property of the Laplacian matrix in Eq. (3.8), we have

$$\begin{aligned}(\mathcal{L} \otimes I_m)\left(\mathbf{1}_{n\times 1} \otimes \boldsymbol{p}_r\right) = \mathbf{0}_{nm\times 1} \\ (\mathcal{L} \otimes I_m)\left(\mathbf{1}_{n\times 1} \otimes \boldsymbol{v}_r\right) = \mathbf{0}_{nm\times 1}\end{aligned} \tag{4.2}$$

The error dynamics can be written as

$$\dot{\hat{X}} = A\hat{X} + BU \tag{4.3}$$

since the control is zero when the agents reach the rendezvous point. The rendezvous is achieved when the system (4.3) is asymptotically stable.

In this section, the rendezvous problem is formulated as an optimal control problem including three cost function components:

$$\begin{aligned}&Min: \mathcal{J} = \mathcal{J}_1 + \mathcal{J}_2^r + \mathcal{J}_3 \\ &S.t. \quad \dot{\hat{X}} = A\hat{X} + BU\end{aligned} \tag{4.4}$$

$\mathcal{J}_1$ has the quadratic form of

$$\mathcal{J}_1 = \int_0^\infty \hat{X}^T R_1 \hat{X} dt = \int_0^\infty \left\{ \hat{X}^T \left(\begin{bmatrix} w_p{}^2\mathcal{L}^2 & 0_{n\times n} \\ 0_{n\times n} & w_v{}^2\mathcal{L}^2 - 2w_p w_c \mathcal{L} \end{bmatrix} \otimes I_m \right) \hat{X} \right\} dt \tag{4.5}$$

where $\mathcal{L}$ is the symmetric Laplacian matrix established by the undirected and connected communication graph. w_p, w_v and w_c represent the weights on position, velocity, and control effort, respectively. Note that R_1 can be shown to be positive semi-definite by Propositions 2.1 and 3.1.

The second cost function $\mathcal{J}_2^r$ has the form of

$$\mathcal{J}_2^r = \int_0^\infty h(\hat{X}) dt \tag{4.6}$$

where $h(\hat{X})$ contains the obstacle avoidance potential function as well as the rendezvous point tracking cost. $h(\hat{X})$ will be constructed by the inverse optimal control approach in Theorem 4.1.

The control effort cost $\mathcal{J}_3$ has the regular quadratic form:

$$\mathcal{J}_3 = \int_0^\infty U^T R_2 U dt \tag{4.7}$$

where $R_2 = {w_c}^2 I_n \otimes I_m$ is positive definite and w_c is the weighting parameter.

The first main result of this chapter is presented in the following theorem.

Theorem 4.1. *For a multi-agent system (3.2) with Assumptions 3.1–3.3, 4.1, w_p, w_v and w_c can be chosen such that the feedback control law*

$$U = \phi(X) = -\frac{w_p}{w_c}(\mathcal{L} \otimes I_m)\boldsymbol{p} - \frac{w_v}{w_c}(\mathcal{L} \otimes I_m)\boldsymbol{v} - \frac{1}{2w_c^2} g_v{}'(X) \tag{4.8}$$

is an optimal control law for the rendezvous problem (4.4) with

$$\begin{aligned} h(\hat{X}) = {} & \frac{w_p}{w_c} g_v{}'^T(\hat{X})(\mathcal{L} \otimes I_m)\hat{\boldsymbol{p}} + \frac{w_v}{w_c} g_v{}'^T(\hat{X})(\mathcal{L} \otimes I_m)\hat{\boldsymbol{v}} - g_p{}'^T(\hat{X})\hat{\boldsymbol{v}} \\ & + \frac{1}{4{w_c}^2}\left\| g_v{}'(\hat{X}) \right\|^2 \end{aligned} \tag{4.9}$$

in Eq. (4.6). $g_v{}'(X)$, $g_v{}'(\hat{X})$ and $g_p{}'(\hat{X})$ in Eq. (4.8) and Eq. (4.9) denote the partial differentiation of $g(.)$ with respect to the velocity error $\hat{\boldsymbol{v}}$ and the position error $\hat{\boldsymbol{p}}$ respectively, where $g(\hat{X})$ is the potential function,

$$g(\hat{X}) = g_{tr}(\hat{X}) + g_{oa}(\hat{X}) \tag{4.10}$$

with the rendezvous point tracking potential function

$$g_{tr}(\hat{X}) \triangleq \sum_{i=1}^{n} g_{tr}^i(\hat{X}) \tag{4.11}$$

$$g_{tr}^i(\hat{X}) = \begin{cases} \begin{bmatrix} (\boldsymbol{p}_i - \boldsymbol{p}_r)^T & \boldsymbol{v}_i^T \end{bmatrix} \begin{bmatrix} w_{tp}^2 I_m & w_{tp} w_{tv} I_m \\ w_{tp} w_{tv} I_m & w_{tv}^2 I_m \end{bmatrix} \begin{bmatrix} \boldsymbol{p}_i - \boldsymbol{p}_r \\ \boldsymbol{v}_i \end{bmatrix} & \text{if agent } i \text{ knows } \boldsymbol{p}_r \\ 0 & \text{if not} \end{cases}, \tag{4.12}$$

where w_{tp}, w_{tv} are tunable weights to control the tracking speed, and the obstacle avoidance potential function

$$g_{oa}(\hat{X}) \triangleq M(\boldsymbol{p})\hat{\boldsymbol{v}} \tag{4.13}$$

with $M(\boldsymbol{p}) = \left[m_r^T(\boldsymbol{p}_1),\quad m_r^T(\boldsymbol{p}_2),\quad \cdots,\quad m_r^T(\boldsymbol{p}_n)\right]$

$$m_r(\boldsymbol{p}_i) = \begin{cases} \mathbf{0} & \boldsymbol{p}_i \in \Pi \\ -\dfrac{\gamma\left(R_{d_j}^2 - \|\boldsymbol{p}_i - \boldsymbol{O}_{bj}\|^2\right)^2}{\left(\|\boldsymbol{p}_i - \boldsymbol{O}_{bj}\|^2 - r_j^2\right)^2}(\boldsymbol{p}_i - \boldsymbol{O}_{bj}) & \boldsymbol{p}_i \in \Gamma_j \quad i = 1, ..., n \\ \textit{not defined} & \boldsymbol{p}_i \in \Lambda_j \end{cases} \tag{4.14}$$

where γ is an adjustable parameter to control the responsiveness of the obstacle avoidance. In addition, the closed-loop system (4.3) is globally asymptotically stable in the safety region Θ.

Proof. Specific to this optimal rendezvous problem, we will have the following equations corresponding to Lemma 2.3 in Chapter 2:

$$T(\hat{\boldsymbol{X}}, \boldsymbol{U}) = \hat{\boldsymbol{X}}^T R_1 \hat{\boldsymbol{X}} + h(\hat{\boldsymbol{X}}) + \boldsymbol{U}^T R_2 \boldsymbol{U} \tag{4.15}$$

$$f(\hat{\boldsymbol{X}}, \boldsymbol{U}) = A\hat{\boldsymbol{X}} + B\boldsymbol{U} \tag{4.16}$$

A candidate Lyapunov function $V(\hat{\boldsymbol{X}})$ is chosen to be

$$V(\hat{\boldsymbol{X}}) = \hat{\boldsymbol{X}}^T P \hat{\boldsymbol{X}} + g(\hat{\boldsymbol{X}}) \tag{4.17}$$

where P is the solution of a Riccati equation, which will be shown afterwards.

In order for the function $V(\hat{\boldsymbol{X}})$ in Eq. (4.17) to be a valid Lyapunov function, it must be continuously differentiable with respect to $\hat{\boldsymbol{X}}$ or equivalently $g(\hat{\boldsymbol{X}})$ must be continuously differentiable with respect to $\hat{\boldsymbol{X}}$. From Eqs. (4.10), (4.12) and (4.14), it suffices to show that $m_r(\boldsymbol{p}_i)$ is continuously differentiable in Θ. In fact, this is true if $m_r(\boldsymbol{p}_i)$ and $\frac{dm_r(\boldsymbol{p}_i)}{d\boldsymbol{p}_i}$ is continuous at the boundary of Ψ_j. Since Eq. (4.14) implies that $\lim\limits_{\|\boldsymbol{p}_i - \boldsymbol{O}_{bj}\| \to R_{d_j}^-} m_r(\boldsymbol{p}_i) = \mathbf{0}_{m\times 1} = \lim\limits_{\|\boldsymbol{p}_i - \boldsymbol{O}_{bj}\| \to R_{d_j}^+} m_r(\boldsymbol{p}_i)$, $m_r(\boldsymbol{p}_i)$ is continuous at the boundary of Ψ_j by the definition of continuity and thus continuous over Θ. In addition,

$$\frac{dm_r(\boldsymbol{p}_i)}{d\boldsymbol{p}_i} = \begin{cases} \mathbf{0}_{m\times m} & \boldsymbol{p}_i \in \Pi \\ \dfrac{4\gamma(R_{d_j}^2 - r_j^2)(R_{d_j}^2 - \|\boldsymbol{p}_i - \boldsymbol{O}_{bj}\|^2)(\boldsymbol{p}_i - \boldsymbol{O}_{bj})(\boldsymbol{p}_i - \boldsymbol{O}_{bj})^T}{\left(\|\boldsymbol{p}_i - \boldsymbol{O}_{bj}\|^2 - r_j^2\right)^3} & \\ -\dfrac{\gamma(R_{d_j}^2 - \|\boldsymbol{p}_i - \boldsymbol{O}_{bj}\|^2)^2 \cdot \boldsymbol{I}_{m\times m}}{\left(\|\boldsymbol{p}_i - \boldsymbol{O}_{bj}\|^2 - r_j^2\right)^2} & \boldsymbol{p}_i \in \Gamma_j \\ \text{not defined} & \boldsymbol{p}_i \in \Lambda_j \end{cases}, \tag{4.18}$$

and it is easy to see that $\lim\limits_{\|\boldsymbol{p}_i - \boldsymbol{O}_{bj}\| \to R_{d_j}^-} \frac{dm_r(\boldsymbol{p}_i)}{d\boldsymbol{p}_i} = \mathbf{0}_{m\times m} = \lim\limits_{\|\boldsymbol{p}_i - \boldsymbol{O}_{bj}\| \to R_{d_j}^+} \frac{dm_r(\boldsymbol{p}_i)}{d\boldsymbol{p}_i}$.

Hence, $\frac{dm_r(\boldsymbol{p}_i)}{d\boldsymbol{p}_i}$ is continuous at the boundary of Ψ_j and thus continuous in Θ. Therefore, $g(\hat{X})$ and the Lyapunov function $V(\hat{X})$ are continuously differentiable with respect to $\hat{X}$ in Θ.

After showing $V(\hat{X})$ is continuously differentiable, the Hamiltonian for this optimal rendezvous problem can be written as

$$\begin{aligned} H(\hat{X}, U, V'^T(\hat{X})) &= T(\hat{X}, U) + V'^T(\hat{X}) f(\hat{X}, U) \\ &= \hat{X}^T R_1 \hat{X} + h(\hat{X}) + U^T R_2 U + [2\hat{X}^T P + g'^T(\hat{X})][A\hat{X} + BU] \end{aligned} \tag{4.19}$$

Setting $(\partial/\partial U) H(\hat{X}, U, V'^T(\hat{X})) = 0$ yields the optimal control law

$$U = \phi(\hat{X}) = -\frac{1}{2} R_2^{-1} B^T V'(\hat{X}) = -R_2^{-1} B^T P \hat{X} - \frac{1}{2} R_2^{-1} B^T g'(\hat{X}) \tag{4.20}$$

With Eq. (4.20), it follows that

$$\begin{aligned} V'^T(\hat{X}) f(\hat{X}, \phi(\hat{X})) &= \hat{X}^T (A^T P + PA - 2PSP)\hat{X} - \hat{X}^T PSg'(\hat{X}) \\ &+ g'^T(\hat{X})(A - SP)\hat{X} - \frac{1}{2} g'^T(\hat{X}) S g'(\hat{X}) \end{aligned} \tag{4.21}$$

where $S \triangleq BR_2^{-1}B^T$. Using Eq. (4.20) and Eq. (4.21) along with Eq. (4.19) yields

$$\begin{aligned} H(\hat{X}, \phi(\hat{X}), V'^T(\hat{X})) &= \hat{X}^T (A^T P + PA + R_1 - PSP)\hat{X} \\ &+ g'^T(\hat{X})(A - SP)\hat{X} + h(\hat{X}) - \frac{1}{4} g'^T(\hat{X}) S g'(\hat{X}) \end{aligned} \tag{4.22}$$

In order to prove that the control law (4.20) is an optimal solution for the rendezvous problem (4.3) using the Lemma 2.3 in Chapter 2, the conditions (2.4)–(2.9) need to be verified.

To satisfy the condition (2.8) in Lemma 2.3 in Chapter 2 or let Eq. (4.22) be zero, we can let

$$A^T P + PA + R_1 - PSP = 0 \tag{4.23}$$

and require that

$$g'^T(\hat{X})(A - SP)\hat{X} + h(\hat{X}) - \frac{1}{4} g'^T(\hat{X}) S g'(\hat{X}) = 0 \tag{4.24}$$

Similarly as Eq. (3.66), it can be shown that

$$H(\hat{X}, U, V'^T(\hat{X})) = [U - \phi(\hat{X})]^T R_2 [U - \phi(\hat{X})] \geq 0 \tag{4.25}$$

Therefore, the condition (2.9) is satisfied.

Substituting A, B, R_1, R_2 in Eq. (4.23) and assuming $P = \begin{bmatrix} P_1 & P_2 \\ P_2 & P_3 \end{bmatrix} \otimes I_m$ yields

$$\begin{bmatrix} -\frac{1}{w_c^2} P_2^2 & P_1 - \frac{1}{w_c^2} P_2 P_3 \\ P_1 - \frac{1}{w_c^2} P_3 P_2 & 2P_2 - \frac{1}{w_c^2} P_3^2 \end{bmatrix} + \begin{bmatrix} w_p{}^2 \mathcal{L}^2 & 0_{n\times n} \\ 0_{n\times n} & w_v{}^2 \mathcal{L}^2 - 2w_p w_c L \end{bmatrix} = 0 \tag{4.26}$$

Then, P can be solved in the analytical form

$$P = \begin{bmatrix} w_p w_v \mathcal{L}^2 & w_p w_c \mathcal{L} \\ w_p w_c \mathcal{L} & w_c w_v \mathcal{L} \end{bmatrix} \otimes I_m \tag{4.27}$$

Note that the motivation of introducing the term $-2w_p w_c L$ in R_1 (see Eq. (4.5)) is to let it cancel $2P_2$ when solving P_3 from Eq. (4.26) in order to make P_3 a linear function of the Laplacian matrix $\mathcal{L}$.

Next, the cost function term $h(\hat{X})$ in J_2^r is constructed from solving Eq. (4.24) and using Eq. (4.27):

$$\begin{aligned} h(\hat{X}) = & \frac{w_p}{w_c} {g_v}'^T(\hat{X})(\mathcal{L} \otimes I_m)\hat{p} + \frac{w_v}{w_c} {g_v}'^T(\hat{X})(\mathcal{L} \otimes I_m)\hat{v} \\ & - {g_p}'^T(\hat{X})\hat{v} + \frac{1}{4{w_c}^2} \left\| {g_v}'(\hat{X}) \right\|^2 \end{aligned} \tag{4.28}$$

which turns out to be Eq. (4.9).

Using Eqs. (4.23), (4.24) and (4.20) becomes

$$V'^T(\hat{X}) f(\hat{X}, \phi(\hat{X})) = -[\hat{X}^T R_1 \hat{X} + h(\hat{X}) + (\hat{X}^T P + \frac{1}{2} g'^T(\hat{X})) S (P\hat{X} + \frac{1}{2} g'(\hat{X}))] \tag{4.29}$$

It can be seen from Eq. (4.29) that the condition (2.7) can be met when $h(\hat{X}) \geq 0$ since $\hat{X}^T R_1 \hat{X}$ is positive semi-definite and $(\hat{X}^T P + \frac{1}{2} g'^T(\hat{X})) S (P\hat{X} + \frac{1}{2} g'(\hat{X}))$ is positive definite. By selecting proper values of the weights w_p, w_v and w_c in Eq. (4.28), one can always ensure $h(\hat{X}) > 0$. Specifically, it can be achieved by choosing a small enough w_c for given w_p and w_v such that the positive term $\frac{1}{4{w_c}^2} \left\| {g_v}'(\hat{X}) \right\|^2$ in Eq. (4.28) is always greater than the other sign-indefinite terms.

Next we will verify the condition (2.5). First note that the matrix

$$\begin{bmatrix} w_{tp}^2 I_m & w_{tp} w_{tv} I_m \\ w_{tp} w_{tv} I_m & w_{tv}^2 I_m \end{bmatrix}$$

in tracking cost (4.12) is positive semi-definite. To see this, we can inspect the eigenvalues of this matrix by

$$\det\left(\lambda I - \begin{bmatrix} w_{tp}^2 I_m & w_{tp} w_{tv} I_m \\ w_{tp} w_{tv} I_m & w_{tv}^2 I_m \end{bmatrix}\right) = \det\left(\begin{bmatrix} (\lambda - w_{tp}^2) I_m & -w_{tp} w_{tv} I_m \\ -w_{tp} w_{tv} I_m & (\lambda - w_{tv}^2) I_m \end{bmatrix}\right) \tag{4.30}$$

Since $(\lambda - w_{tp}^2) I_m$ and $-w_{tp} w_{tv} I_m$ commute, we have

$$\begin{aligned} & \det\left(\begin{bmatrix} (\lambda - w_{tp}^2) I_m & -w_{tp} w_{tv} I_m \\ -w_{tp} w_{tv} I_m & (\lambda - w_{tv}^2) I_m \end{bmatrix}\right) \\ & = \det\left((\lambda - w_{tp}^2)(\lambda - w_{tv}^2) I_m - w_{tp}^2 w_{tv}^2 I_m\right) = \det\left(\lambda^2 I_m - (w_{tp}^2 + w_{tv}^2)\lambda I_m\right) \end{aligned} \tag{4.31}$$

The eigenvalues are $\lambda = \underbrace{0, \cdots, 0}_{m}, \underbrace{w_{tp}^2 + w_{tv}^2, \cdots w_{tp}^2 + w_{tv}^2}_{m} \geq 0$ and thus $g_{tr}(\hat{X}) \geq 0$. Moreover, from the solution of P in Eq. (4.27), the Lyapunov function turns out to be

$$V(\hat{X}) = \hat{X}^T P \hat{X} + g(\hat{X})$$
$$= \begin{cases} w_p w_v \boldsymbol{p}^T(\mathcal{L}^2 \otimes I_m)\boldsymbol{p} + w_c w_v \boldsymbol{v}^T(\mathcal{L} \otimes I_m)\boldsymbol{v} + 2w_p w_c \boldsymbol{p}^T(\mathcal{L} \otimes I_m)\boldsymbol{v} & \boldsymbol{p}_i \in \Pi \\ w_p w_v \boldsymbol{p}^T(\mathcal{L}^2 \otimes I_m)\boldsymbol{p} + w_c w_v \boldsymbol{v}^T(\mathcal{L} \otimes I_m)\boldsymbol{v} + 2w_p w_c \boldsymbol{p}^T(\mathcal{L} \otimes I_m)\boldsymbol{v} & \\ \quad + g_{tr}(\hat{X}) + g_{oa}(\hat{X}) & \boldsymbol{p}_i \in \Gamma_j \\ \text{not defined} & \boldsymbol{p}_i \in \Lambda_j \end{cases} \tag{4.32}$$

Note that the property (4.2) is used in Eq. (4.32) so that the first three terms in $V(\hat{X})$ only contain $\boldsymbol{p}$ and $\boldsymbol{v}$. $V(\hat{X}) > 0$ when $\hat{X} \neq 0$ can be guaranteed by selecting a large enough w_v for given w_p and w_c such that the positive term $w_p w_v \boldsymbol{p}^T(\mathcal{L}^2 \otimes I_m)\boldsymbol{p} + w_c w_v \boldsymbol{v}^T(\mathcal{L} \otimes I_m)\boldsymbol{v}$ and $g_{tr}(\hat{X})$ are always greater than the sign-indefinite terms and thus the condition (2.5) can be satisfied.

To recapitulate the selection of the weight parameters, one can choose a large w_v given w_p for Eq. (4.32) to meet the condition (2.5). Then with the above chosen w_v and w_p, one can choose a small enough w_c for Eq. (4.28) to meet the condition (2.7). In addition, this rule of weight selection also applies for a meaningful cost functional (4.5) by means of satisfying the condition (3.46). Meanwhile, the tracking weights w_{tp}, w_{tv} are chosen to obtain a proper tracking speed, and the parameter γ can be used to adjust the obstacle avoidance responsiveness.

Substituting Eq. (4.27) and $g'(\hat{X})$ into Eq. (4.20) leads to the optimal control law

$$\phi(\hat{X}) = -\frac{w_p}{w_c}(\mathcal{L} \otimes I_m)\hat{\boldsymbol{p}} - \frac{w_v}{w_c}(\mathcal{L} \otimes I_m)\hat{\boldsymbol{v}} - \frac{1}{2w_c^2} g_v{}'(\hat{X}) \tag{4.33}$$

Substituting $\hat{\boldsymbol{p}} = \boldsymbol{p} - \mathbf{1}_{n\times 1} \otimes \boldsymbol{p}_r$ and $\hat{\boldsymbol{v}} = \boldsymbol{v} - \mathbf{0}_{nm\times 1}$ into Eq. (4.33) and using Eq. (4.2), the optimal control law (4.33) becomes

$$\boldsymbol{U} = \phi(X) = -\frac{w_p}{w_c}(\mathcal{L} \otimes I_m)\boldsymbol{p} - \frac{w_v}{w_c}(\mathcal{L} \otimes I_m)\boldsymbol{v} - \frac{1}{2w_c^2} g_v{}'(\hat{X})$$

Also note that since $\boldsymbol{p}_r$ and $\boldsymbol{v}_r$ are known to agent i,

$$g_v{}'(\hat{X}) = g_{tr_v}{}'(X) + M^T(\boldsymbol{p}) = g_{tr_v}{}'(X) + M^T(\boldsymbol{p}) = g_v{}'(X) \tag{4.34}$$

where $g_{tr_v}{}'(\hat{X}) = \frac{\partial g_{tr}(\hat{X})}{\partial \hat{\boldsymbol{v}}}$. Thus, the optimal control law turns out to be Eq. (4.8)

$$\boldsymbol{U} = \phi(X) = -\frac{w_p}{w_c}(\mathcal{L} \otimes I_m)\boldsymbol{p} - \frac{w_v}{w_c}(\mathcal{L} \otimes I_m)\boldsymbol{v} - \frac{1}{2w_c^2} g_v{}'(X)$$

Note that Assumption 4.1 guarantees the conditions (2.4) and (2.6) since both $g(\hat{X})$ in Eq. (4.17) and $g_v{}'(\hat{X})$, which is the same as $g_v{}'(X)$ in Eq. (4.8), are equal to zero when $\hat{X} = \mathbf{0}$.

Now, all the conditions (2.4)–(2.9) in Lemma 2.3 in Chapter 2 can be satisfied. Therefore, according to Lemma 2.3 in Chapter 2, the control law (4.8) is an optimal control law for the problem (4.3) in the sense of Eq. (2.11) and Eq. (2.12), and the closed-loop system is asymptotically stable. It implies $\boldsymbol{X} = \boldsymbol{X}_r$ and the rendezvous is achieved.

In addition, it can be easily seen from Eq. (4.17) that $V(\hat{\boldsymbol{X}}) \to \infty$ as $\|\boldsymbol{X}\| \to \infty$. Therefore, the closed-loop system is globally asymptotically stable. Note that the globally asymptotic stability region excludes the undefined area Θ^c, which is physically meaningful because no agent can start from inside the obstacle. □

As can be seen from the proof of Theorem 4.1, the potential function $g(\hat{\boldsymbol{X}})$ provides the rendezvous point tracking and obstacle avoidance capabilities. The inverse optimal control approach facilitates the integration of this potential function into the cost function $h(\hat{\boldsymbol{X}})$ such that the analytical optimal control law to achieve these capabilities can be obtained.

Remark 4.1. The obstacle avoidance control in this section only requires the relative position information between the agent and the obstacle, rather than both relative position and velocity information as in Section 3.3 of Chapter 3. This makes the algorithm more intuitive in the common sense of obstacle avoidance.

Remark 4.2. It can be noticed that the form of the penalty function $M(\boldsymbol{p})$ does not affect the proof, which provides more flexibility of using other avoidance penalty functions with different behaviors.

Remark 4.3. The optimal control law (4.8) of each agent only requires the local information from its neighbors since it is a linear function of the Laplacian matrix $\mathcal{L}$. This feature offers a great implementation advantage.

4.1.3 Numerical examples

In this section, the effectiveness of the proposed optimal rendezvous control law given by Theorem 4.1 for 4 mobile agents is illustrated. The scenario is shown in Fig. 4.1. The problem is considered in a 2D environment, i.e. $m = 2$. The Laplacian matrix can be set up by the definition in Chapter 2

$$\mathcal{L} = \begin{bmatrix} 2 & -1 & 0 & -1 \\ -1 & 2 & -1 & 0 \\ 0 & -1 & 2 & -1 \\ -1 & 0 & -1 & 2 \end{bmatrix}. \tag{4.35}$$

Also, only agent 1 knows the rendezvous point at $(7.5, 0)$ m. The initial positions of the four agents are assumed to be $(-5, -5)$ m, $(5, -5)$ m, $(5, 5)$ m, and $(-5, 5)$ m, respectively. The initial velocities of the four agents are assumed to be $(0.25, 0.2)$ m/s, $(-0.15, 0.2)$ m/s, $(-0.125, -0.1)$ m/s, and $(0.2, -0.1)$ m/s, respectively. The associated weights in the control law are set to $w_p = 0.05$, $w_v = 1.5$, $w_c = 0.5$, $\gamma = 0.01$, $w_{tp} = 0.2\sqrt{3}$, $w_{tv} = 2\sqrt{3}$.

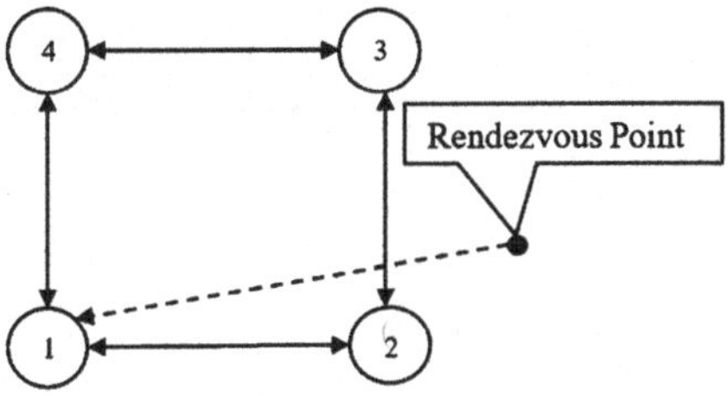

Figure 4.2 Partial access to the reference.

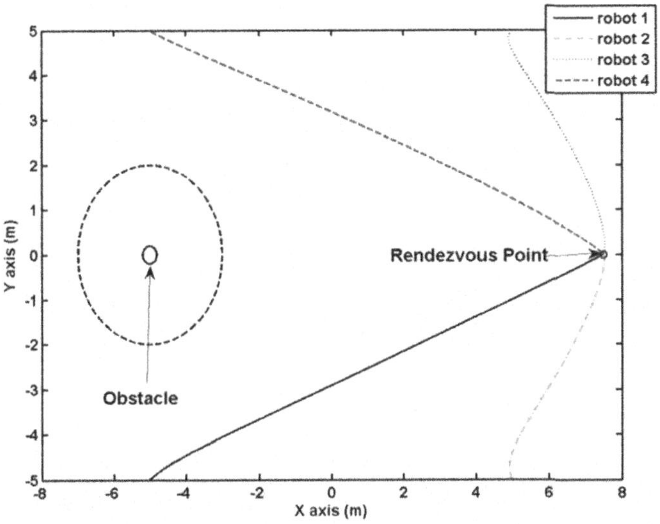

Figure 4.3 Rendezvous without concern of obstacles.

A. Rendezvous without obstacles on the trajectories of agents

In this scenario, an obstacle is assumed to appear on $(-5, 0)$ m, which is not on the trajectory of any agent. The radii of the obstacle and the detection region are set to $r_1 = 0.2$ m and $R_{d_1} = 2$ m. The simulation results of the four agents' motion under the proposed optimal control law are shown in Fig. 4.3. It can be seen that the obstacle avoidance does not take effect since no agent steps into the detection region. Fig. 4.4 presents the time histories of the agents' positions and velocities, respectively. It shows that the agents arrive at the rendezvous point simultaneously. The optimal control inputs are given in Fig. 4.5.

B. Rendezvous with two obstacles on the trajectory of agents

In this scenario, the first obstacle is assumed to appear on $(-2, 3.9)$ m, which is on the trajectory of agent 4. The radii of the obstacle and the detection region are set to $r_1 = 0.2$ m and $R_{d_1} = 2$ m, respectively. The other obstacle is assumed to appear on $(0, -2)$ m, which is on the trajectory of agent 1. The radii of the obstacle and the detection region are set to $r_2 = 0.2$ m and $R_{d_2} = 1$ m, respectively. The simulation results are given in Figs. 4.6–4.8.

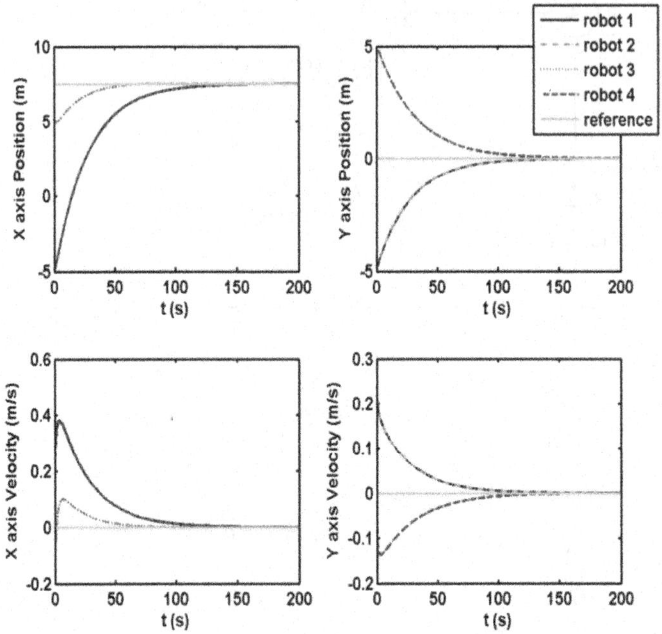

Figure 4.4 Time histories of four agents' positions and velocities.

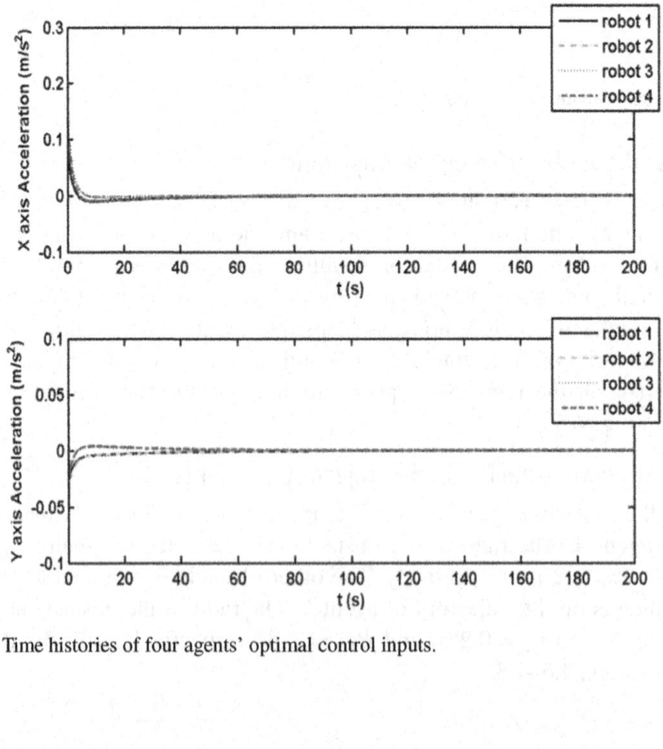

Figure 4.5 Time histories of four agents' optimal control inputs.

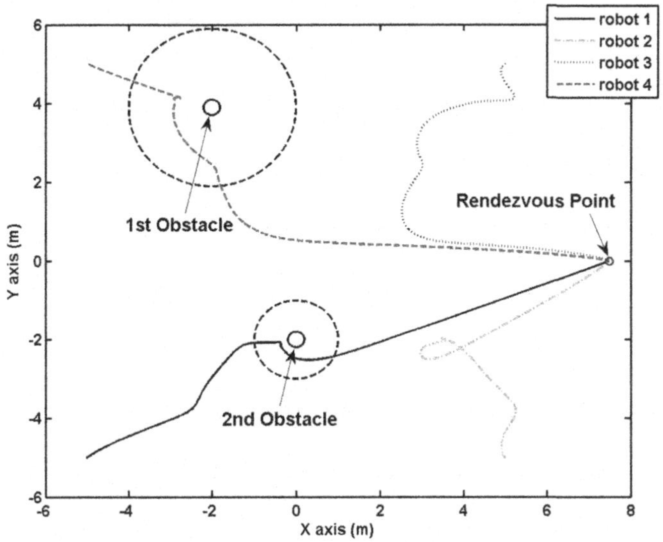

Figure 4.6 Rendezvous with two obstacles.

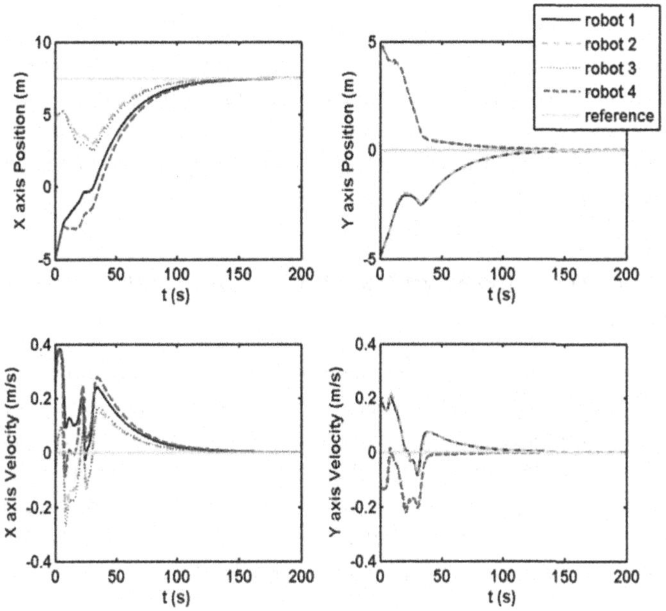

Figure 4.7 Time histories of four agents' positions and velocities.

Fig. 4.6 demonstrates that all agents are able to avoid the multiple obstacles on their trajectories and reach the rendezvous point. Fig. 4.7 presents the time histories of the agents' positions and velocities, respectively. It shows that the agents arrive at

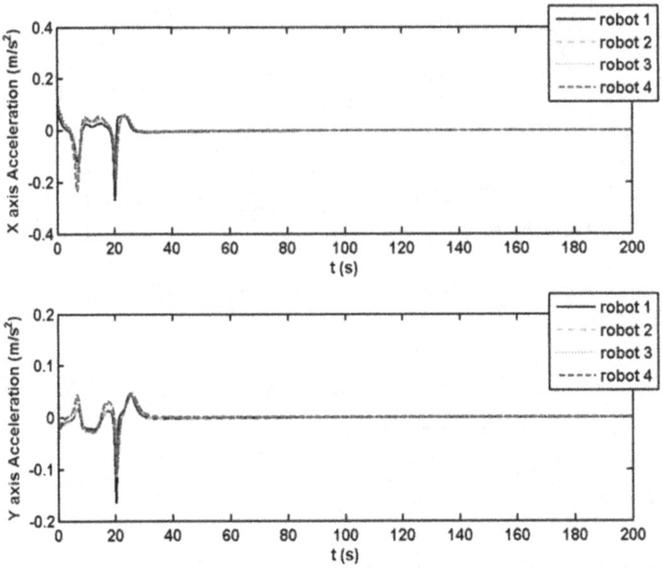

Figure 4.8 Time histories of four agents' optimal control inputs.

the rendezvous point simultaneously. The optimal control inputs are given in Fig. 4.8, which shows the maneuvers when agent 1 and 4 avoid the obstacles during $[5, 10]$ s and $[15, 25]$ s, respectively. These simulation results demonstrate that the proposed optimal cooperative control algorithm is capable of achieving rendezvous under different obstacle scenarios.

4.1.4 Extension to cooperative tracking problem with obstacle avoidance

In this section, we will extend the rendezvous problem to the cooperative tracking problem. The rendezvous problem can be treated as a special case of cooperative tracking problem when the reference is a stationary point. In other words, the cooperative tracking problem will consider the case that all the agents are cooperatively controlled to track a dynamic reference simultaneously. We will first present the cooperative tracking algorithm comparing with the rendezvous problem in Section 4.1.4.1, and then give the simulation results for validation in Section 4.1.4.2.

4.1.4.1 Cooperative tracking algorithm with obstacle avoidance

For the convenience of formulation, we define the error state

$$\hat{X} = \left[\hat{p}^T \quad \hat{v}^T\right]^T \triangleq X - X_r \tag{4.36}$$

where $\boldsymbol{X}_r = \left[\mathbf{1}_{1\times n} \otimes \boldsymbol{p}_r^T \quad \mathbf{1}_{1\times n} \otimes \boldsymbol{v}_r^T\right]^T$ is the reference state. By the property of the Laplacian matrix (3.8), we have

$$\begin{aligned}(\mathcal{L} \otimes I_m)\left(\mathbf{1}_{n\times 1} \otimes \boldsymbol{p}_r\right) &= \mathbf{0}_{nm\times 1}\\ (\mathcal{L} \otimes I_m)\left(\mathbf{1}_{n\times 1} \otimes \boldsymbol{v}_r\right) &= \mathbf{0}_{nm\times 1}\end{aligned} \tag{4.37}$$

The error dynamics can be written as

$$\dot{\hat{\boldsymbol{X}}} = A\hat{\boldsymbol{X}} + B\hat{\boldsymbol{U}} = A\hat{\boldsymbol{X}} + B\left(\boldsymbol{U} - \boldsymbol{U}_r\right) \tag{4.38}$$

where $\boldsymbol{U}_r = \mathbf{1}_{n\times 1} \otimes \boldsymbol{u}_r$ is the control input of the reference trajectory and $\hat{\boldsymbol{U}} = \boldsymbol{U} - \boldsymbol{U}_r$. The tracking is achieved when the system (4.38) is asymptotically stable.

In this section, the cooperative tracking problem is formulated as an optimal control problem exactly in the same way as the rendezvous problem. The main result of this section is presented in the following theorem.

Theorem 4.2. *For a multi-agent system (3.2) with Assumptions 3.1–3.3, 4.1, w_p, w_v and w_c can be chosen such that the feedback control law*

$$\boldsymbol{U} = \boldsymbol{U}_r - \frac{w_p}{w_c}(\mathcal{L} \otimes I_m)\boldsymbol{p} - \frac{w_v}{w_c}(\mathcal{L} \otimes I_m)\boldsymbol{v} - \frac{1}{2w_c^2} g_v{}'(\boldsymbol{X}) \tag{4.39}$$

is an optimal control law for the cooperative tracking problem and the closed-loop system (4.38) is globally asymptotically stable in the safety region Θ. In addition, $h(\hat{\boldsymbol{X}})$ in the cost function $\mathcal{J}_2^r$ in Eq. (4.6) is

$$\begin{aligned}h(\hat{\boldsymbol{X}}) = {} & \frac{w_p}{w_c} g_v{}'^T(\hat{\boldsymbol{X}})(\mathcal{L} \otimes I_m)\hat{\boldsymbol{p}} + \frac{w_v}{w_c} g_v{}'^T(\hat{\boldsymbol{X}})(\mathcal{L} \otimes I_m)\hat{\boldsymbol{v}}\\ & - g_p{}'^T(\hat{\boldsymbol{X}})\hat{\boldsymbol{v}} + \frac{1}{4{w_c}^2}\left\| g_v{}'(\hat{\boldsymbol{X}})\right\|^2\end{aligned} \tag{4.40}$$

where $g_v{}'(\boldsymbol{X})$, $g_v{}'(\hat{\boldsymbol{X}})$ and $g_p{}'(\hat{\boldsymbol{X}})$ in Eqs. (4.39) and (4.40) denote the partial differentiation of $g(\hat{\boldsymbol{X}})$ with respect to the velocity error $\hat{\boldsymbol{v}}$ and the position error $\hat{\boldsymbol{p}}$ respectively. $g(\hat{\boldsymbol{X}})$ is the penalty function defined by

$$g(\hat{\boldsymbol{X}}) = g_{tr}(\hat{\boldsymbol{X}}) + g_{oa}(\hat{\boldsymbol{X}}) \tag{4.41}$$

with the first component $g_{tr}(\hat{\boldsymbol{X}})$ being the tracking penalty function,

$$g_{tr}(\hat{\boldsymbol{X}}) = \sum_{i=1}^{n} g_{tr}^i(\hat{\boldsymbol{X}}) \tag{4.42}$$

$$g_{tr}^i(\hat{\boldsymbol{X}}) \triangleq \begin{cases} \left[(\boldsymbol{p}_i - \boldsymbol{p}_r)^T \quad (\boldsymbol{v}_i - \boldsymbol{v}_r)^T\right]\begin{bmatrix} w_{tp}^2 I_m & w_{tp}w_{tv}I_m \\ w_{tp}w_{tv}I_m & w_{tv}^2 I_m\end{bmatrix}\begin{bmatrix}\boldsymbol{p}_i - \boldsymbol{p}_r \\ \boldsymbol{v}_i - \boldsymbol{v}_r\end{bmatrix}, & \\ \qquad\qquad \textit{if agent } i \textit{ has access to the reference} & \\ 0 \qquad \textit{if not} & \end{cases} \tag{4.43}$$

where w_{tp} and w_{tv} are tunable weights to control the tracking speed, and the second component being the obstacle avoidance penalty function

$$g_{oa}(\hat{\boldsymbol{X}}) = \boldsymbol{M}(\boldsymbol{p})\hat{\boldsymbol{v}} \tag{4.44}$$

with $\boldsymbol{M}(\boldsymbol{p}) = \begin{bmatrix} \boldsymbol{m}_r^T(\boldsymbol{p}_1) & \boldsymbol{m}_r^T(\boldsymbol{p}_2) & \ldots & \boldsymbol{m}_r^T(\boldsymbol{p}_n) \end{bmatrix}$

$$\boldsymbol{m}_r(\boldsymbol{p}_i) \triangleq \begin{cases} \boldsymbol{0}_{m\times 1} & \boldsymbol{p}_i \in \Pi \\ -\dfrac{\gamma\left(R_{d_j}^2 - \|\boldsymbol{p}_i - \boldsymbol{O}_{bj}\|^2\right)^2}{\left(\|\boldsymbol{p}_i - \boldsymbol{O}_{bj}\|^2 - r_j^2\right)^2}\left(\boldsymbol{p}_i - \boldsymbol{O}_{bj}\right) & \boldsymbol{p}_i \in \Gamma_j \\ \text{not defined} & \boldsymbol{p}_i \in \Lambda_j \end{cases} \quad i = 1, ..., n \tag{4.45}$$

where γ is an adjustable parameter to control the responsiveness of the obstacle avoidance.

Proof. For details, please refer the proof of Theorem 4.1. □

The potential function $g(\hat{\boldsymbol{X}})$ provides the reference tracking and obstacle avoidance capabilities. The inverse optimal control approach facilitates the integration of this potential function into the cost function $h(\hat{\boldsymbol{X}})$ such that an analytical distributed optimal control law to achieve these capabilities can be obtained.

Remark 4.4. The optimal control law (4.39) can be rewritten as

$$\boldsymbol{U} = \boldsymbol{U}_r - \frac{1}{2w_c^2} g_{tr_v}{}'(\hat{\boldsymbol{X}}) - \frac{w_p}{w_c}(\mathcal{L} \otimes I_m)\boldsymbol{p} - \frac{w_v}{w_c}(\mathcal{L} \otimes I_m)\boldsymbol{v} - \frac{1}{2w_c^2} g_{oa_v}{}'(\hat{\boldsymbol{X}}) \tag{4.46}$$

If the trajectory to track is a rendezvous point or a straight line with constant velocity, the control input of the desired trajectory will become zero, i.e. $\boldsymbol{u}_r = \boldsymbol{0}_{m\times 1}$ and $\boldsymbol{U}_r = \boldsymbol{0}_{nm\times 1}$, and thus the optimal control law (4.46) does not need to know $\boldsymbol{U}_r$. In this case, only partial access to the reference is sufficient to guarantee the tracking convergence (see the definition of $g_{tr}(\hat{\boldsymbol{X}})$ in Eq. (4.43)). In other words, only some of the agents need to know the reference position and velocity. In the case of tracking a reference with nonzero $\boldsymbol{U}_r$, full access to the reference $\boldsymbol{U}_r$ is required as shown in the control law (4.46). Note that the first two terms $\boldsymbol{U}_r$ and $g_{tr_v}{}'(\hat{\boldsymbol{X}})$ in Eq. (4.46) are the tracking control law. It only requires the agent's own information and does not need other agents' information. It may need to know the reference depending on the reference characteristics and whether this agent has access to the reference as discussed above. The next two terms $-\frac{w_p}{w_c}(\mathcal{L} \otimes I_m)\boldsymbol{p}$ and $-\frac{w_p}{w_c}(\mathcal{L} \otimes I_m)\boldsymbol{v}$ in Eq. (4.46) are the optimal consensus laws to guarantee that all agents reach the reference simultaneously. They are linear functions of $(\mathcal{L} \otimes I_m)\boldsymbol{p}$ and $(\mathcal{L} \otimes I_m)\boldsymbol{v}$ and thus only require the local information from its neighbors based on the communication topology. The last term is the obstacle avoidance law to provide the obstacle avoidance capability. Note that the obstacle avoidance law of each agent only requires the relative position information of the agent with respect to the obstacle and does not need other robots' information. In summary, the proposed distributed optimal control law of each agent only depends on the local information from the neighbors, and the requirement to have access to the reference depends on the acceleration characteristics of the reference.

Remark 4.5. The selection rule of the weight parameters can be referred to Theorem 4.1. Meanwhile, the tracking weights w_{tp} and w_{tv} can be tuned to obtain a proper tracking speed, and the parameter γ can be used to adjust the obstacle avoidance responsiveness.

4.1.4.2 Numerical examples

In this section, the effectiveness of the proposed optimal cooperative tracking law given by Theorem 4.2 is illustrated. Cooperative control of four mobile agents is considered in a 2D environment, i.e. $m = 2$. The communication topology is shown in Fig. 4.1 and the Laplacian matrix is given in Eq. (4.35). The initial positions of the four agents and associated weights are exactly the same as in Section 4.1.3.

A. Cooperative tracking of a reference with constant velocity

In this scenario, the multi-agent cooperative tracking of a reference trajectory with constant velocity is illustrated. In this case, $\boldsymbol{u}_r = \mathbf{0}_{m\times 1}$. Thus, as discussed in Remark 4.4, partial access to the reference is sufficient to guarantee the cooperative tracking behavior. We still assume that only agent 1 has access to the reference as shown in Fig. 4.2. The initial position and velocity of the reference trajectory are $(7.5, 0)$ m and $(0.1, 0.1)$ m/s, respectively.

First, an obstacle is assumed to appear at $(-5, 0)$ m, which is not on the trajectory of any agent. The radii of the obstacle and the detection region are set to $r_1 = 0.2$ m and $R_{d_1} = 2$ m, respectively. The four agents' motions are shown in Figs. 4.9–4.11. The obstacle avoidance does not take effect since no agent steps into the detection

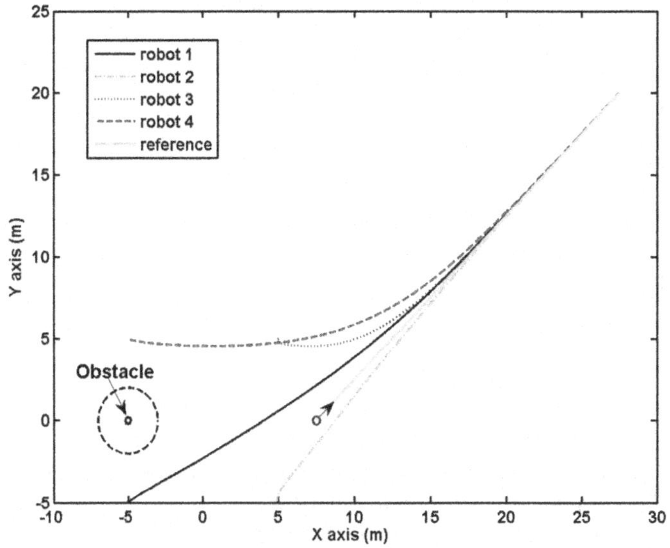

Figure 4.9 Multi-agent cooperative tracking of a reference trajectory with constant velocity without obstacle on the trajectory.

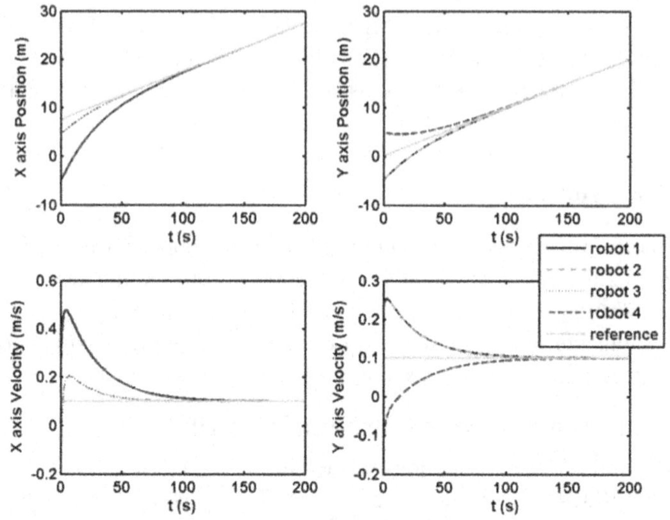

Figure 4.10 Time histories of four agents' positions and velocities.

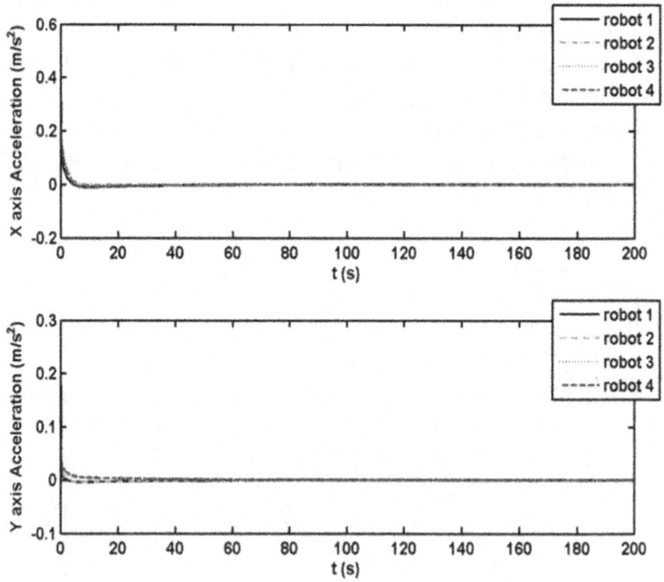

Figure 4.11 Time histories of four agents' optimal control inputs.

region. The four agents reach the desired reference trajectory simultaneously by the proposed optimal cooperative tracking control algorithm.

In the next scenario, one obstacle is assumed to appear at $(-2, 4.57)$ m, which is on the trajectory of agent 4. The radii of the obstacle and the detection region are set to $r_1 = 0.2$ m and $R_{d_1} = 2$ m, respectively. The other obstacle is assumed to appear

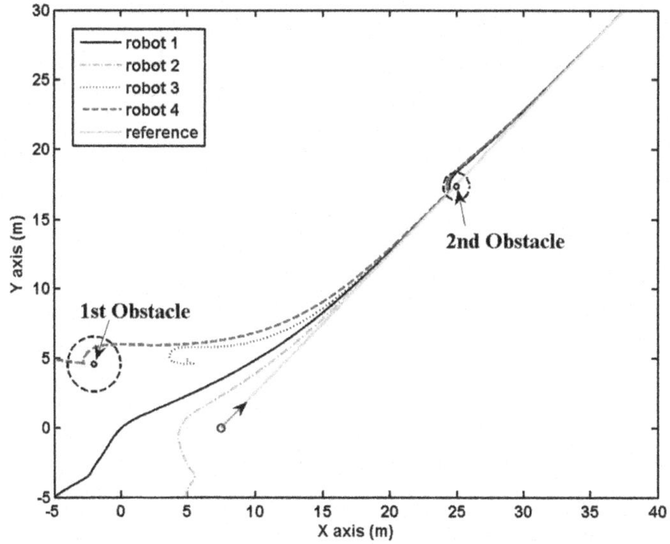

Figure 4.12 Multi-agent cooperative tracking of a reference trajectory with constant velocity and two obstacles on the trajectories.

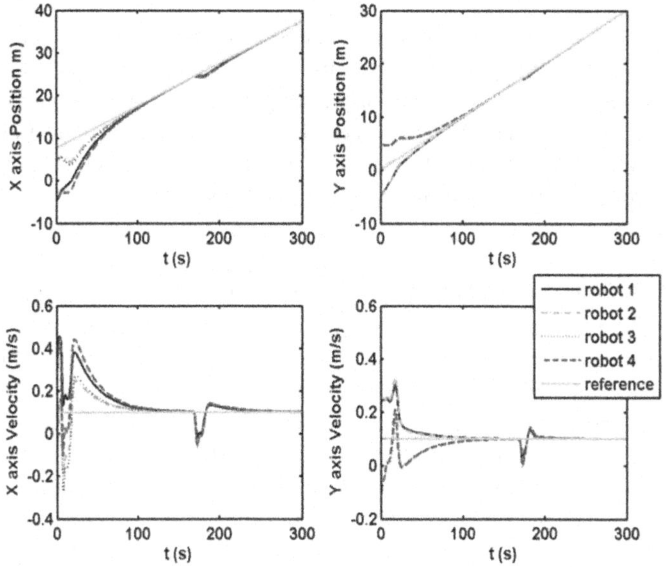

Figure 4.13 Time histories of four agents' positions and velocities.

at (25, 17.4) m, which is on the trajectories of all the agents when they nearly complete the tracking. The radii of the second obstacle and its detection region are set to $r_2 = 0.2$ m and $R_{d_2} = 1$ m, respectively. The four agents' motions are shown in Figs. 4.12–4.14. Fig. 4.12 demonstrates that all agents are able to avoid the multiple

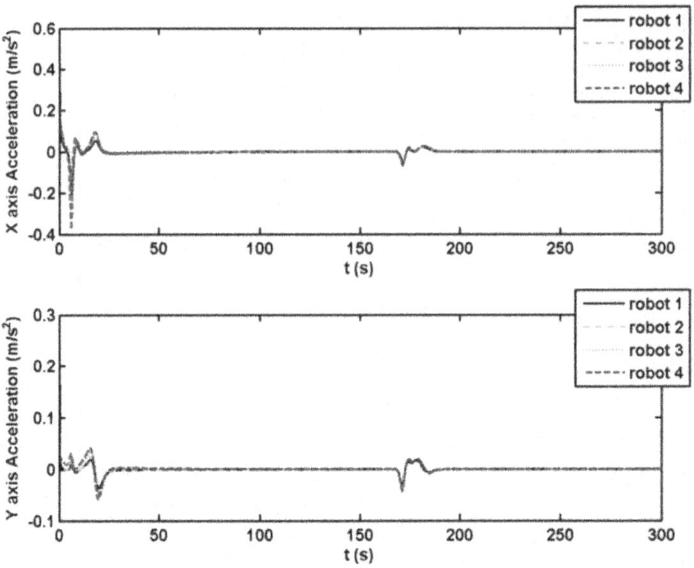

Figure 4.14 Time histories of four agents' optimal control inputs.

obstacles on their trajectories and reach the desired reference trajectory. Fig. 4.13 presents the time histories of the agents' positions and velocities, which indicates that they reach the desired reference trajectory at the same time. The optimal control inputs are shown in Fig. 4.14 and it can be seen that the obstacle avoidance control occurs in the intervals of [2, 15] s and [165, 185] s. Furthermore, the control response shows that the optimal obstacle avoidance control law does not require large control effort.

B. Cooperative tracking of a dynamic reference trajectory

In this scenario, the multi-agent cooperative tracking of a dynamic reference trajectory is illustrated. Since the reference velocity is not constant, $\boldsymbol{u}_r \neq \mathbf{0}_{m\times 1}$. Thus, as discussed in Remark 4.4, full access to the reference is required to guarantee the cooperative behavior, which is shown in Fig. 4.15. The desired reference trajectory is assumed to be a circular motion with initial conditions of $(0, 3)$ m, $(0.15, 0)$ m/s, and the angular velocity of 0.05 rad/s. The initial conditions of the four agents are the same as in the previous scenarios.

First, an obstacle is assumed to appear at $(-5, 0)$ m, which is not on the trajectory of any agent. The radii of the obstacle and the detection region are set to $r_1 = 0.2$ m and $R_{d_1} = 1$ m, respectively. The four agents' motions under the proposed optimal cooperative control law are shown in Figs. 4.16–4.18. It can be seen that the obstacle avoidance does not take effect since no agent steps into the detection region. The desired cooperative behavior is achieved by the proposed optimal control algorithm.

Then in the next scenario, one obstacle is assumed to appear at $(-2, 4)$ m, which is on the trajectory of agent 4. The radii of the obstacle and the detection region

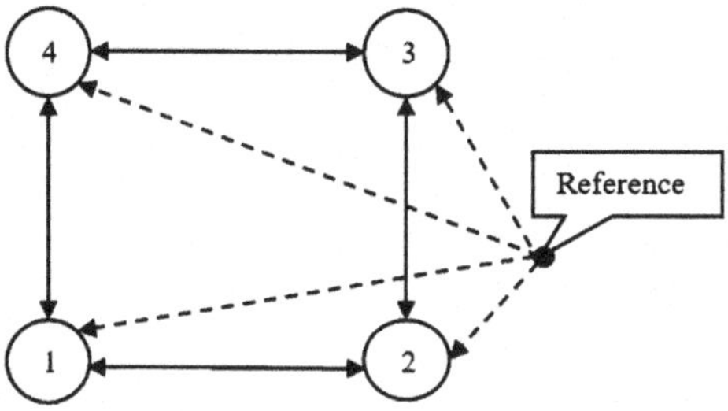

Figure 4.15 Full access to the reference.

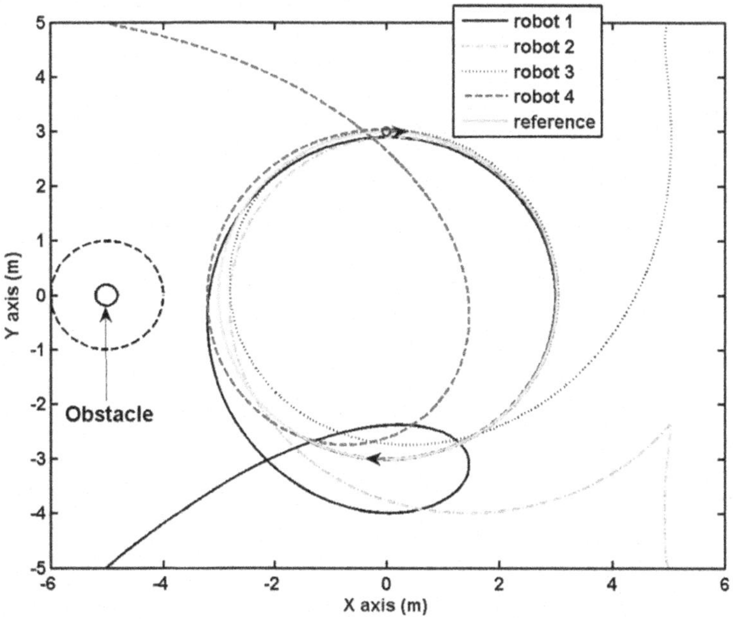

Figure 4.16 Multi-agent cooperative tracking of a circular reference trajectory without obstacle on the trajectory.

are set to $r_1 = 0.2$ m and $R_{d_1} = 1$ m, respectively. The other obstacle is assumed to appear at $(1.5, 2.6)$ m, which is on the trajectories of all the agents. The radii of the obstacle and the detection region are set to $r_2 = 0.1$ m and $R_{d_2} = 0.5$ m, respectively. The four agents' motions under the proposed optimal control law are shown in Figs. 4.19–4.21. Fig. 4.19 demonstrates that all agents are able to avoid the multiple obstacles and track the desired reference trajectory. Fig. 4.20 presents the time histories of the agents' positions and velocities, which indicates that they reach the desired reference trajectory simultaneously. The optimal control inputs are shown in Fig. 4.21 and it can be seen that the obstacle avoidance control occurs in the interval of $[5, 25]$ s and $[130, 145]$ s. If the simulation runs longer, the agents will meet and avoid the second obstacle again. For better viewing of the

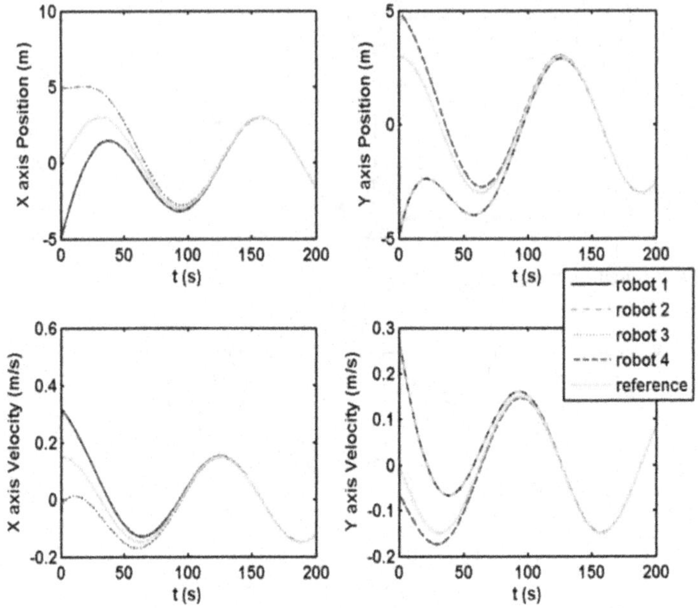

Figure 4.17 Time histories of four agents' positions and velocities.

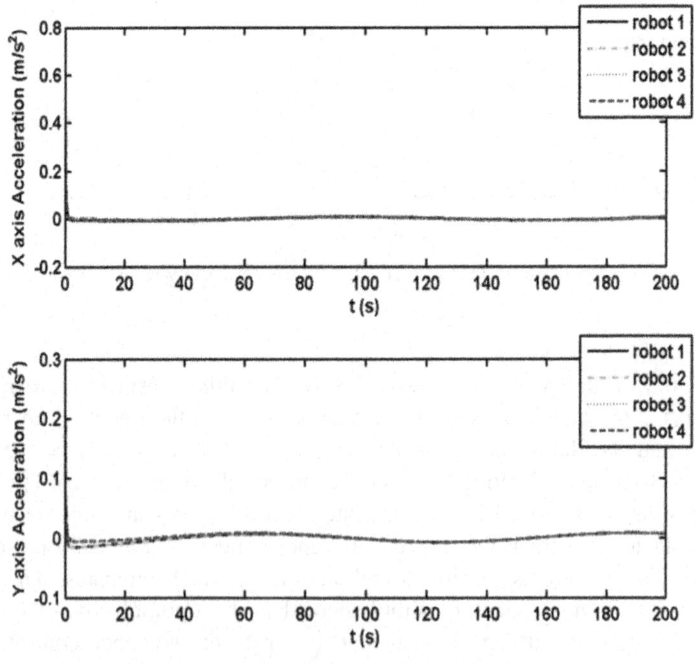

Figure 4.18 Time histories of four agents' optimal control inputs.

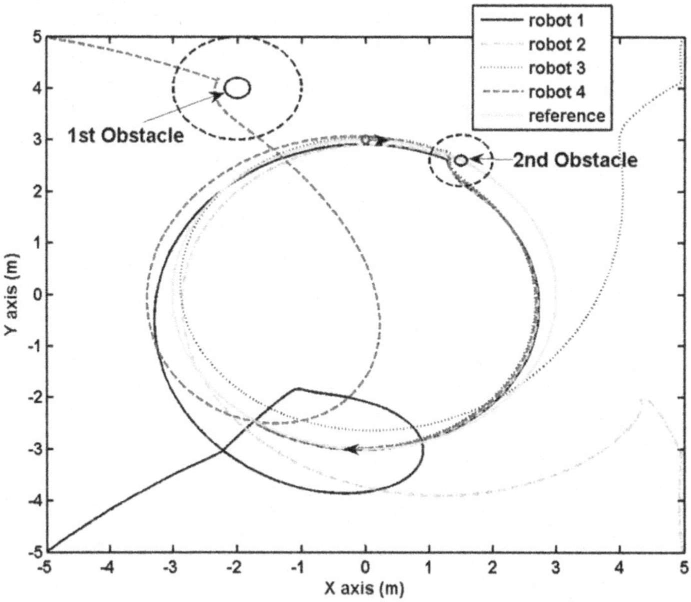

Figure 4.19 Multi-agent cooperative tracking to a circular reference trajectory with two obstacles on the trajectories.

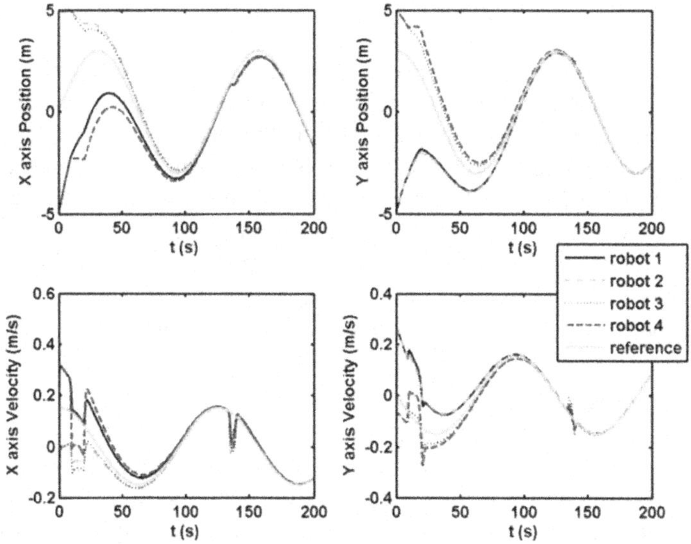

Figure 4.20 Time histories of four agents' positions and velocities.

results, we just provide the results in the first period. Furthermore, the control response shows that the optimal obstacle avoidance control law does not demand large control effort.

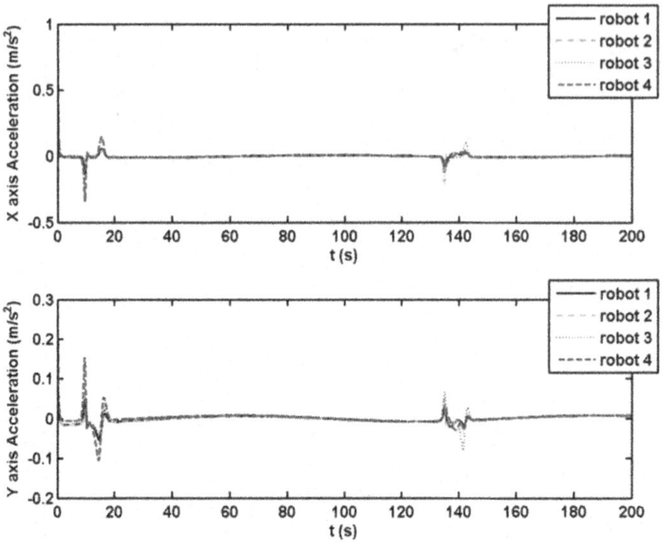

Figure 4.21 Time histories of four agents' optimal control inputs.

4.2 Optimal flocking control design with obstacle avoidance

Flocking problem is initially motivated by the bioscience, which studies how birds flock together. The similar behaviors in the biological world are swarms of bees, a herd of land animals and school of fish. Since some similar behaviors can be employed in the cooperative control of multi-agent, flocking has become another important cooperative mission. The engineering applications of flocking include some military missions such as reconnaissance, surveillance, and combat using cooperative unmanned aerial vehicles (UAVs). The flocking problem is also a direct application of consensus. The variable of interest for a flocking problem is the velocity of the multi-agent systems, which is usually required to be the same desired value among agents.

4.2.1 Problem formulation

The flocking problem in this chapter is to design a distributed optimal control law based on local information such that proposed flocking behaviors including velocity alignment, navigation (desired trajectory and velocity tracking), cohesion, and collision/obstacle avoidance are all achieved.

For the convenience of formulation, the agents and obstacles are modeled as circle-shaped objects in Fig. 4.22. R denotes the radius of the collision/obstacle avoidance region; r denotes the radius of each agent. Without loss of generality, we assume that each agent has the same r and R. r_k denotes the radius of the kth obstacle, where $k = 1, \ldots, q$ and q is the number of obstacles; $\boldsymbol{O}_k$ denotes the location of the center

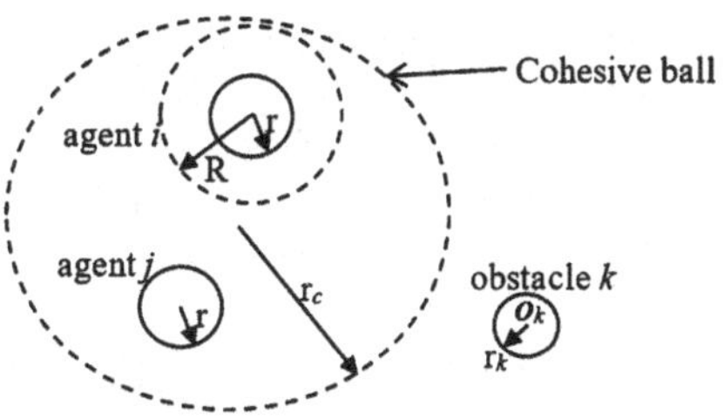

Figure 4.22 Illustration of agents and obstacles.

of the kth obstacle; r_c denotes the radius of the cohesive ball. The following regions can be defined.

Collision region for the ith agent:

$$\Phi_i \triangleq \left\{ \boldsymbol{x} \left| \boldsymbol{x} \in \mathbb{R}^m, \left\| \boldsymbol{x} - \boldsymbol{p}_i \right\| \leq r + \bar{r} \right. \right\}$$

Reaction region for the ith agent:

$$\Sigma_i \triangleq \left\{ \boldsymbol{x} \left| \boldsymbol{x} \in \mathbb{R}^m, r + \bar{r} < \left\| \boldsymbol{x} - \boldsymbol{p}_i \right\| \leq R + \bar{r} \right. \right\}$$

Safety region for the ith agent:

$$\Xi_i \triangleq \left\{ \boldsymbol{x} \left| \boldsymbol{x} \in \mathbb{R}^m, \left\| \boldsymbol{x} - \boldsymbol{p}_i \right\| > R + \bar{r} \right. \right\}$$

where $\boldsymbol{p}_i$ is the position of the colliding object, and

$$\bar{r} = \begin{cases} r & \text{the colliding object is the other agent} \\ r_k & \text{the colliding object is the } k^{th} \text{ obstacle} \end{cases}$$

Global cohesive ball:

$$\Upsilon \triangleq \left\{ \boldsymbol{x} \left| \boldsymbol{x} \in \mathbb{R}^m, \left\| \boldsymbol{x} - \boldsymbol{c} \right\| \leq r_c \right. \right\}$$

where $\boldsymbol{c} = \frac{1}{n} \sum\limits_{i=1}^{n} \boldsymbol{p}_i$.

Local cohesive ball for the ith agent:

$$\Delta_i \triangleq \left\{ \boldsymbol{x} \left| \boldsymbol{x} \in \mathbb{R}^m, \left\| \boldsymbol{x} - \boldsymbol{c}_i \right\| \leq r_{ci} \right. \right\}$$

where $\boldsymbol{c}_i = \frac{1}{|N_i|} \sum\limits_{i \in N_i} \boldsymbol{p}_i$. N_i denotes the closed neighborhood of agent i, which includes the agent i and all the neighboring nodes (agents) that have communication connections with agent i defined by the adjacency matrix $\mathcal{A}$. $|N_i|$ denotes the cardinality of N_i. r_{ci} denotes the radius of the local cohesive ball for agent i. Note that if each agent is cohesive in its local cohesive ball and the communication topology is connected, a global cohesive ball Υ exists.

4.2.2 Optimal flocking control algorithm design

For the convenience of formulation, we define the error state

$$\hat{X} = [\hat{p}^T \quad \hat{v}^T]^T \triangleq X - X_f \tag{4.47}$$

where $X_f = [p_f^T \quad \mathbf{1}_{1\times n} \otimes v_r^T]^T$; $p_f \in \mathbb{R}^{nm}$ is the final position and $v_r \in \mathbb{R}^m$ is the desired flocking velocity. Note that for the flocking problem, all agents will reach the same velocity v_r, while the position of each agent should be confined in the cohesive ball but may not be specified.

By the property of the Laplacian matrix (3.8), we have

$$(\mathcal{L} \otimes I_m)(\mathbf{1}_{n\times 1} \otimes v_r) = \mathbf{0}_{nm\times 1} \tag{4.48}$$

The reference state X_f satisfies

$$\dot{X}_f = AX_f + BU_f \tag{4.49}$$

where $U_f = \mathbf{1}_{n\times 1} \otimes u_f$ is the control input of the reference trajectory. In the scope of this chapter, we assume that the desired flocking velocity v_r is a constant. Thus, $U_r = \mathbf{0}_{nm\times 1}$. From Eqs. (3.37) and (4.49), the error dynamics can be written as

$$\dot{\hat{X}} = A\hat{X} + B\hat{U} = A\hat{X} + BU \tag{4.50}$$

where $\hat{U} = U - U_f = U$. The flocking is achieved when the system (4.50) is asymptotically stable.

In this section, the flocking problem is formulated as a unified optimal control problem including three cost function components:

$$\begin{aligned} &Min: \mathcal{J}_f = \mathcal{J}_{f1} + \mathcal{J}_{f2} + \mathcal{J}_{f3} \\ &S.t. \quad \dot{\hat{X}} = A\hat{X} + BU \end{aligned} \tag{4.51}$$

$\mathcal{J}_{f1}$ is the cost that ensures the alignment of velocities.

$$J_{f1} = \int_0^\infty \hat{X}^T R_{f1} \hat{X} dt = \int_0^\infty \left\{ \hat{X}^T \left(\begin{bmatrix} 0_{n\times n} & 0_{n\times n} \\ 0_{n\times n} & w_v^2 \mathcal{L}^2 \end{bmatrix} \otimes I_m \right) \hat{X} \right\} dt \tag{4.52}$$

where $\mathcal{L}$ is the symmetric Laplacian matrix established by the undirected and connected communication graph, w_v is the weight on the velocity. Note that in the cost function $\mathcal{J}_{f1}$, velocity alignment is the objective and thus the weight on the position error is zero. The weight matrix $w_v^2\mathcal{L}^2$ for the velocity error is used to guarantee that the analytical solution of the Riccati equation for the optimal control law is a linear function of the Laplacian matrix $\mathcal{L}$, and is thus completely dependent on the local communication topology, which will be shown in the proof of Theorem 4.3. R_{f1} can be shown to be positive semi-definite by Proposition 2.1.

The second cost function $\mathcal{J}_{f_2}$ has the form of

$$\mathcal{J}_{f_2} = \int_0^\infty h_f(\hat{\boldsymbol{X}})dt \tag{4.53}$$

where $h_f(\hat{\boldsymbol{X}})$ denotes the penalty function of navigation, cohesion, and collision/obstacle avoidance. It will be constructed from the inverse optimal control approach in Theorem 4.3.

The control effort cost $\mathcal{J}_{f_3}$ has the regular quadratic form

$$\mathcal{J}_{f_3} = \int_0^\infty \boldsymbol{U}^T R_{f_2} \boldsymbol{U} dt \tag{4.54}$$

where $R_{f_2} = w_c^2 I_n \otimes \boldsymbol{I}_m$ is positive definite and w_c is the weighting parameter.

Before presenting the main result, we first define the penalty function $g(\hat{\boldsymbol{X}})$,

$$g(\hat{\boldsymbol{X}}) = g_t(\hat{\boldsymbol{X}}) + g_c(\hat{\boldsymbol{X}}) + g_a(\hat{\boldsymbol{X}}) \tag{4.55}$$

to address the three flocking characteristics including navigation, cohesiveness, and obstacle/collision avoidance. The first component $g_t(\hat{\boldsymbol{X}})$ in Eq. (4.55) is the navigation penalty function for the entire flock to track a desired trajectory:

$$g_t(\hat{\boldsymbol{X}}) \triangleq \sum_{i=1}^{n} g_{t_i}(\hat{\boldsymbol{X}}) \tag{4.56}$$

$$g_{t_i}(\hat{\boldsymbol{X}}) \triangleq \begin{cases} \begin{bmatrix} (\boldsymbol{p}_i - \boldsymbol{p}_r)^T & (\boldsymbol{v}_i - \boldsymbol{v}_r)^T \end{bmatrix} \begin{bmatrix} w_{tp}^2 \boldsymbol{I}_m & \alpha w_{tp} w_{tv} \boldsymbol{I}_m \\ \alpha w_{tp} w_{tv} \boldsymbol{I}_m & w_{tv}^2 \boldsymbol{I}_m \end{bmatrix} & \\ \quad \begin{bmatrix} \boldsymbol{p}_i - \boldsymbol{p}_r \\ \boldsymbol{v}_i - \boldsymbol{v}_r \end{bmatrix}, & \text{if agent } i \text{ has access to the reference} \\ \quad 0, & \text{if not} \end{cases} \quad i = 1, \ldots, n \tag{4.57}$$

where w_{tp} and w_{tv} are tunable weights to control the tracking speed. α is chosen between $(-1, 1)$ to guarantee $g_t(\hat{\boldsymbol{X}}) > 0$, which will be shown in the proof of Theorem 4.3. $\boldsymbol{p}_r$ and $\boldsymbol{v}_r$ are the position and velocity of the reference.

It is worth noting that the navigation penalty function (4.56) differs from the cost function $\mathcal{J}_{f_1}$ in Eq. (4.52) because $\mathcal{J}_{f_1}$ is used to guarantee that all agents reach the same velocity cooperatively whereas the penalty function (4.56) is used for agent i to track the desired trajectory such that all other agents can flock around this desired trajectory. In addition, $\mathcal{J}_{f_1}$ is to be minimized for all agents whereas the penalty function (4.56) exists only for some of the agents that have access to the information of the reference as shown in Eq. (4.57). Also note that in Eq. (4.47), the ith component of $\boldsymbol{p}_f$ will be specified as $\boldsymbol{p}_r$ if agent i has the access to the reference. For the agents that do not have access to the reference, the corresponding components of $\boldsymbol{p}_f$ will not be specified but should be confined in the cohesive ball such that the flock can move around the trajectory of reference.

The second component $g_c(\hat{X})$ in Eq. (4.55) is the cohesiveness penalty function defined as

$$g_c(\hat{X}) \triangleq M_c(p)\hat{v} \tag{4.58}$$

with $M_c(p) = \begin{bmatrix} m_c^T(p_1) & m_c^T(p_2) & \cdots & m_c^T(p_n) \end{bmatrix}$

$$m_c(p_i) \triangleq \begin{cases} 0_{m\times 1} & p_i \in \Delta_i \\ \left(r_{ci}^2 - \|p_i - c_i\|^2\right)^2 (p_i - c_i) & p_i \notin \Delta_i \end{cases}, \quad i = 1, ..., n \tag{4.59}$$

The third component $g_a(\hat{X})$ in Eq. (4.55) is the obstacle/collision avoidance penalty function defined as

$$g_a(\hat{X}) \triangleq M_a(p)\hat{v} \tag{4.60}$$

(6.14) with $M_a(p) = \left[\left(\sum_{\Omega_1} m_{l_1}(p_1)\right)^T \quad \left(\sum_{\Omega_2} m_{l_2}(p_2)\right)^T \quad \cdots \quad \left(\sum_{\Omega_n} m_{l_n}(p_n)\right)^T \right]$

$$m_{l_i}(p_i) \triangleq \begin{cases} 0_{m\times 1} & p_{l_i} \in \Xi_i \\ -\dfrac{\left((R+\bar{r})^2 - \left\|p_i - p_{l_i}\right\|^2\right)^2}{\left(\left\|p_i - p_{l_i}\right\|^2 - (r+\bar{r})^2\right)^2} (p_i - p_{l_i}) & p_{l_i} \in \Sigma_i \\ \text{not defined} & p_{l_i} \in \Phi_i \end{cases} \quad i = 1, ..., n \tag{4.61}$$

where $p_{l_i} \in \Omega_i$ and Ω_i denotes the set of obstacles and for all the agents excluding agent i:

$$\Omega_i = \{p_1, \cdots, p_j, \cdots, p_n, O_1, \cdots O_k, \cdots, O_q\} \setminus \{p_i\}$$

The main result of this section is presented in the following theorem.

Theorem 4.3. *For a multi-agent system (3.2) with undirected and connected communication topology, w_v, w_c, w_{tp} and w_{tv} can be chosen such that the feedback control law*

$$U = -\frac{w_v}{w_c}(\mathcal{L} \otimes I_m)v - \frac{1}{2w_c^2} g_v{}'(X) \tag{4.62}$$

is an optimal control law for the flocking problem (4.51) and the closed-loop system (4.50) is globally asymptotically stable. In addition, $h_f(\hat{X})$ in the cost function $\mathcal{J}_{f_2}$ in Eq. (4.53) is

$$h_f(\hat{X}) = \frac{w_v}{w_c} g_v{}'^T(\hat{X})(\mathcal{L} \otimes I_m)\hat{v} - g_p{}'^T(\hat{X})\hat{v} + \frac{1}{4w_c{}^2}\left\|g_v{}'(\hat{X})\right\|^2 \tag{4.63}$$

where $g_v{}'(\boldsymbol{X})$, $g_v{}'(\hat{\boldsymbol{X}})$ *and* $g_p{}'(\hat{\boldsymbol{X}})$ *in Eq. (4.62) and Eq. (4.63) denote the partial differentiation of* $g(\hat{\boldsymbol{X}})$ *with respect to the velocity error* $\hat{\boldsymbol{v}}$ *and the position error* $\hat{\boldsymbol{p}}$ *respectively.*

Proof. Specific to this optimal flocking problem, we have the following equations corresponding to the equations in Lemma 2.3 in Chapter 2:

$$T(\hat{\boldsymbol{X}}, \boldsymbol{U}) = \hat{\boldsymbol{X}}^T R_{f_1} \hat{\boldsymbol{X}} + h_f(\hat{\boldsymbol{X}}) + \boldsymbol{U}^T R_{f_2} \boldsymbol{U} \tag{4.64}$$

$$\boldsymbol{f}(\hat{\boldsymbol{X}}, \boldsymbol{U}) = A\hat{\boldsymbol{X}} + B\boldsymbol{U} \tag{4.65}$$

A candidate Lyapunov function $V(\hat{\boldsymbol{X}})$ is chosen to be

$$V(\hat{\boldsymbol{X}}) = \hat{\boldsymbol{X}}^T P \hat{\boldsymbol{X}} + g(\hat{\boldsymbol{X}}) \tag{4.66}$$

where P is the solution of a Riccati equation, which will be shown afterwards.

In order for the function $V(\hat{\boldsymbol{X}})$ to be a valid Lyapunov function, it must be continuously differentiable with respect to $\hat{\boldsymbol{X}}$ or equivalently $g(\hat{\boldsymbol{X}})$ must be continuously differentiable with respect to $\hat{\boldsymbol{X}}$. From Eqs. (4.55)–(4.61), it suffices to show that $\boldsymbol{m}_c(\boldsymbol{p}_i)$ and $\boldsymbol{m}_{l_i}(\boldsymbol{p}_i)$ are continuously differentiable in the defined regions. In fact, this is true if the function itself and its derivative are continuous at the boundary of the defined regions. It is obvious that $\boldsymbol{m}_c(\boldsymbol{p}_i)$ and $\frac{d\boldsymbol{m}_c(\boldsymbol{p}_i)}{d\boldsymbol{p}_i}$ are continuous at the boundary of Δ_i by the definition of continuity. Moreover, since Eq. (4.61) implies that $\lim_{\|p_i - p_{l_i}\| \to (R+\bar{r})^-} \boldsymbol{m}_{l_i}(\boldsymbol{p}_i) = \boldsymbol{0}_{m\times 1} = \lim_{\left\|\boldsymbol{p}_i - \boldsymbol{p}_{l_i}\right\| \to (R+\bar{r})^+} \boldsymbol{m}_{l_i}(\boldsymbol{p}_i)$, $\boldsymbol{m}_{l_i}(\boldsymbol{p}_i)$ is continuous at the outer boundary of Σ_i by the definition of continuity. In addition,

$$\frac{d\boldsymbol{m}_{l_i}(\boldsymbol{p}_i)}{d\boldsymbol{p}_i} = \begin{cases} \boldsymbol{0}_{m\times m} & \boldsymbol{p}_{l_i} \in \Xi_i \\ \dfrac{4((R+\bar{r})^2-(r+\bar{r})^2)((R+\bar{r})^2-\left\|\boldsymbol{p}_i-\boldsymbol{p}_{l_i}\right\|^2)\left(\boldsymbol{p}_i-\boldsymbol{p}_{l_i}\right)\left(\boldsymbol{p}_i-\boldsymbol{p}_{l_i}\right)^T}{\left(\left\|\boldsymbol{p}_i-\boldsymbol{p}_{l_i}\right\|^2-(r+\bar{r})^2\right)^3} & \\ \quad -\dfrac{((R+\bar{r})^2-\left\|\boldsymbol{p}_i-\boldsymbol{p}_{l_i}\right\|^2)^2\cdot \boldsymbol{I}_{m\times m}}{\left(\left\|\boldsymbol{p}_i-\boldsymbol{p}_{l_i}\right\|^2-(r+\bar{r})^2\right)^2} & \boldsymbol{p}_{l_i} \in \Sigma_i \\ \text{not defined} & \boldsymbol{p}_{l_i} \in \Phi_i \end{cases} \tag{4.67}$$

It is easy to see that $\lim_{\left\|\boldsymbol{p}_i - \boldsymbol{p}_{l_i}\right\| \to (R+\bar{r})^-} \frac{d\boldsymbol{m}_{l_i}(\boldsymbol{p}_i)}{d\boldsymbol{p}_i} = \boldsymbol{0}_{m\times m} = \lim_{\left\|\boldsymbol{p}_i - \boldsymbol{p}_{l_i}\right\| \to (R+\bar{r})^+} \frac{d\boldsymbol{m}_{l_i}(\boldsymbol{p}_i)}{d\boldsymbol{p}_i}$. Hence, $\frac{d\boldsymbol{m}_{l_i}(\boldsymbol{p}_i)}{d\boldsymbol{p}_i}$ is continuous at the outer boundary of Σ_i. Therefore, $g(\hat{\boldsymbol{X}})$ and the Lyapunov function $V(\hat{\boldsymbol{X}})$ are continuously differentiable with respect to $\hat{\boldsymbol{X}}$.

The Hamiltonian for this optimal flocking problem can be written as

$$\begin{aligned} H(\hat{X}, U, V'^T(\hat{X})) &= T(\hat{X}, U) + V'^T(\hat{X}) f(\hat{X}, U) \\ &= \hat{X}^T R_{f_1} \hat{X} + h_f(\hat{X}) + U^T R_{f_2} U \\ &\quad + [2\hat{X}^T P + g'^T(\hat{X})][A\hat{X} + BU] \end{aligned} \tag{4.68}$$

Setting $(\partial/\partial U) H(\hat{X}, U, V'^T(\hat{X})) = \mathbf{0}$ yields the optimal control law

$$U^* = \phi(\hat{X}) = -\frac{1}{2} R_{f_2}^{-1} B^T V'(\hat{X}) = -R_{f_2}^{-1} B^T P \hat{X} - \frac{1}{2} R_{f_2}^{-1} B^T g'(\hat{X}) \tag{4.69}$$

With Eq. (4.69), it follows that

$$\begin{aligned} V'^T(\hat{X}) f(\hat{X}, \phi(\hat{X})) &= \hat{X}^T (A^T P + PA - 2PSP)\hat{X} \\ &\quad - \hat{X}^T PSg'(\hat{X}) + g'^T(\hat{X})(A - SP)\hat{X} - \frac{1}{2} g'^T(\hat{X}) Sg'(\hat{X}) \end{aligned} \tag{4.70}$$

where $S \triangleq BR_{f_2}^{-1} B^T$. Using Eq. (4.69) and Eq. (4.70) along with Eq. (4.68) yields

$$\begin{aligned} H(\hat{X}, \phi(\hat{X}), V'^T(\hat{X})) &= \hat{X}^T (A^T P + PA + R_{f_1} - PSP)\hat{X} \\ &\quad + g'^T(\hat{X})(A - SP)\hat{X} + h_f(\hat{X}) - \frac{1}{4} g'^T(\hat{X}) Sg'(\hat{X}) \end{aligned} \tag{4.71}$$

In order to prove that the control law (4.69) is an optimal solution for the flocking problem (4.51) using Lemma 2.3 in Chapter 2, the conditions (2.4)–(2.9) need to be verified.

To satisfy the condition (2.8) in Lemma 2.3 in Chapter 2 or let Eq. (4.71) be zero, we can let

$$A^T P + PA + R_{f_1} - PSP = 0 \tag{4.72}$$

and

$$g'^T(\hat{X})(A - SP)\hat{X} + h_f(\hat{X}) - \frac{1}{4} g'^T(\hat{X}) Sg'(\hat{X}) = 0 \tag{4.73}$$

Similar to Eq. (3.66), it can be shown that

$$H(\hat{X}, U, V'^T(\hat{X})) = [U - \phi(\hat{X})]^T R_{f_2} [U - \phi(\hat{X})] \geq 0 \tag{4.74}$$

Therefore, the condition (2.9) is satisfied.

Substituting A, B, R_{f_1}, R_{f_2} in Eq. (4.72) and assuming $P = \begin{bmatrix} P_1 & P_2 \\ P_2 & P_3 \end{bmatrix} \otimes I_m$ yield

$$\begin{bmatrix} -\frac{1}{w_c{}^2} P_2^2 & P_1 - \frac{1}{w_c{}^2} P_2 P_3 \\ P_1 - \frac{1}{w_c{}^2} P_3 P_2 & 2P_2 - \frac{1}{w_c{}^2} P_3^2 \end{bmatrix} + \begin{bmatrix} 0_{n\times n} & 0_{n\times n} \\ 0_{n\times n} & w_v{}^2 \mathcal{L}^2 \end{bmatrix} = 0 \tag{4.75}$$

Then, P can be solved in the analytical form

$$P = \begin{bmatrix} 0_{n\times n} & 0_{n\times n} \\ 0_{n\times n} & w_c w_v L \end{bmatrix} \otimes I_m \tag{4.76}$$

Now we will verify the condition (2.5). First note that the matrix

$$\begin{bmatrix} w_{tp}^2 I_m & \alpha w_{tp} w_{tv} I_m \\ \alpha w_{tp} w_{tv} I_m & w_{tv}^2 I_m \end{bmatrix}$$

in the tracking potential function (4.57) is positive definite. To see this, we can inspect the eigenvalues of this matrix by

$$\det\left(\lambda I - \begin{bmatrix} w_{tp}^2 I_m & \alpha w_{tp} w_{tv} I_m \\ \alpha w_{tp} w_{tv} I_m & w_{tv}^2 I_m \end{bmatrix}\right) = \det\left(\begin{bmatrix} (\lambda - w_{tp}^2) I_m & -\alpha w_{tp} w_{tv} I_m \\ -\alpha w_{tp} w_{tv} I_m & (\lambda - w_{tv}^2) I_m \end{bmatrix}\right) \tag{4.77}$$

Since $(\lambda - w_{tp}^2) I_m$ and $-\alpha w_{tp} w_{tv} I_m$ commute [146], i.e.

$$\left[(\lambda - w_{tp}^2) I_m\right]\left[-\alpha w_{tp} w_{tv} I_m\right] = \left[-\alpha w_{tp} w_{tv} I_m\right]\left[(\lambda - w_{tp}^2) I_m\right],$$

we have

$$\begin{aligned} &\det\left(\begin{bmatrix} (\lambda - w_{tp}^2) I_m & -\alpha w_{tp} w_{tv} I_m \\ -\alpha w_{tp} w_{tv} I_m & (\lambda - w_{tv}^2) I_m \end{bmatrix}\right) \\ &= \det\left((\lambda - w_{tp}^2)(\lambda - w_{tv}^2) I_m - \alpha^2 w_{tp}^2 w_{tv}^2 I_m\right) \\ &= \det\left(\lambda^2 I_m - (w_{tp}^2 + w_{tv}^2)\lambda I_m + (1-\alpha^2) w_{tp}^2 w_{tv}^2 I_m\right) \end{aligned} \tag{4.78}$$

The eigenvalues are

$$\begin{aligned} \lambda = &\underbrace{\frac{w_{tp}^2 + w_{tv}^2 + \sqrt{\left(w_{tp}^2 + w_{tv}^2\right)^2 - 4(1-\alpha^2) w_{tp}^2 w_{tv}^2}}{2}, \cdots,}_{m} \\ &\underbrace{\frac{w_{tp}^2 + w_{tv}^2 - \sqrt{\left(w_{tp}^2 + w_{tv}^2\right)^2 - 4(1-\alpha^2) w_{tp}^2 w_{tv}^2}}{2}, \cdots}_{m} \end{aligned} \tag{4.79}$$

It can be seen that all the eigenvalues are positive when $\alpha \in (-1, 1)$, which implies that the weighting matrix in $g_t(\hat{X})$ is positive definite and $g_t(\hat{X}) > 0$ for $\hat{X} \neq 0$. Moreover, from the solution of P in Eq. (4.76), the Lyapunov function becomes

$$V(\hat{X}) = \hat{X}^T P \hat{X} + g(\hat{X}) = w_c w_v \boldsymbol{v}^T (\mathcal{L} \otimes I_m) \boldsymbol{v} + g_t(\hat{X}) + g_c(\hat{X}) + g_a(\hat{X}) \tag{4.80}$$

Note that the property (4.48) is used in Eq. (4.80) so that the first term in $V(\hat{\boldsymbol{X}})$ only contains $\boldsymbol{v}$. We have shown $w_c w_v \boldsymbol{v}^T(\mathcal{L}\otimes I_m)\boldsymbol{v} > 0$ and $g_t(\hat{\boldsymbol{X}}) > 0$ for $\hat{\boldsymbol{X}} \neq 0$. Therefore, $V(\hat{\boldsymbol{X}}) > 0$ ($\hat{\boldsymbol{X}} \neq 0$) can be guaranteed by selecting a large enough w_{tp} (or w_{tv}) in $g_t(\hat{\boldsymbol{X}})$ such that the positive terms $g_t(\hat{\boldsymbol{X}})$ and $w_c w_v \boldsymbol{v}^T(\mathcal{L}\otimes I_m)\boldsymbol{v}$ are always greater than the last two sign-indefinite terms and thus the condition (2.5) can be satisfied.

Next, the cost function term $h_f(\hat{\boldsymbol{X}})$ in J_{f_2} (Eq. (4.53)) is constructed from solving Eq. (4.73) and using Eq. (4.76):

$$h_f(\hat{\boldsymbol{X}}) = \frac{w_v}{w_c} {g_v}'^T(\hat{\boldsymbol{X}})(\mathcal{L}\otimes I_m)\hat{\boldsymbol{v}} - {g_p}'^T(\hat{\boldsymbol{X}})\hat{\boldsymbol{v}} + \frac{1}{4{w_c}^2}\left\| {g_v}'(\hat{\boldsymbol{X}})\right\|^2 \tag{4.81}$$

which turns out to be Eq. (4.63).

Using Eq. (4.72) and Eq. (4.73), Eq. (4.70) becomes

$$V'^T(\hat{\boldsymbol{X}})f(\hat{\boldsymbol{X}},\phi(\hat{\boldsymbol{X}})) = -[\hat{\boldsymbol{X}}^T R_{f_1}\hat{\boldsymbol{X}} + h_f(\hat{\boldsymbol{X}}) + (\hat{\boldsymbol{X}}^T P + \frac{1}{2}g'^T(\hat{\boldsymbol{X}}))S(P\hat{\boldsymbol{X}} + \frac{1}{2}g'(\hat{\boldsymbol{X}}))] \tag{4.82}$$

Since $\hat{\boldsymbol{X}}^T R_{f_1}\hat{\boldsymbol{X}} \geq 0$ and $(\hat{\boldsymbol{X}}^T P + \frac{1}{2}g'^T(\hat{\boldsymbol{X}}))S(P\hat{\boldsymbol{X}} + \frac{1}{2}g'(\hat{\boldsymbol{X}})) > 0$ when $\hat{\boldsymbol{X}} \neq \boldsymbol{0}$, it can be seen from Eq. (4.82) that the condition (2.7) can be met if $h_f(\hat{\boldsymbol{X}}) \geq 0$. By selecting proper values of the weights w_v and w_c in Eq. (4.81), one can always ensure $h_f(\hat{\boldsymbol{X}}) \geq 0$. Specifically, it can be achieved by choosing a small enough w_c for given w_{tp} (or w_{tv}) and w_v such that the positive term $\frac{1}{4{w_c}^2}\left\| {g_v}'(\hat{\boldsymbol{X}})\right\|^2$ in Eq. (4.81) is always greater than the first two sign-indefinite terms.

Substituting P and $g'(\hat{\boldsymbol{X}})$ into Eq. (4.69) leads to the optimal control law

$$\phi(\hat{\boldsymbol{X}}) = -\frac{w_v}{w_c}(\mathcal{L}\otimes I_m)\hat{\boldsymbol{v}} - \frac{1}{2w_c^2}{g_v}'(\hat{\boldsymbol{X}}) \tag{4.83}$$

Substituting $\hat{\boldsymbol{v}} = \boldsymbol{v} - \boldsymbol{1}_{n\times 1}\otimes \boldsymbol{v}_r$ into Eq. (4.83) and using Eq. (4.48), the optimal control law (4.83) becomes

$$\hat{\boldsymbol{U}} = \phi(\boldsymbol{X}) = -\frac{w_v}{w_c}(\mathcal{L}\otimes I_m)\boldsymbol{v} - \frac{1}{2w_c^2}{g_v}'(\hat{\boldsymbol{X}}) \tag{4.84}$$

Also note that

$$\begin{aligned} {g_v}'(\hat{\boldsymbol{X}}) &= {g_{t_v}}'(\hat{\boldsymbol{X}}) + {g_{c_v}}'(\hat{\boldsymbol{X}}) + {g_{a_v}}'(\hat{\boldsymbol{X}}) \\ &\triangleq \frac{\partial g_t(\hat{\boldsymbol{X}})}{\partial \hat{\boldsymbol{v}}} + \frac{\partial g_c(\hat{\boldsymbol{X}})}{\partial \hat{\boldsymbol{v}}} + \frac{\partial g_a(\hat{\boldsymbol{X}})}{\partial \hat{\boldsymbol{v}}} = \frac{\partial g_t(\hat{\boldsymbol{X}})}{\partial \hat{\boldsymbol{v}}} + \boldsymbol{M}_c^T(\boldsymbol{p}) + \boldsymbol{M}_a^T(\boldsymbol{p}) \end{aligned} \tag{4.85}$$

Since $\boldsymbol{p}_r$ and $\boldsymbol{v}_r$ are known to agent i if it has access to the reference, ${g_{t_v}}'(\hat{\boldsymbol{X}}) = {g_{t_v}}'(\boldsymbol{X})$ and thus ${g_v}'(\hat{\boldsymbol{X}}) = {g_v}'(\boldsymbol{X})$. Therefore, using ${g_v}'(\hat{\boldsymbol{X}}) = {g_v}'(\boldsymbol{X})$ in Eq. (4.84), the optimal control law turns out to be Eq. (4.62), $\boldsymbol{U} = -\frac{w_v}{w_c}(\mathcal{L}\otimes I_m)\boldsymbol{v} - \frac{1}{2w_c^2}{g_v}'(\boldsymbol{X})$

It remains to verify the conditions (2.4) and (2.6). It can be seen from Eqs. (4.66) and (4.69) that the conditions (2.4) and (2.6) are met if $g(\hat{\boldsymbol{X}}) = 0$ and $g'(\hat{\boldsymbol{X}}) = \boldsymbol{0}$ when $\hat{\boldsymbol{X}} = \boldsymbol{0}$. From the definitions of the penalty functions (4.55)–(4.60), one can see that $g(\hat{\boldsymbol{X}}) = 0$ when $\hat{\boldsymbol{X}} = \boldsymbol{0}$. Eqs. (4.84), (4.85), (4.59) and (4.61) imply that if one flocking agent encounters an obstacle or the other agent, the avoidance force is not zero and the agent will be driven away from the obstacle or the other agent until a new flocking is reached. Therefore, when the agents cohesively flock along a reference and have no collision, we have $g_v{}'(\boldsymbol{X}) = \boldsymbol{0}$, which validates the conditions (2.4) and (2.6).

Now, all the conditions (2.4)–(2.9) in Lemma 2.3 in Chapter 2 can be satisfied. Therefore, according to Lemma 2.3 in Chapter 2, the control law (4.62) is an optimal control law for the problem (4.51) in the sense of Eq. (2.11) and Eq. (2.12), and the closed-loop system is asymptotically stable. It implies $\boldsymbol{X} = \boldsymbol{X}_f$ and the flocking is achieved.

In addition, it can be easily seen from Eq. (4.66) that $V(\hat{\boldsymbol{X}}) \to \infty$ as $\left\| \hat{\boldsymbol{X}} \right\| \to \infty$. Therefore, the closed-loop system is globally asymptotically stable. Note that the global asymptotic stability region excludes the undefined area Φ_i, which is physically meaningful because no agent can start from inside the obstacle or the other agent. □

As can be seen from the proof of Theorem 4.3, the potential function $g(\hat{\boldsymbol{X}})$ provides the navigation/reference tracking, cohesion, and obstacle/collision avoidance capabilities. The inverse optimal control approach facilitates the integration of this potential function into the cost function $h_f(\hat{\boldsymbol{X}})$ such that an analytical distributed optimal control law to achieve these capabilities can be obtained.

Remark 4.6. The optimal control law (4.62) can be rewritten as

$$\boldsymbol{U} = -\frac{w_v}{w_c}(\mathcal{L} \otimes I_m)\boldsymbol{v} - \frac{1}{2w_c^2} g_{t_v}{}'(\boldsymbol{X}) - \frac{1}{2w_c^2} g_{c_v}{}'(\boldsymbol{X}) - \frac{1}{2w_c^2} g_{a_v}{}'(\boldsymbol{X}) \tag{4.86}$$

Note that the first term $-\frac{w_v}{w_c}(\mathcal{L} \otimes I_m)\boldsymbol{v}$ in Eq. (4.86) is the velocity alignment control. It is a linear function of $(\mathcal{L} \otimes I_m)\boldsymbol{v}$ and thus only requires the local information from the agent's neighbors based on the communication topology. The second term in Eq. (4.86) is the tracking control law existing only for the agents that have the access to the reference trajectory information. It only requires the agent's own information and the reference information. The third term is the cohesive control, which takes effect when the agent is out of the cohesive ball. It can be seen from Eq. (4.59) that the cohesive control only requires the local information (local cohesion center $\boldsymbol{c}_i$) from its neighbors. The last term in Eq. (4.86) is to provide the obstacle/collision avoidance capability. Note that the obstacle/collision avoidance law of each agent requires the relative position information of the agent with respect to the obstacle or the colliding agent from the agent's onboard sensors, which is also local because it is only needed when the obstacle/colliding agent enters the reaction region that is in the local agent's sensor detection region. In summary, the proposed distributed optimal flocking law of each agent only depends on the local information.

Remark 4.7. To recapitulate the selection of the weight parameters, one can choose a large w_{tp} (or w_{tv}) for Eq. (4.80) to meet the condition (2.5). Then with the above chosen w_{tp} (or w_{tv}), one can choose a small enough w_c for Eq. (4.81) to meet the condition (2.7). Meanwhile, the parameter α is selected to guarantee $g_t(\hat{X}) > 0$ for $\hat{X} \neq \mathbf{0}$. Moreover, w_v can be tuned to adjust how fast the velocity alignment is achieved.

Remark 4.8. For the flocking problem, most of the literature only concerns velocity alignment, whereas the proposed flocking algorithm in this chapter is able to achieve tracking of the desired reference trajectory as well as the velocity alignment. Note that $\boldsymbol{v}_r$ is the desired final flocking velocity. The velocity of agent i (there may be more than one agent) that has access to the reference is controlled to track the desired $\boldsymbol{v}_r$ by virtue of the penalty function $g_t(\hat{X})$ and the velocities of all the other agents are cooperatively controlled to align with the velocity of agent i by minimizing the cost function J_{f_1}. Additionally, agent i can track the desired reference trajectory as well so that the whole flock can move closely along the reference.

4.2.3 Numerical examples

In this section, simulation results are shown to demonstrate the effectiveness of the proposed optimal flocking law given by Theorem 4.3. Flocking control of four mobile agents is considered in a 2D environment, i.e. $m = 2$. The communication topology is given in Fig. 4.23. The Laplacian matrix can be set up by the definition

$$\mathcal{L} = \begin{bmatrix} 2 & -1 & 0 & -1 \\ -1 & 2 & -1 & 0 \\ 0 & -1 & 2 & -1 \\ -1 & 0 & -1 & 2 \end{bmatrix} \tag{4.87}$$

Assume that agent 1 has access to the reference, which is a rectangular trajectory with the initial position being [5, 5] m and the initial velocity being [0, −0.2] m/s. At the instant of 50 s, the reference velocity changes to [−0.2, 0] m/s. At the instant of 100 s, the reference velocity changes to [0, 0.2] m/s and then changes to [0.2, 0] m/s at the instant of 150 s. Finally, the reference will return to the starting point.

The initial positions of the four agents are assumed to be (2, 2) m, (2, −2) m, (−2, −2) m, and (−2, 2) m, respectively. The initial velocities of the four agents are assumed to be (0.2, −0.4) m/s, (−0.4, 0.6) m/s, (0.6, −0.2) m/s, and (−0.2, −0.8)

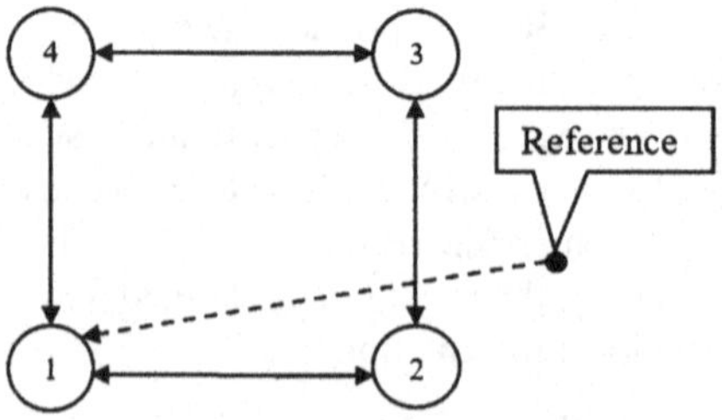

Figure 4.23 Communication topology and reference access.

m/s, respectively. The radius of each agent is $r = 0.2$ m. The radius of the collision/obstacle avoidance region is $R = 0.5$ m. The radii of local cohesive balls are all set to $r_{ci} = 0.5$ m to guarantee a global cohesive ball with the radius of $r_c = 1.2$ m. α is chosen to be 0.9. The weights in the control law are set to $w_v = 2$, $w_c = 1$, $w_{tp} = 5.5$ and $w_{tv} = 1$.

The next three subsections will demonstrate the flocking behaviors successively including velocity alignment, navigation, cohesion, and obstacle/collision avoidance.

A. Flocking with velocity alignment and navigation

In this scenario, the multi-agent flocking with velocity alignment and navigation capabilities is demonstrated. An obstacle is placed at (8, 0) m with the radius of $r_1 = 0.4$ m, which is not on the trajectory of any agent. The flocking results using the proposed optimal control law can be seen from Fig. 4.24 and Fig. 4.25.

Fig. 4.24 demonstrates that all agents are able to achieve flocking from the initial positions, which is marked by '$t = 0$ s'. The whole flock travels along the reference (solid green line; mid gray in print version) as the time evolves since agent 1 (solid black line) tracks the reference and the other agents keep the same velocity with agent 1. Fig. 4.25 presents the time histories of the agents' positions and velocities, which indicate the alignment of velocities and the tracking of reference. Note that agents 2, 3, and 4 are free to move in terms of positions with the only constraint of velocity alignment with agent 1 because they don't have access to the reference and don't track a given trajectory.

The four circles in Fig. 4.24 denote the desired cohesive ball and it can be seen that all the agents fail to stay inside the cohesive ball since we do not take the cohesion into consideration in the optimal control law design. In addition, since the obstacle is not on

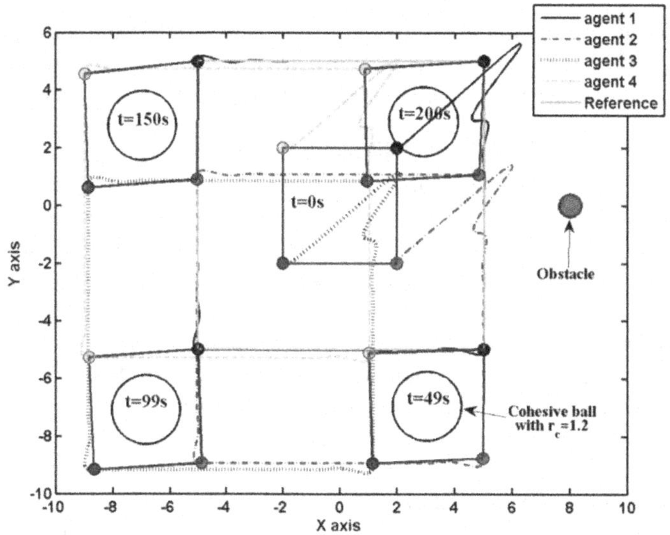

Figure 4.24 Flocking demonstration with velocity alignment and navigation.

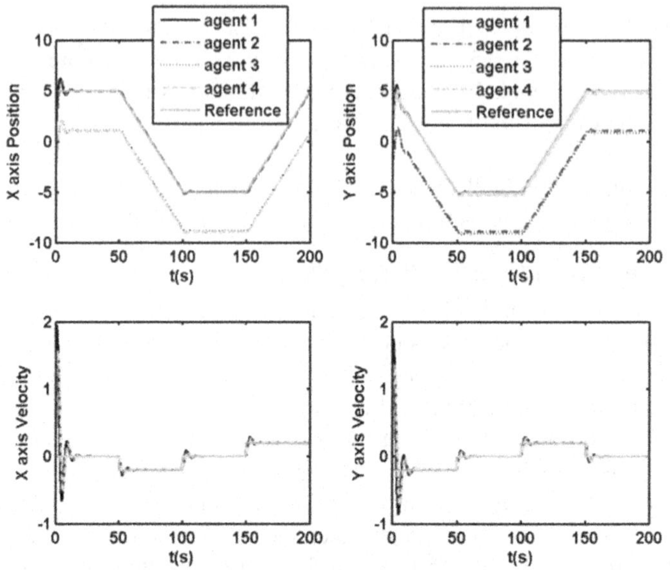

Figure 4.25 Time histories of positions and velocities.

the trajectory of any agent, the obstacle avoidance penalty function is zero. Therefore, the penalty function $g(\hat{X})$ only contains the tracking penalty function $g_t(\hat{X})$. The proposed optimal control law reduces to

$$\phi(X) = -\frac{w_v}{w_c}(\mathcal{L} \otimes I_m)v - \frac{1}{2w_c^2} g_{t_v}{}'(X) \tag{4.88}$$

where the first term ensures the coordination of velocities and is the same as the conventional flocking control in the literature except the weighting parameters. The second term ensures the agent's tracking to the reference.

B. Flocking with velocity alignment, navigation, and cohesion

In this scenario, the multi-agent flocking with cohesion capability is added to the previous optimal control law to demonstrate the velocity alignment, navigation, and cohesion. The obstacle is the same as that for the scenario A. In this case, the penalty function $g(\hat{X})$ contains the tracking penalty function $g_t(\hat{X})$ and the cohesiveness penalty function $g_c(\hat{X})$ since the obstacle is not on the trajectory of any agent. Fig. 4.26 demonstrates that all agents are able to flock along the reference and stay inside the cohesive ball all the time. Compared with Fig. 4.24, it validates the effectiveness of the cohesive component in the optimal control law.

Fig. 4.27 presents the time histories of the agents' positions and velocities, which indicates the alignment of velocities and the tracking of the reference. It is worth noting that the motions of agents 2, 3, and 4 in this scenario are not as free as the motions in the previous scenario A because they must be confined in the cohesive ball.

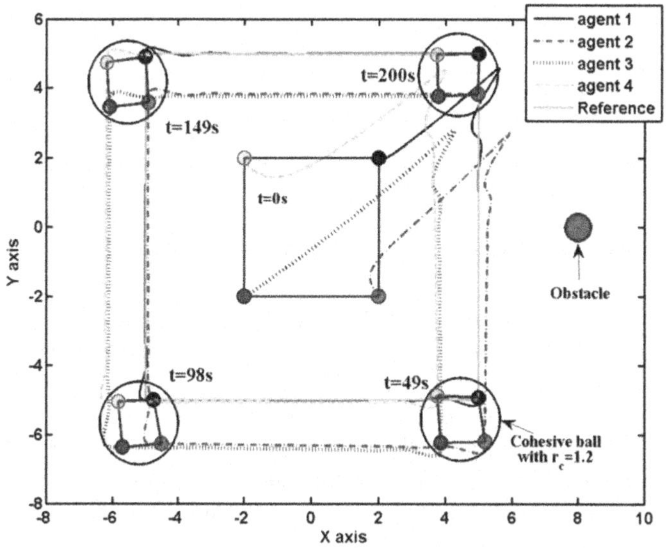

Figure 4.26 Flocking demonstration with velocity alignment, navigation, and cohesion.

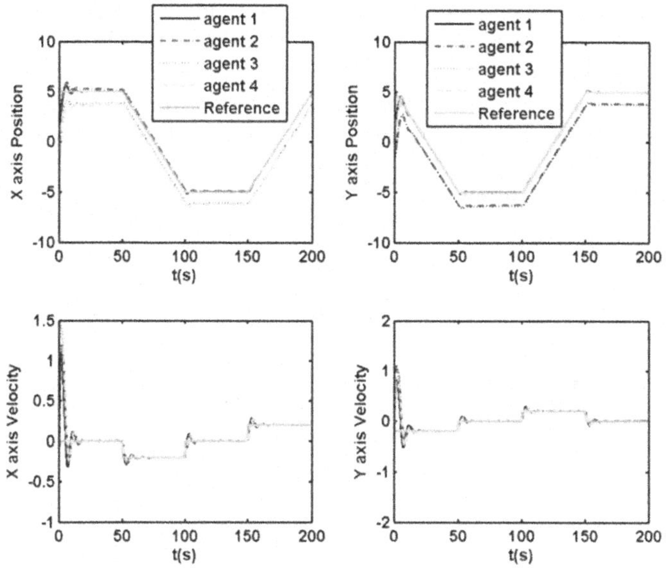

Figure 4.27 Time histories of positions and velocities.

Note that the global cohesion is achieved by implementing the local cohesive control since only local information is available to each agent. The radius of the global cohesive ball r_c is predefined to be 1.2 m. Then, the radius of the local cohesive ball r_{ci}, which is actually used in the controller, can be chosen to be 0.5 m. The proposed optimal control law turns out to be

$$\phi(X) = -\frac{w_v}{w_c}(\mathcal{L} \otimes I_m)v - \frac{1}{2w_c^2} g_{t_v}{}'(X) - \frac{1}{2w_c^2} g_{c_v}{}'(X) \tag{4.89}$$

where the first two terms are the same as for the scenario A and the last term ensures the cohesion.

C. Flocking with velocity alignment, navigation, cohesion, and obstacle/ collision avoidance

In this scenario, all the characteristics of flocking will be illustrated. The penalty function $g(\hat{X})$ contains the tracking penalty function $g_t(\hat{X})$, the cohesiveness penalty function $g_c(\hat{X})$, and the obstacle/collision avoidance penalty function $g_a(\hat{X})$. The optimal control law will be Eq. (4.86). The first obstacle is placed at $(5.3, 0)$ m with the radius of $r_1 = 0.4$ m, which is on the trajectories of agent 1 and agent 2. The second obstacle is placed at $(-5, -2)$ m with the radius of $r_2 = 0.3$ m, which is on the trajectories of agent 1, agent 2, and agent 3. The simulation results are shown in Figs. 4.28–4.30.

Fig. 4.28 demonstrates that all agents are able to achieve flocking with all the desired behaviors including velocity alignment, navigation, cohesion, and obstacle/collision avoidance. It can be seen that the trajectories are different from those in the scenario B due to the obstacle/collision avoidance. The obstacle avoidance takes effect when agents detect the obstacle, which can be seen around the obstacle 1 and 2. The rectangle region shows the collision avoidance whenever the distance between two agents are less than $R + r$. Meanwhile, all the agents still travel along the ref-

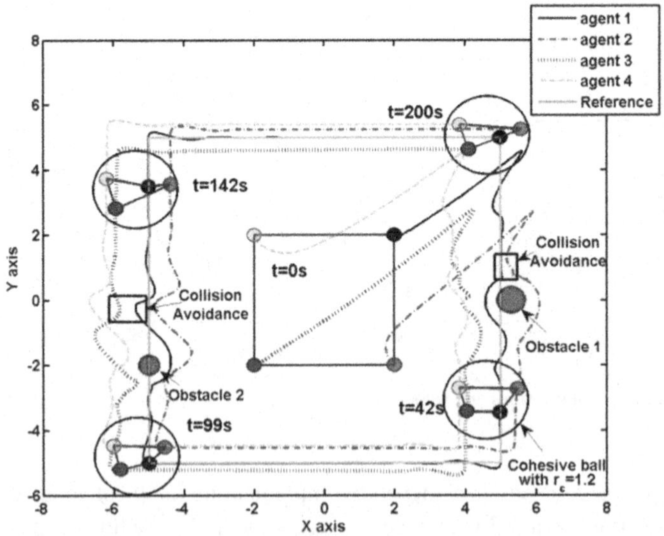

Figure 4.28 Flocking demonstration with velocity alignment, navigation, cohesion, and obstacle/collision avoidance.

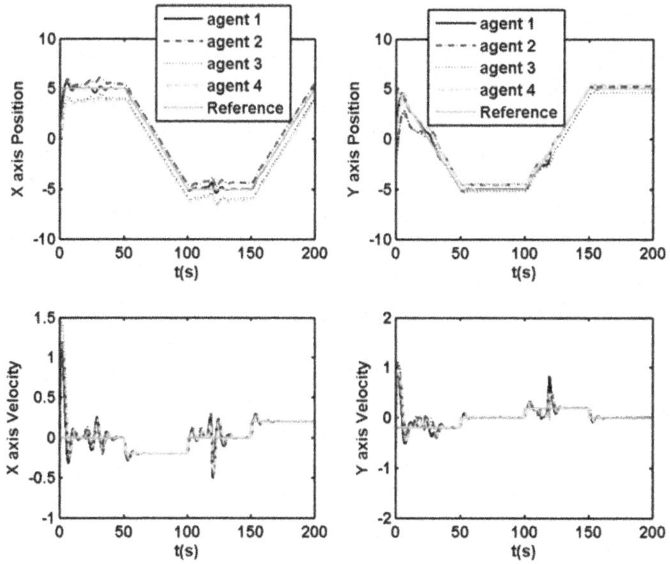

Figure 4.29 Time histories of positions and velocities.

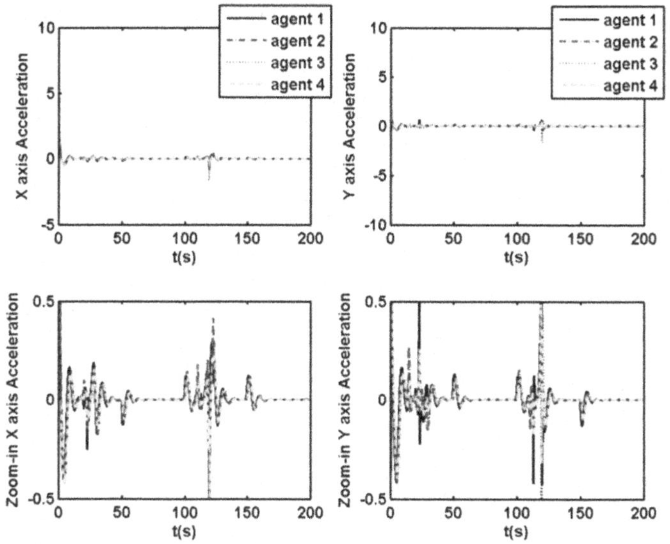

Figure 4.30 Time histories of optimal control inputs.

erence within the cohesive ball. Fig. 4.29 presents the time histories of the agents' positions and velocities, which indicate the alignment of velocities and the tracking of reference. The optimal control inputs are shown in Fig. 4.30. Some large variations occurred when agents do the obstacle/collision avoidance and it can be seen that the optimal flocking does not require a large control effort.

4.3 Conclusion remarks

In this chapter, we have investigated the multi-robot rendezvous problem with obstacle avoidance and then the results are extended to the multi-robot cooperative tracking problem. The stability and optimality of the closed systems have been proved. Simulations are conducted to validate the proposed optimal rendezvous algorithm or cooperative tracking control law. The simulations results show the capabilities of rendezvous and cooperative tracking as well as the obstacle avoidance. Besides, we have investigated the multi-agent flocking algorithm with the characteristics of velocity alignment, navigation, cohesion, and obstacle/collision avoidance. The stability and optimality of the closed system have been proved. Simulations have been conducted to validate the proposed optimal flocking algorithm and the simulations results show that all the desired flocking characteristics have been accomplished.

Optimal formation control of multiple UAVs

5

In this chapter, the kinematic and dynamic models of the UAV are first described. A feedback linearization technique is then employed to obtain a reduced double-integrator model. The formation control problem with obstacle avoidance will be defined based on the reduced model. The original control can be obtained via nonlinear transformations.

5.1 Problem formulation

Point-mass aircraft model [115] is used to describe the motion of formation of flying UAVs in the present research. The related variables are defined with respect to the inertial coordinate frame $(\hat{x}, \hat{y}, \hat{h})$ and are shown in Fig. 5.1.

The model assumes that the aircraft thrust is directed along the velocity vector and that the aircraft always performs coordinated maneuvers. It is also assumed that the Earth is flat and the fuel expenditure is negligible (i.e., the center of mass is time-invariant) [197]. Under these assumptions, the kinematic and dynamic UAV equations of motion can be described as follows.

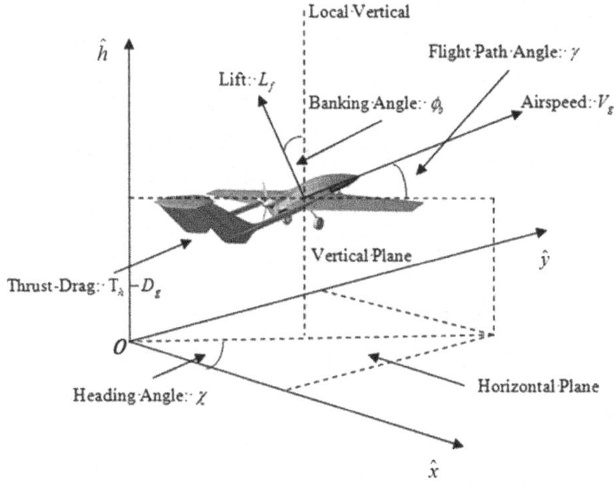

Figure 5.1 UAV models.

Cooperative Control of Multi-Agent Systems. https://doi.org/10.1016/B978-0-12-820118-3.00015-6

Kinematics:

$$\begin{aligned}\dot{x}_i &= V_{g_i} \cos\gamma_i \cos\chi_i \\ \dot{y}_i &= V_{g_i} \cos\gamma_i \sin\chi_i \\ \dot{h}_i &= V_{g_i} \sin\gamma_i\end{aligned} \tag{5.1}$$

where $i = 1, ..., n$ is the index of multiple UAVs under consideration. n is the number of UAVs. For UAV i, x_i is the down-range displacement; y_i is the cross-range displacement; h_i is the altitude; V_{g_i} is the ground speed; γ_i is the flight-path angle and χ_i is the heading angle.

Dynamics:

$$\begin{aligned}\dot{V}_{g_i} &= \frac{T_{h_i} - D_{g_i}}{m_i} - g_a \sin\gamma_i \\ \dot{\gamma}_i &= \frac{g_a}{V_{g_i}}\left(n_{g_i} \cos\phi_{b_i} - \cos\gamma_i\right) \\ \dot{\chi}_i &= \frac{L_{f_i} \sin\phi_{b_i}}{m_i V_{g_i} \cos\gamma_i}\end{aligned} \tag{5.2}$$

where T_{h_i} is the engine thrust; D_{g_i} is the drag; m_i is the mass; g_a is the acceleration due to gravity; L_{f_i} is the vehicle lift; ϕ_{b_i} is the banking angle. The control variables in the UAVs are the g-load $n_{g_i} = \frac{L_{f_i}}{g_a m_i}$ controlled by the elevator, the banking angle ϕ_{b_i} controlled by the combination of rudder and ailerons, and the engine thrust T_{h_i} controlled by the throttle. Throughout the formation control process, the control variables will be constrained to remain within their respective limits.

The nonlinear UAV model can be transformed into a linear time-invariant form by the feedback linearization [114]. Specifically, one can differentiate the kinematic equation (5.1) once with respect to time and then substitute the dynamic equation (5.2) to obtain a double-integrator model:

$$\ddot{x}_i = a_{x_i}, \qquad \ddot{y}_i = a_{y_i} \qquad \ddot{h}_i = a_{h_i} \tag{5.3}$$

where a_{x_i}, a_{y_i}, and a_{h_i} are the new control variables in the point-mass model. The relationships between these control variables and the actual control variables are given by Eq. (5.4) [114]:

$$\begin{aligned}\phi_{b_i} &= \tan^{-1}\left[\frac{a_{y_i} \cos\chi_i - a_{x_i} \sin\chi_i}{\cos\gamma_i (a_{h_i} + g_a) - \sin\gamma_i (a_{x_i} \cos\chi_i + a_{y_i} \sin\chi_i)}\right] \\ n_{g_i} &= \frac{\cos\gamma_i (a_{h_i} + g_a) - \sin\gamma_i (a_{x_i} \cos\chi_i + a_{y_i} \sin\chi_i)}{g_a \cos\phi_{b_i}} \\ T_{h_i} &= [\sin\gamma_i (a_{h_i} + g_a) + \cos\gamma_i (a_{x_i} \cos\chi_i + a_{y_i} \sin\chi_i)] m_i + D_{g_i}\end{aligned} \tag{5.4}$$

The heading angle χ_i and the flight-path angle γ_i are computed as

$$\tan\chi_i = \frac{\dot{y}_i}{\dot{x}_i}, \qquad \sin\gamma_i = \frac{\dot{h}_i}{V_{g_i}} \tag{5.5}$$

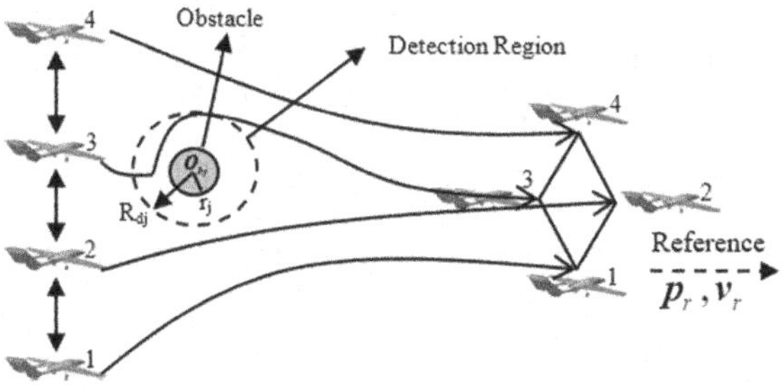

Figure 5.2 Formation flying scenario with obstacle avoidance.

The reduced UAV models can also be expressed as

$$\begin{cases} \dot{\boldsymbol{p}}_i = \boldsymbol{v}_i \\ \dot{\boldsymbol{v}}_i = \boldsymbol{a}_i \end{cases}, \quad i = 1, \ldots, n \tag{5.6}$$

where $\boldsymbol{p}_i = [x_i \quad y_i \quad h_i]^T$, $\boldsymbol{v}_i = [\dot{x}_i \quad \dot{y}_i \quad \dot{h}_i]^T$, $\boldsymbol{a}_i = [a_{x_i} \quad a_{y_i} \quad a_{h_i}]^T$ are the position, velocity, and acceleration, respectively. Eq. (5.6) can be rewritten in a matrix form:

$$\dot{\boldsymbol{X}} = A\boldsymbol{X} + B\boldsymbol{U} \tag{5.7}$$

with

$$\begin{gathered} A = \begin{bmatrix} 0_{n\times n} & I_n \\ 0_{n\times n} & 0_{n\times n} \end{bmatrix} \otimes I_3, \; B = \begin{bmatrix} 0_{n\times n} \\ I_n \end{bmatrix} \otimes I_3 \\ \boldsymbol{X} = [\underbrace{\boldsymbol{p}_1{}^T, \ldots, \boldsymbol{p}_n{}^T}_{\boldsymbol{p}^T}, \underbrace{\boldsymbol{v}_1{}^T, \ldots, \boldsymbol{v}_n{}^T}_{\boldsymbol{v}^T}]^T, \boldsymbol{U} = [\boldsymbol{a}_1{}^T, \ldots, \boldsymbol{a}_n{}^T]^T \end{gathered} \tag{5.8}$$

where $\boldsymbol{X} \in \mathbb{R}^{6n}$ and $\boldsymbol{U} \in \mathbb{R}^{3n}$ are the aggregate state and aggregate control input of all UAVs, respectively.

The formation control problem in this chapter is to design a distributed optimal control law $\boldsymbol{a}_i(t)$ based on the information flow topology, and thereby the actual control laws ϕ_{b_i}, n_{g_i}, and T_{h_i} such that all the UAVs achieve the desired formation and the formation follows a prescribed reference trajectory. In the meantime, each UAV can avoid the obstacle along its trajectory. Fig. 5.2 illustrates the scenario, where R_{dj} denotes the radius of the detection region, r_j denotes the radius of the obstacle, and $\boldsymbol{O}_{bj}$ is the center of the obstacle. $\boldsymbol{p}_r$ and $\boldsymbol{v}_r$ represent the reference trajectory.

For the convenience of formulation, the following regions can be defined:

Collision region for the jth obstacle: $\Lambda_j \triangleq \left\{\boldsymbol{x} \,\middle|\, \boldsymbol{x} \in \mathbb{R}^3, \left\|\boldsymbol{x} - \boldsymbol{O}_{bj}\right\| \le r_j \right\}$

Detection region for the jth obstacle: $\Psi_j \triangleq \left\{\boldsymbol{x} \,\middle|\, \boldsymbol{x} \in \mathbb{R}^3, \left\|\boldsymbol{x} - \boldsymbol{O}_{bj}\right\| < R_{d_j} \right\}$

Reaction region for the jth obstacle: $\Gamma_j \triangleq \left\{\boldsymbol{x} \,\middle|\, \boldsymbol{x} \in \mathbb{R}^3, r_j < \left\|\boldsymbol{x} - \boldsymbol{O}_{bj}\right\| < R_{d_j} \right\}$

The entire safety region can be denoted as $\Theta = \left(\bigcup_j \Lambda_j\right)^c$, and the entire detection-free region can be denoted as $\Pi = \left(\bigcup_j \Psi_j\right)^c$. The symbol '$\bigcup$' denotes the union of sets and the superscript 'c' denotes the complement of a set. In this chapter, we make Assumptions 3.1–3.3, 4.1.

5.2 Integrated optimal control approach to formation control problem

In this section, we propose an integrated optimal control approach to address the formation control problem with trajectory tracking and obstacle avoidance capabilities. In order to formulate the formation control problem, the following definition is needed.

Definition 5.1 ([82]). n vehicles in a formation can be defined by a constant offset vector:

$$\boldsymbol{\sigma} = [\underbrace{\boldsymbol{\sigma}_{p_1}{}^T, \ldots, \boldsymbol{\sigma}_{p_n}{}^T}_{\sigma_p^T}, \mathbf{0}^T, \ldots, \mathbf{0}^T]^T \tag{5.9}$$

The n vehicles are said to be in formation σ at time t if there exist $\mathbb{R}^3$ valued vectors $\boldsymbol{p}_{cs}$ and $\boldsymbol{v}_{cs}$ such that $\boldsymbol{p}_i - \boldsymbol{\sigma}_{p_i} = \boldsymbol{p}_{cs}$ and $\boldsymbol{v}_i = \boldsymbol{v}_{cs}$ for $i = 1, 2, \ldots, n$. $\boldsymbol{p}_{cs}$ and $\boldsymbol{v}_{cs}$ represent the consensus position and velocity, respectively. The vehicles converge to the formation $\boldsymbol{\sigma}$ if there exist real valued functions $\boldsymbol{p}_{cs}(t)$ and $\boldsymbol{v}_{cs}(t)$ such that $\boldsymbol{p}_i(t) - \boldsymbol{\sigma}_{p_i} - \boldsymbol{p}_{cs}(t) \to \mathbf{0}$ and $\boldsymbol{v}_i(t) - \boldsymbol{v}_{cs}(t) \to \mathbf{0}$, as $t \to \infty$ for $i = 1, 2, \ldots, n$.

Fig. 5.3 illustrates the interpretation of the vectors in the above definition using four UAVs. The offset vector $\boldsymbol{\sigma}$ is a constant vector that defines the desired formation pattern with $\boldsymbol{\sigma}_{p_i}$ being the location of vehicle i in the formation. If $\boldsymbol{p}_i - \boldsymbol{\sigma}_{p_i} = \boldsymbol{p}_{cs}$, it implies that $\boldsymbol{p}_i - \boldsymbol{p}_j = \boldsymbol{\sigma}_{p_i} - \boldsymbol{\sigma}_{p_j}$ and thus the formation is achieved. The final velocities of all UAVs should reach the same consensus velocity $\boldsymbol{v}_{cs}$.

Note that in the formulation of [82], a desired formation can be achieved with unpredictable trajectory and velocity, whereas the formulation in this section will be able to guarantee the formation to follow a desired trajectory $\boldsymbol{p}_{ref}$ and desired formation velocity $\boldsymbol{v}_{ref}$, which will be shown in the following optimal control formulation.

Define a formation vector

$$\bar{\boldsymbol{X}} = \left[\bar{\boldsymbol{p}}^T \quad \bar{\boldsymbol{v}}^T\right]^T = \boldsymbol{X} - \boldsymbol{\sigma} = \left[(\boldsymbol{p} - \boldsymbol{\sigma}_p)^T \quad (\boldsymbol{v} - \mathbf{0})^T\right]^T \tag{5.10}$$

Then,

$$\dot{\bar{\boldsymbol{X}}} = \dot{\boldsymbol{X}} - \dot{\boldsymbol{\sigma}} = A\boldsymbol{X} + B\boldsymbol{U} = A(\bar{\boldsymbol{X}} + \boldsymbol{\sigma}) + B\boldsymbol{U} = A\bar{\boldsymbol{X}} + B\boldsymbol{U} \tag{5.11}$$

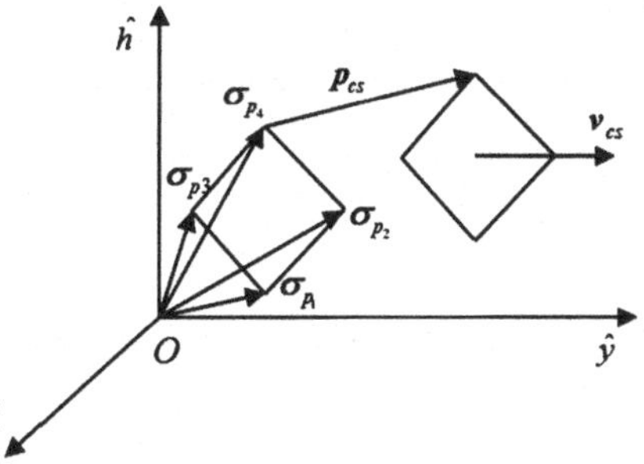

Figure 5.3 UAVs in formation.

since $A\boldsymbol{\sigma} = \mathbf{0}$ and $\dot{\boldsymbol{\sigma}} = \mathbf{0}$. Define a formation error vector

$$\hat{\boldsymbol{X}} = \begin{bmatrix} \hat{\boldsymbol{p}}^T & \hat{\boldsymbol{v}}^T \end{bmatrix}^T \triangleq \bar{\boldsymbol{X}} - \bar{\boldsymbol{X}}_{fs} \tag{5.12}$$

where

$$\begin{aligned} \bar{\boldsymbol{X}}_{fs} &= [\underbrace{\boldsymbol{p}_{cs}{}^T, \ldots, \boldsymbol{p}_{cs}{}^T}_{\bar{\boldsymbol{p}}_{cs}^T}, \underbrace{\boldsymbol{v}_{cs}{}^T, \ldots, \boldsymbol{v}_{cs}{}^T}_{\bar{\boldsymbol{v}}_{cs}^T}]^T \\ &= \begin{bmatrix} \mathbf{1}_{1\times n} \otimes \begin{bmatrix} \alpha_x & \alpha_y & \alpha_h \end{bmatrix} & \mathbf{1}_{1\times n} \otimes \begin{bmatrix} \beta_x & \beta_y & \beta_h \end{bmatrix} \end{bmatrix}^T \end{aligned} \tag{5.13}$$

is the final consensus state. $\alpha_x, \alpha_y, \alpha_h$ are the final consensus positions along $\hat{x}$ axis, $\hat{y}$ axis, and $\hat{h}$ axis, respectively. $\beta_x, \beta_y, \beta_h$ are the final consensus velocities along $\hat{x}$ axis, $\hat{y}$ axis, and $\hat{h}$ axis, respectively. Note that $\bar{\boldsymbol{p}}_{cs}$ and $\bar{\boldsymbol{v}}_{cs}$ are not known a priori.

According to the property of the Laplacian $\mathcal{L}$ in Eq. (3.8), if the undirected communication graph is connected, when the UAVs reach formation,

$$\begin{aligned} (\mathcal{L} \otimes I_3)\bar{\boldsymbol{p}}_{cs} &= \mathbf{0}_{3n\times 1} \\ (\mathcal{L} \otimes I_3)\bar{\boldsymbol{v}}_{cs} &= \mathbf{0}_{3n\times 1} \end{aligned} \tag{5.14}$$

The final consensus state satisfies the dynamic equation

$$\dot{\bar{\boldsymbol{X}}}_{fs} = A\bar{\boldsymbol{X}}_{fs} + B\bar{\boldsymbol{U}}_{fs} = A\bar{\boldsymbol{X}}_{fs} \tag{5.15}$$

since $\bar{\boldsymbol{U}}_{fs} = \mathbf{0}_{3n\times 1}$ when the UAVs reach consensus. From Eqs. (5.11) and (5.15), the error dynamics becomes

$$\dot{\hat{\boldsymbol{X}}} = \dot{\bar{\boldsymbol{X}}} - \dot{\bar{\boldsymbol{X}}}_{fs} = A\bar{\boldsymbol{X}} + B\boldsymbol{U} - A\bar{\boldsymbol{X}}_{fs} = A\hat{\boldsymbol{X}} + B\boldsymbol{U} \tag{5.16}$$

The formation is achieved when the system (5.16) is asymptotically stable.

In this section, the formation control problem is formulated in a unified optimal control framework including three cost function components:

$$\begin{aligned} &Min : \mathcal{J}_u = \mathcal{J}_{u_1} + \mathcal{J}_{u_2} + \mathcal{J}_{u_3} \\ &S.t. \quad \dot{\hat{X}} = A\hat{X} + BU \end{aligned} \tag{5.17}$$

where $\mathcal{J}_{u_1}, \mathcal{J}_{u_2}, \mathcal{J}_{u_3}$ represent formation cost, obstacle avoidance and tracking cost, and control effort cost, respectively.

$\mathcal{J}_{u_1}$ has the form of

$$\mathcal{J}_{u_1} = \int_0^\infty \hat{X}^T R_{u_1} \hat{X} dt = \int_0^\infty \left\{ \hat{X}^T \left(\begin{bmatrix} w_p^2 \mathcal{L}^2 & 0_{n\times n} \\ 0_{n\times n} & {w_v}^2 \mathcal{L}^2 - 2 w_p w_c \mathcal{L} \end{bmatrix} \otimes I_3 \right) \hat{X} \right\} dt \tag{5.18}$$

where $\mathcal{L}$ is the symmetric Laplacian matrix established by the undirected and connected information exchange graph. w_p, w_v, and w_c represent the weights on position, velocity, and control effort, respectively. Note that R_{u_1} can be shown to be positive semi-definite by Propositions 2.1 and 3.1.

$\mathcal{J}_{u_2}$ has the form of

$$\mathcal{J}_{u_2} = \int_0^\infty h_u(\hat{X}) dt \tag{5.19}$$

where $h_u(\hat{X})$ contains the tracking penalty function as well as the obstacle avoidance penalty function. It will be constructed from the inverse optimal control approach in Theorem 5.1. $\mathcal{J}_{u_3}$ has the regular quadratic form of

$$\mathcal{J}_{u_3} = \int_0^\infty U^T R_{u_2} U dt \tag{5.20}$$

where $R_{u_2} = {w_c}^2 I_n \otimes I_3$ is positive definite and w_c is the weighting parameter.

Before presenting the main result, we first define the penalty function $g(\hat{X})$:

$$g(\hat{X}) = g_{tr}(\hat{X}) + g_{oa}(\hat{X}) \tag{5.21}$$

The first component is the tracking penalty function

$$g_{tr}(\hat{X}) \triangleq \sum_{i=1}^{n} g_{tr_i}(\hat{X}) \tag{5.22}$$

$$g_{tr_i}(\hat{X}) = \begin{cases} \begin{bmatrix} (\boldsymbol{p}_i - \boldsymbol{p}_r)^T & (\boldsymbol{v}_i - \boldsymbol{v}_r)^T \end{bmatrix} \begin{bmatrix} w_{tp}^2 I_3 & w_{tp} w_{tv} I_3 \\ w_{tp} w_{tv} I_3 & w_{tv}^2 I_3 \end{bmatrix} \\ \quad \begin{bmatrix} \boldsymbol{p}_i - \boldsymbol{p}_r \\ \boldsymbol{v}_i - \boldsymbol{v}_r \end{bmatrix}, \quad \text{if agent } i \text{ has access to the reference} \\ \quad 0, \qquad \text{if not} \end{cases} \tag{5.23}$$

where $i = 1, \cdots, n$, w_{tp}, w_{tv} are tunable weights to control the tracking speed. $\boldsymbol{p}_r$ and $\boldsymbol{v}_r$ are the position and velocity of the reference.

Remark 5.1. The UAV that has access to the reference will track the desired trajectory by means of minimizing the cost $\mathcal{J}_{u_2}$ that contains $g_{tr}(\hat{\boldsymbol{X}})$, which will be shown in Theorem 5.1.

Note that minimizing the cost function $\mathcal{J}_{u_1}$ guarantees that all the UAVs' positions and velocities are coordinated to achieve the desired formation and reach the consensus velocity v_{cs} synchronously. But where the formation goes cannot be specified as desired by merely minimizing $\mathcal{J}_{u_1}$ because the final UAV's velocity depends on each UAV's initial velocity and results from the negotiation among UAVs during the flight via the optimal consensus algorithm, i.e. minimizing $\mathcal{J}_{u_1}$. Therefore, combining the tracking penalty function $g_{tr}(\hat{\boldsymbol{X}})$ with the formation cost $\mathcal{J}_{u_1}$, the formation is able to follow a desired trajectory $\boldsymbol{p}_r$ with the desired formation velocity $\boldsymbol{v}_r$, i.e. $\boldsymbol{v}_{cs} = \boldsymbol{v}_r$. It is also worth noting that only one UAV having access to the reference is sufficient to guarantee the entire formation to track the desired trajectory.

The second component in $g(\hat{\boldsymbol{X}})$ is the obstacle avoidance penalty function

$$g_{oa}(\hat{\boldsymbol{X}}) = \sum_{i=1}^{n} g_{oa_i}(\hat{\boldsymbol{X}}) \tag{5.24}$$

$$\boldsymbol{g}_{oa_i}(\hat{\boldsymbol{X}}) \triangleq \begin{cases} 0 & \boldsymbol{p}_i \in \Pi \\ \dfrac{\left(R_{d_j}^2 - \|\boldsymbol{p}_i - \boldsymbol{O}_{bj}\|^2\right)^2}{\left(\|\boldsymbol{p}_i - \boldsymbol{O}_{bj}\|^2 - r_j^2\right)^2} (\boldsymbol{v}_i^{x,y} - \boldsymbol{v}_{oa})^T (\boldsymbol{v}_i^{x,y} - \boldsymbol{v}_{oa}) & \boldsymbol{p}_i \in \Gamma_j \quad i = 1, ..., n \\ \text{not defined} & \boldsymbol{p}_i \in \Lambda_j \end{cases} \tag{5.25}$$

where $\boldsymbol{v}_i^{x,y} \in \mathbb{R}^2$ is the projection of the UAV's velocity vector $\boldsymbol{v}_i$ in the $\hat{x} - \hat{y}$ plane; $\boldsymbol{v}_{oa} \in \mathbb{R}^2$ is the prescribed obstacle avoidance velocity. There are many ways to avoid the obstacle from different directions. Since UAVs move primarily along the downrange and cross-range directions, we design the avoidance strategy in the $(\hat{x}, \hat{y})$ plane, which is shown in Fig. 5.4, where r_j denotes the radius of obstacle j and r_{safe} is defined as the safe distance between the UAV and the obstacle. The coordinate system $\hat{x} - O - \hat{y}$ is the horizontal plane of $(\hat{x}, \hat{y}, \hat{h})$. The coordinate system $\hat{x}' - UAV - \hat{y}'$ is co-planar to $\hat{x} - O - \hat{y}$ with the origin at the center of the UAV and the $\hat{x}'$ axis along the tangential direction from the UAV to the dashed circle centered at the obstacle's center with the radius of $r_j + r_{\text{safe}}$. The $\boldsymbol{v}_{oa}$ is chosen to be the projection of $\boldsymbol{v}_i^{x,y}$ on the direction of $\hat{x}'$. Note that the obstacle is avoided if the UAV is driven to the direction of $\boldsymbol{v}_{oa}$.

$\boldsymbol{v}_{oa}$ can be calculated as follows: The rotation angle β between the two coordinate frames $\hat{x} - O - \hat{y}$ and $\hat{x}' - UAV - \hat{y}'$ can be determined from the geometry when the current UAV's position and the obstacle's position are known.

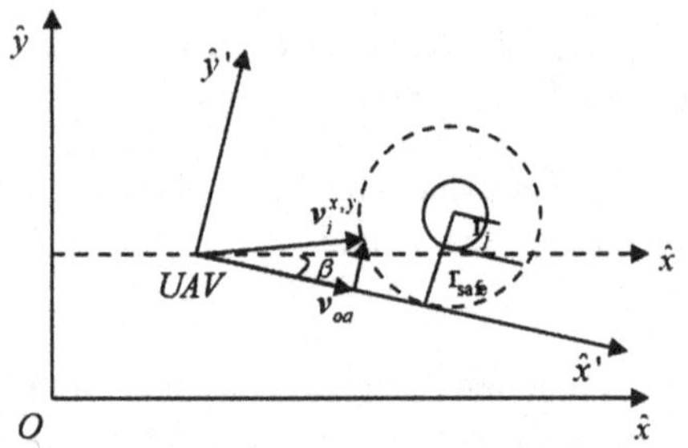

Figure 5.4 Illustration of obstacle avoidance.

Then, the velocity vector $v_i^{x,y}$ expressed in the $\hat{x}' - UAV - \hat{y}'$ frame denoted as $v_i^{x',y'}$ can be calculated by

$$v_i^{x',y'} = \begin{bmatrix} v_i^{x'} \\ v_i^{y'} \end{bmatrix} = C(\beta) \cdot v_i^{x,y} \tag{5.26}$$

where $C(\beta) = \begin{bmatrix} \cos\beta & -\sin\beta \\ \sin\beta & \cos\beta \end{bmatrix}$ is the rotation matrix.

The prescribed obstacle avoidance velocity v_{oa} is obtained by

$$v_{oa} = C^{-1}(\beta) \begin{bmatrix} v_i^{x'} \\ 0 \end{bmatrix} \tag{5.27}$$

Remark 5.2. The obstacle avoidance penalty function is designed to penalize the error between the current velocity $v_i^{x,y}$ and the avoidance velocity v_{oa}. The term $\frac{\left(R_{d_j}^2 - \|p_i - O_{bj}\|^2\right)^2}{\left(\|p_i - O_{bj}\|^2 - r_j^2\right)^2}$ in $g_{oa_i}(\hat{X})$ is designed to guarantee the continuity of the penalty function on the boundary of the detection region, which will be shown in the proof of Theorem 5.1.

The main result of this chapter is presented in the following theorem.

Theorem 5.1. *For a multi-UAV system Eq. (5.1) and Eq. (5.2) with Assumptions 3.1–3.3, 4.1, one can always find proper weights w_p, w_v, and w_c such that the distributed feedback control law*

$$U = \phi(X) = -\frac{w_p}{w_c}(\mathcal{L} \otimes I_3)\left(p - \sigma_p\right) - \frac{w_v}{w_c}(\mathcal{L} \otimes I_3)v - \frac{1}{2w_c^2} g_v'(X) \tag{5.28}$$

is an optimal control for the formation control problem (5.17) with the cost function term $h_u(\hat{X})$ in $\mathcal{J}_{u_2}$ being

$$h_u(\hat{X}) = \frac{w_p}{w_c} g_v{}'^T(\hat{X})(\mathcal{L} \otimes I_3)\hat{p} + \frac{w_v}{w_c} g_v{}'^T(\hat{X})(\mathcal{L} \otimes I_3)\hat{v} - g_p{}'^T(\hat{X})\hat{v} + \frac{1}{4w_c^2}\left\| g_v{}'(\hat{X}) \right\|^2 \tag{5.29}$$

where $g_v{}'(X)$, $g_v{}'(\hat{X})$ and $g_p{}'(\hat{X})$ in Eq. (5.28) and Eq. (5.29) denote the partial differentiation of $g(\hat{X})$ with respect to the velocity error $\hat{v}$ and the position error $\hat{p}$ respectively. In addition, the closed-loop system is globally asymptotically stable.

Proof. Pertaining to this optimal formation control problem, we have the following equations corresponding to Lemma 2.3 in Chapter 2:

$$T(\hat{X}, U) = \hat{X}^T R_{u_1} \hat{X} + h_u(\hat{X}) + U^T R_{u_2} U \tag{5.30}$$

$$f(\hat{X}, U) = A\hat{X} + BU \tag{5.31}$$

A candidate Lyapunov function $V(\hat{X})$ is chosen to be

$$V(\hat{X}) = \hat{X}^T P \hat{X} + g(\hat{X}) \tag{5.32}$$

where P is the solution of a Riccati equation, which will be shown afterwards.

In order for the function $V(\hat{X})$ in Eq. (5.32) to be a valid Lyapunov function, it must be continuously differentiable with respect to $\hat{X}$ or equivalently $g(\hat{X})$ must be continuously differentiable with respect to $\hat{X}$. From Eqs. (5.23) and (5.25), it suffices to show that $g_{oa_i}(\hat{X})$ is continuously differentiable in the safety region Θ. In fact, this is true if $g_{oa_i}(\hat{X})$ and $\frac{d g_{oa_i}(\hat{X})}{d p_i}$ are continuous at the boundary of Ψ_j. Since Eq. (5.25) implies that $\lim_{\|p_i - O_{bj}\| \to R_{d_j}^-} g_{oa_i}(\hat{X}) = 0 = \lim_{\|p_i - O_{bj}\| \to R_{d_j}^+} g_{oa_i}(\hat{X})$, $g_{oa_i}(\hat{X})$ is continuous at the boundary of Ψ_j by the definition of continuity and thus continuous over the safety region Θ. In addition,

$$\frac{d g_{oa_i}(\hat{X})}{d p_i} = \begin{cases} \mathbf{0} & p_i \in \Pi \\ \frac{-4(R_{d_j}^2 - r_j^2)(R_{d_j}^2 - \|p_i - O_{bj}\|^2)}{\left(\|p_i - O_{bj}\|^2 - r_j^2\right)^3} (v_i^{x,y} - v_{oa})^T (v_i^{x,y} - v_{oa}) & p_i \in \Gamma_j \\ \text{not defined} & p_i \in \Lambda_j \end{cases}, \tag{5.33}$$

it is easy to see that $\lim_{\|p_i - O_{bj}\| \to R_{d_j}^-} \frac{d g_{oa_i}(\hat{X})}{d p_i} = \mathbf{0}_{3\times 1} = \lim_{\|p_i - O_{bj}\| \to R_{d_j}^+} \frac{d g_{oa_i}(\hat{X})}{d p_i}$. Hence, $\frac{d g_{oa_i}(\hat{X})}{d p_i}$ is continuous at the boundary of Ψ_j and thus continuous over the safety region Θ. Therefore, $g(\hat{X})$ and the Lyapunov function $V(\hat{X})$ are continuously differentiable with respect to $\hat{X}$ in the safety region Θ.

After showing $V(\hat{X})$ is continuously differentiable, the Hamiltonian for this optimal control problem can be written as

$$\begin{aligned} H(\hat{X}, U, V'^T(\hat{X})) &= T(\hat{X}, U) + V'^T(\hat{X}) f(\hat{X}, U) \\ &= \hat{X}^T R_{u_1} \hat{X} + h_u(\hat{X}) + U^T R_{u_2} U + [2\hat{X}^T P + g'^T(\hat{X})][A\hat{X} + BU] \end{aligned} \tag{5.34}$$

Setting $(\partial/\partial U) H(\hat{X}, U, V'^T(\hat{X})) = 0$ yields the optimal control law

$$U = \phi(\hat{X}) = -\frac{1}{2} R_{u_2}^{-1} B^T V'(\hat{X}) = -R_{u_2}^{-1} B^T P \hat{X} - \frac{1}{2} R_{u_2}^{-1} B^T g'(\hat{X}) \tag{5.35}$$

With Eq. (5.35) it follows that

$$\begin{aligned} V'^T(\hat{X}) f(\hat{X}, \phi(\hat{X})) &= \hat{X}^T (A^T P + PA - 2PSP)\hat{X} - \hat{X}^T PSg'(\hat{X}) \\ &\quad + g'^T(\hat{X})(A - SP)\hat{X} - \frac{1}{2} g'^T(\hat{X}) S g'(\hat{X}) \end{aligned} \tag{5.36}$$

where $S \triangleq B R_{u_2}^{-1} B^T$. Using Eq. (5.35) and Eq. (5.36) along with Eq. (5.34) yields

$$\begin{aligned} H(\hat{X}, \phi(\hat{X}), V'^T(\hat{X})) &= \hat{X}^T (A^T P + PA + R_{u_1} - PSP)\hat{X} \\ &\quad + g'^T(\hat{X})(A - SP)\hat{X} + h_u(\hat{X}) - \frac{1}{4} g'^T(\hat{X}) S g'(\hat{X}) \end{aligned} \tag{5.37}$$

In order to prove that the control law (5.35) is an optimal solution for the formation flying problem (5.17) using Lemma 2.3 in Chapter 2, the conditions (2.4)–(2.9) need to be verified.

To satisfy the condition (2.8) in Lemma 2.3 in Chapter 2 or let Eq. (5.37) to be zero, we can let

$$A^T P + PA + R_{u_1} - PSP = 0 \tag{5.38}$$

and require that

$$g'^T(\hat{X})(A - SP)\hat{X} + h_u(\hat{X}) - \frac{1}{4} g'^T(\hat{X}) S g'(\hat{X}) = 0 \tag{5.39}$$

Similar to Eq. (3.66), it can be shown that

$$H(\hat{X}, U, V'^T(\hat{X})) = [U - \phi(\hat{X})]^T R_{u_2} [U - \phi(\hat{X})] \geq 0 \tag{5.40}$$

Therefore, the condition (2.9) is satisfied.

Next, substituting A, B, R_{u_1}, R_{u_2} in Eq. (5.38) and assuming $P = \begin{bmatrix} P_1 & P_2 \\ P_2 & P_3 \end{bmatrix} \otimes I_3$ yields

$$\begin{bmatrix} -\frac{1}{w_c^2}P_2^2 & P_1 - \frac{1}{w_c^2}P_2P_3 \\ P_1 - \frac{1}{w_c^2}P_3P_2 & 2P_2 - \frac{1}{w_c^2}P_3^2 \end{bmatrix} + \begin{bmatrix} w_p^2\mathcal{L}^2 & 0_{n\times n} \\ 0_{n\times n} & w_v^2\mathcal{L}^2 - 2w_pw_c\mathcal{L} \end{bmatrix} = 0 \quad (5.41)$$

Then, P can be solved in the analytical form

$$P = \begin{bmatrix} w_pw_v\mathcal{L}^2 & w_pw_c\mathcal{L} \\ w_pw_c\mathcal{L} & w_cw_v\mathcal{L} \end{bmatrix} \otimes I_3 \quad (5.42)$$

Note that the purpose of introducing the term $-2w_pw_cL$ in R_{u_1} (see Eq. (5.18)) is to let it cancel $2P_2$ when solving P_3 from Eq. (5.41) in order to make P_3 a linear function of the Laplacian matrix $\mathcal{L}$.

Now we will verify the condition (2.5). First note that the matrix

$$\begin{bmatrix} w_{tp}^2I_3 & w_{tp}w_{tv}I_3 \\ w_{tp}w_{tv}I_3 & w_{tv}^2I_3 \end{bmatrix}$$

in the tracking penalty function (5.23) is positive semi-definite. To see this, we can inspect the eigenvalues of this matrix by

$$\det\left(\lambda I - \begin{bmatrix} w_{tp}^2I_3 & w_{tp}w_{tv}I_3 \\ w_{tp}w_{tv}I_3 & w_{tv}^2I_3 \end{bmatrix}\right) = \det\left(\begin{bmatrix} (\lambda - w_{tp}^2)I_3 & -w_{tp}w_{tv}I_3 \\ -w_{tp}w_{tv}I_3 & (\lambda - w_{tv}^2)I_3 \end{bmatrix}\right) \quad (5.43)$$

Since $(\lambda - w_{tp}^2)I_3$ and $-w_{tp}w_{tv}I_3$ commute [147], i.e. $\left[(\lambda - w_{tp}^2)I_3\right]\left[-w_{tp}w_{tv}I_3\right] = \left[-w_{tp}w_{tv}I_3\right]\left[(\lambda - w_{tp}^2)I_3\right]$, we have

$$\begin{aligned} &\det\left(\begin{bmatrix} (\lambda - w_{tp}^2)I_3 & -w_{tp}w_{tv}I_3 \\ -w_{tp}w_{tv}I_3 & (\lambda - w_{tv}^2)I_3 \end{bmatrix}\right) \\ &= \det\left((\lambda - w_{tp}^2)(\lambda - w_{tv}^2)I_3 - w_{tp}^2w_{tv}^2I_3\right) \\ &= \det\left(\lambda^2I_3 - (w_{tp}^2 + w_{tv}^2)\lambda I_3\right) \end{aligned} \quad (5.44)$$

The eigenvalues are $\lambda = 0, 0, 0, w_{tp}^2 + w_{tv}^2, w_{tp}^2 + w_{tv}^2, w_{tp}^2 + w_{tv}^2 \geq 0$, which implies that the weighting matrix in $g_{tr}(\hat{X})$ is positive semi-definite and $g_{tr}(\hat{X}) \geq 0$.

Moreover, from the solution of P in Eq. (5.42), the Lyapunov function becomes

$$V(\hat{\boldsymbol{X}}) = \hat{\boldsymbol{X}}^T P\hat{\boldsymbol{X}} + g(\hat{\boldsymbol{X}}) = \begin{cases} w_p w_v \bar{\boldsymbol{p}}^T(\mathcal{L}^2 \otimes I_3)\bar{\boldsymbol{p}} + w_c w_v \boldsymbol{v}^T(\mathcal{L} \otimes I_3)\boldsymbol{v} + 2w_p w_c \bar{\boldsymbol{p}}^T(\mathcal{L} \otimes I_3)\boldsymbol{v} + g_{tr}(\hat{\boldsymbol{X}}) & \boldsymbol{p}_i \in \Pi \\ w_p w_v \bar{\boldsymbol{p}}^T(\mathcal{L}^2 \otimes I_3)\bar{\boldsymbol{p}} + w_c w_v \boldsymbol{v}^T(\mathcal{L} \otimes I_3)\boldsymbol{v} + 2w_p w_c \bar{\boldsymbol{p}}^T(\mathcal{L} \otimes I_3)\boldsymbol{v} + g_{tr}(\hat{\boldsymbol{X}}) + g_{oa}(\hat{\boldsymbol{X}}) & \boldsymbol{p}_i \in \Gamma_j \\ \text{not defined} & \boldsymbol{p}_i \in \Lambda_j \end{cases} \tag{5.45}$$

Note that the property (5.14) is used in Eq. (5.45) so that the first three terms in $V(\hat{\boldsymbol{X}})$ only contains $\bar{\boldsymbol{p}}$ and $\boldsymbol{v}$. $V(\hat{\boldsymbol{X}}) > 0$ when $\hat{\boldsymbol{X}} \neq \boldsymbol{0}$ can be guaranteed by selecting a large enough w_v for given w_p such that the positive terms $w_p w_v \bar{\boldsymbol{p}}^T(\mathcal{L}^2 \otimes I_3)\bar{\boldsymbol{p}}$, $w_c w_v \boldsymbol{v}^T(\mathcal{L} \otimes I_3)\boldsymbol{v}$, and $g_{tr}(\hat{\boldsymbol{X}})$ are always greater than the sign-indefinite terms and thus the condition (2.5) can be satisfied.

Next, the cost function term $h_u(\hat{\boldsymbol{X}})$ in $\mathcal{J}_{u_2}$ (Eq. (5.20)) is constructed from solving Eq. (5.39) and using Eq. (5.42):

$$\begin{aligned} h_u(\hat{\boldsymbol{X}}) = {} & \frac{w_p}{w_c} {g_v}'^T(\hat{\boldsymbol{X}})(\mathcal{L} \otimes I_3)\hat{\boldsymbol{p}} + \frac{w_v}{w_c} {g_v}'^T(\hat{\boldsymbol{X}})(\mathcal{L} \otimes I_3)\hat{\boldsymbol{v}} \\ & - {g_p}'^T(\hat{\boldsymbol{X}})\hat{v} + \frac{1}{4w_c^2}\left\| {g_v}'(\hat{\boldsymbol{X}}) \right\|^2 \end{aligned} \tag{5.46}$$

which turns out to be Eq. (5.29). Note that the cost $h_u(\hat{\boldsymbol{X}})$ is obtained from the inverse optimal control strategy since it is not given a priori but is constructed from the optimality condition (5.39).

Using Eq. (5.38) and Eq. (5.39), Eq. (5.36) becomes

$$V'^T(\hat{\boldsymbol{X}})\boldsymbol{f}(\hat{\boldsymbol{X}}, \phi(\hat{\boldsymbol{X}})) = -[\hat{\boldsymbol{X}}^T R_{u_1}\hat{\boldsymbol{X}} + h_u(\hat{\boldsymbol{X}}) + (\hat{\boldsymbol{X}}^T P + \frac{1}{2}g'^T(\hat{\boldsymbol{X}}))S(P\hat{\boldsymbol{X}} + \frac{1}{2}g'(\hat{\boldsymbol{X}}))] \tag{5.47}$$

It can be seen from Eq. (5.47) that the condition (2.7) can be met when $h_u(\hat{\boldsymbol{X}}) \geq 0$ since $\hat{\boldsymbol{X}}^T R_{u_1}\hat{\boldsymbol{X}}$ is positive semi-definite and $(\hat{\boldsymbol{X}}^T P + \frac{1}{2}g'^T(\hat{\boldsymbol{X}}))S(P\hat{\boldsymbol{X}} + \frac{1}{2}g'(\hat{\boldsymbol{X}}))$ is positive definite. By selecting proper values of the weights w_p, w_v, and w_c in Eq. (5.46), one can always ensure $h_u(\hat{\boldsymbol{X}}) \geq 0$. Specifically, it can be achieved by choosing a small enough w_c for given w_p and w_v such that the positive term $\frac{1}{4{w_c}^2}\left\| {g_v}'(\hat{\boldsymbol{X}}) \right\|^2$ in Eq. (5.46) is always greater than the other sign-indefinite terms.

Substituting P and $g'(\hat{\boldsymbol{X}})$ into Eq. (5.35) leads to the optimal control law

$$\phi(\hat{\boldsymbol{X}}) = -\frac{w_p}{w_c}(\mathcal{L} \otimes I_3)\hat{\boldsymbol{p}} - \frac{w_v}{w_c}(\mathcal{L} \otimes I_3)\hat{\boldsymbol{v}} - \frac{1}{2w_c^2}{g_v}'(\hat{\boldsymbol{X}}) \tag{5.48}$$

Substituting $\hat{\boldsymbol{p}} = \bar{\boldsymbol{p}} - \bar{\boldsymbol{p}}_{cs}$ and $\hat{\boldsymbol{v}} = \bar{\boldsymbol{v}} - \bar{\boldsymbol{v}}_{cs}$ into Eq. (5.48) and using Eq. (5.14), the optimal control law (5.48) becomes

$$\phi(\boldsymbol{X}) = -\frac{w_p}{w_c}(\mathcal{L} \otimes I_3)\bar{\boldsymbol{p}} - \frac{w_v}{w_c}(\mathcal{L} \otimes I_3)\bar{\boldsymbol{v}} - \frac{1}{2w_c^2} g_v{}'(\hat{\boldsymbol{X}}) \tag{5.49}$$

Also note that

$$g_v{}'(\hat{\boldsymbol{X}}) = g_{tr_v}{}'(\hat{\boldsymbol{X}}) + g_{oa_v}{}'(\hat{\boldsymbol{X}}) \triangleq \frac{\partial g_{tr}(\hat{\boldsymbol{X}})}{\partial \hat{\boldsymbol{v}}} + \frac{\partial g_{oa}(\hat{\boldsymbol{X}})}{\partial \hat{\boldsymbol{v}}} \tag{5.50}$$

Since $\boldsymbol{p}_r$ and $\boldsymbol{v}_r$ are known to UAV i if it has access to the reference, $g_{tr_v}{}'(\hat{\boldsymbol{X}}) = g_{tr_v}{}'(\boldsymbol{X})$. Also, since $\boldsymbol{v}_{oa}$ is determined by the current position and velocity of UAV i and the obstacle position, i.e. $\boldsymbol{p}_i$, $\boldsymbol{v}_i$, and $\boldsymbol{O}_{bj}$ if UAV i enters the detection region, we can see from Eq. (5.25) that $g_{oa_v}{}'(\hat{\boldsymbol{X}}) = g_{oa_v}{}'(\boldsymbol{X})$ and thus $g_v{}'(\hat{\boldsymbol{X}}) = g_v{}'(\boldsymbol{X})$.

Therefore, using $g_v{}'(\hat{\boldsymbol{X}}) = g_v{}'(\boldsymbol{X})$ in Eq. (5.49) and noting $\bar{\boldsymbol{v}} = \boldsymbol{v}$, the optimal control law turns out to be Eq. (5.28), i.e.

$$\boldsymbol{U} = \phi(\boldsymbol{X}) = -\frac{w_p}{w_c}(\mathcal{L} \otimes I_3)\left(\boldsymbol{p} - \sigma_p\right) - \frac{w_v}{w_c}(\mathcal{L} \otimes I_3)\boldsymbol{v} - \frac{1}{2w_c^2} g_v{}'(\boldsymbol{X})$$

It remains to verify the conditions (2.4) and (2.6). It can be seen from Eq. (5.32) and Eq. (5.35) that the conditions (2.4) and (2.6) are met if $g(\hat{\boldsymbol{X}}) = 0$ and $g'(\hat{\boldsymbol{X}}) = \boldsymbol{0}$ when $\hat{\boldsymbol{X}} = \boldsymbol{0}$. Eq. (5.25) and Eq. (5.39) imply that if the UAV is inside the detection region, the avoidance force is not zero and the UAV will be driven outside the detection region. When the UAV flies outside the detection region and along the reference trajectory, we have $g(\hat{\boldsymbol{X}}) = 0$ and $g'(\boldsymbol{X}) = \boldsymbol{0}$, which validates the conditions (2.4) and (2.6).

Now, all the conditions (2.4)–(2.9) in Lemma 2.3 in Chapter 2 can be satisfied. According to Lemma 2.3 in Chapter 2, the control law (5.28) is an optimal control law for the problem (5.17) in the sense of Eq. (2.11) and Eq. (2.12), and the closed-loop system is asymptotically stable. It implies $\boldsymbol{X} - \boldsymbol{\sigma} = \bar{\boldsymbol{X}}_{fs}$ and the desired formation is achieved.

In addition, it can be easily seen from Eq. (5.35) that $V(\hat{\boldsymbol{X}}) \to \infty$ as $\left\|\hat{\boldsymbol{X}}\right\| \to \infty$. Therefore, the closed-loop system is globally asymptotically stable. Note that the global asymptotic stability region excludes the undefined area Θ^c, which is physically meaningful because no UAV can start from inside the obstacle. □

As can be seen from the proof of Theorem 5.1, the penalty function $g(\hat{\boldsymbol{X}})$ provides the reference tracking and obstacle avoidance capabilities. The inverse optimal control approach facilitates the integration of this penalty function into the cost function $h_u(\hat{\boldsymbol{X}})$ such that an analytical, distributed, and optimal control law to achieve these capabilities can be obtained.

Remark 5.3. To recapitulate the selection of the weight parameters, one can choose a large w_v given w_p for Eq. (5.45) to meet the condition (2.5). Then with the above chosen w_v and w_p, one can choose a small enough w_c for Eq. (5.46) to meet the condition (2.7). In addition, this rule of weight selection also applies to the condition (3.46) for a meaningful formation cost functional (5.18). Meanwhile, the tracking weights w_{tp} and w_{tv} can be tuned to obtain a proper tracking speed.

Remark 5.4. From Eq. (5.28), Eq. (5.23), and Eq. (5.25), it can be seen that the optimal control law of each UAV only requires the local information from its neighbors since the first two terms of the optimal control law is a linear function of the Laplacian matrix $\mathcal{L}$ and the last term only requires the information about the UAV's own position and velocity, the obstacle's position, and the reference if this UAV has access to it.

5.3 Numerical examples

The effectiveness of the proposed optimal formation control law given by Theorem 5.1 is illustrated in this section. In this chapter, all formation-to-go UAVs are assumed to be identical. The drag in the UAV model Eq. (5.2) is calculated by [197]

$$D_{g_i} \triangleq 0.5\rho(V_{g_i} - V_{w_i})^2 A_{rea} C_{D0} + 2k_d k_n^2 n_{g_i}^2 m_i^2 / [\rho(V_{g_i} - V_{w_i})^2 A_{rea}] \quad (5.51)$$

with the wing area $A_{rea} = 37.16\ \mathrm{m}^2$, zero-lift drag coefficient $C_{D0} = 0.02$, load-factor effectiveness $k_n = 1$, induced drag coefficient $k_d = 0.1$, gravitational coefficient $g_a = 9.81\ \mathrm{kg/m}^2$, atmospheric density $\rho = 1.225\ \mathrm{kg/m}^3$, and the weight of the UAV $m_i = 14,515$ kg. The gust model $V_{w_i} = V_{w_i,\text{normal}} + V_{w_i,\tan}$ is scaled based on [121] and varies according to the altitude h_i. In the simulated gust, the normal wind shear is given by

$$V_{w_i,\text{normal}} = 0.215 V_m \log_{10}(h_i) + 0.285 V_m \quad (5.52)$$

where $V_m = 4.0$ m/s is the mean wind speed at an altitude of 80 m, which is the simulated altitude. The turbulence part of the wind gust $V_{m_i,\tan}$ has a Gaussian distribution with a zero mean and a standard derivation of $0.09V_m$. The constraints on the control variables are $T_{h_i} \le 1.13 \times 10^5 N$, $-1.5 \le n_{g_i} \le 3.5$, and $-80° \le \phi_{b_i} \le 80°$.

The undirected and connected communication topology of the four UAVs is illustrated in Fig. 5.2 and can be described by the Laplacian matrix $\mathcal{L}$

$$\mathcal{L} = \begin{bmatrix} 1 & -1 & 0 & 0 \\ -1 & 2 & -1 & 0 \\ 0 & -1 & 2 & -1 \\ 0 & 0 & -1 & 1 \end{bmatrix} \quad (5.53)$$

The initial positions of the four UAVs are given by (0, −1500, 70) m, (0, −500, 80) m, (0, 500, 90) m and (0, 1500, 95) m, respectively and the initial velocities of the four UAVs are assumed to be 100 m/s, 130 m/s, 70 m/s, and 100 m/s, respectively.

All the initial flight-path angles and heading angles are set to be 0 degrees. The UAVs are required to fly in a rhombic-shape formation as shown in Fig. 5.2.

The desired rhombic formation can be described as follows

$$\boldsymbol{\sigma} = [\underbrace{\boldsymbol{\sigma}_{p_1}{}^T, \ldots, \boldsymbol{\sigma}_{p_n}{}^T}_{\boldsymbol{\sigma}_p^T}, \mathbf{0}^T, \ldots, \mathbf{0}^T]^T \tag{5.54}$$

where

$$\boldsymbol{\sigma}_{p_1} = [0; -1000; 1] \text{ m}; \boldsymbol{\sigma}_{p_2} = [1000; 0; 1] \text{ m};$$
$$\boldsymbol{\sigma}_{p_3} = [-1000; 0; 1] \text{ m}; \boldsymbol{\sigma}_{p_4} = [0; 1000; 1] \text{ m}$$

The final velocity and altitude of the formation are desired to be the same as the reference. The reference is a straight line trajectory with the initial position being $[0 \quad -1000 \quad 80]^T$ m and the constant velocity being $[100 \quad 0 \quad 0]^T$ m/s. Assume that only UAV 1 has access to the reference. The associated weights in the control law are set to $w_p = 0.06$, $w_v = 2$, $w_c = 1$, $w_{tp} = 0.03$ and $w_{tv} = 1$. Two simulation scenarios are considered – without obstacles and with obstacles – on the trajectories of the UAVs, respectively.

5.3.1 Formation control without obstacles on the trajectories of UAVs

In this scenario, an obstacle is assumed to appear at $(-1000, 0, 80)$ m, which is not on the trajectory of any UAV. The radius of the obstacle and the detection region are assumed to be $r_1 = 30$ m and $R_{d_1} = 500$ m, respectively. The safe distance of the obstacle is set to 30 m. The simulation results of the four UAVs' motion under the proposed optimal formation control law are shown in Figs. 5.5–5.14. Fig. 5.5

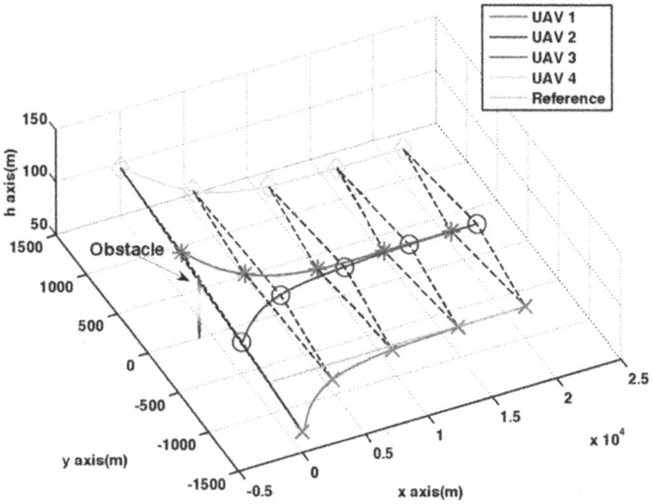

Figure 5.5 3D trajectories of UAVs' formation flying without concern of obstacles.

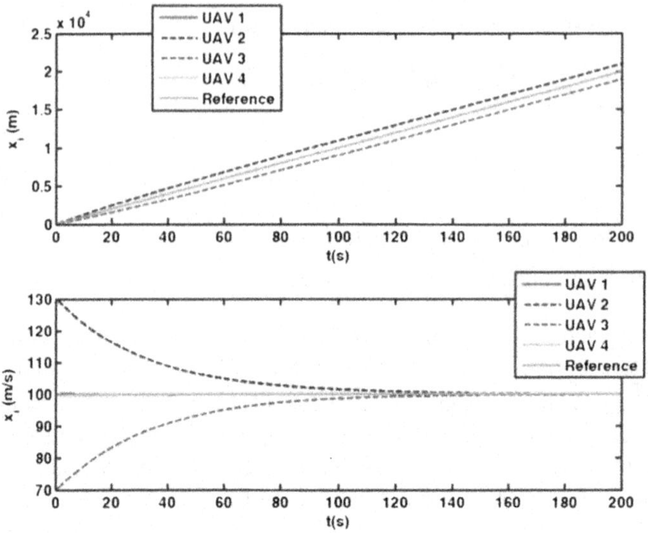

Figure 5.6 Time histories of the positions x_i and velocities $\dot{x}_i$.

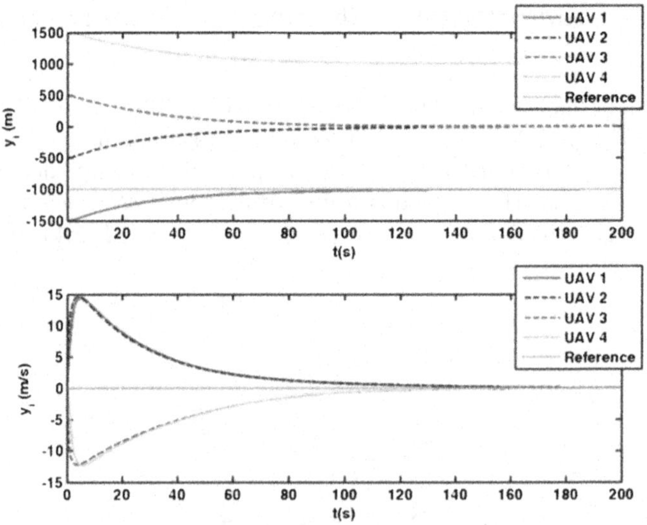

Figure 5.7 Time histories of the positions y_i and velocities $\dot{y}_i$.

demonstrates the 3D trajectories. As can be seen, the obstacle avoidance does not take effect since no UAV enters the detection region of the obstacle. All the UAVs are driven to a rhombic formation and the formation follows the desired reference trajectory. Figs. 5.6–5.11 present the time histories of the UAVs' positions, velocities, flight path angles, and heading angles, respectively.

It can be seen that UAV 1 tracks the reference and all UAVs fly in formation with the desired altitude of 80 m and the desired velocity of $[100 \quad 0 \quad 0]^T$ m/s. This is

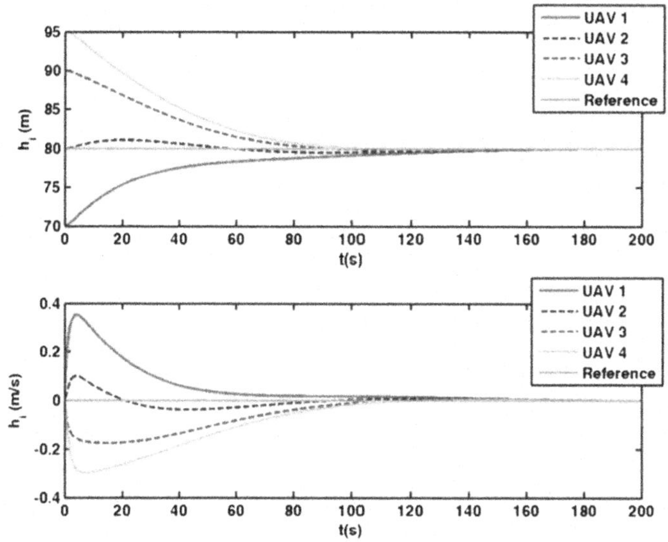

Figure 5.8 Time histories of the positions h_i and velocities $\dot{h}_i$.

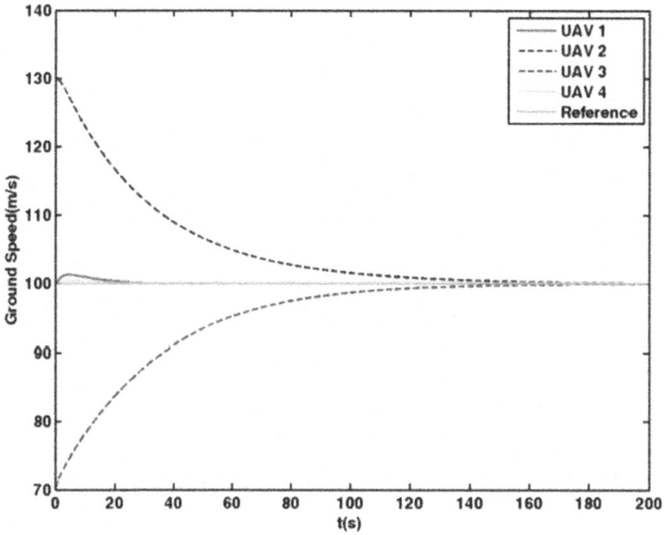

Figure 5.9 Time histories of the ground speed V_{g_i}.

achieved by combining the optimization of the first cost function component J_{u_1} with the optimization of the tracking cost $g_{tr}(\hat{\boldsymbol{X}})$ in J_{u_2}. In addition, it is worth noting that all the UAVs achieve the formation simultaneously, which is guaranteed by formulating the Laplacian matrix $\mathcal{L}$ in the cost function J_{u_1}. Figs. 5.12–5.14 demonstrate the control responses and they are all within the prescribed constraints.

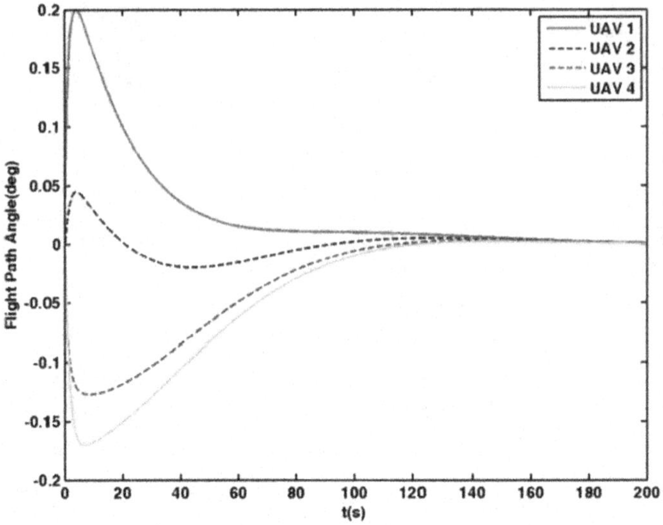

Figure 5.10 Time histories of the flight path angle γ_i.

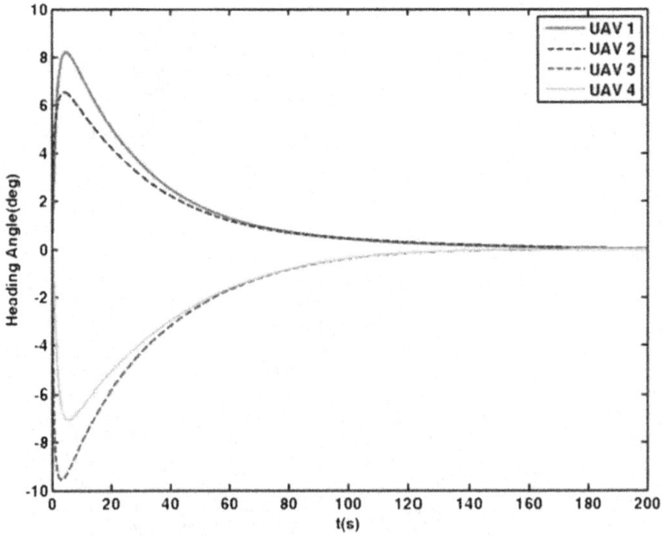

Figure 5.11 Time histories of the heading angle χ_i.

Note that the penalty function $g(\hat{\boldsymbol{X}})$ in this simulation scenario only contains the tracking penalty function $g_{tr}(\hat{\boldsymbol{X}})$ since obstacle avoidance is not needed. The proposed optimal control law reduces to

$$\phi(\hat{\boldsymbol{X}}) = -\frac{w_p}{w_c}(L \otimes I_3)\left(\boldsymbol{p} - \boldsymbol{\sigma}_p\right) - \frac{w_v}{w_c}(L \otimes I_3)\boldsymbol{v} - \frac{1}{2w_c^2} {g_{tr_v}}'(\boldsymbol{X}) \tag{5.55}$$

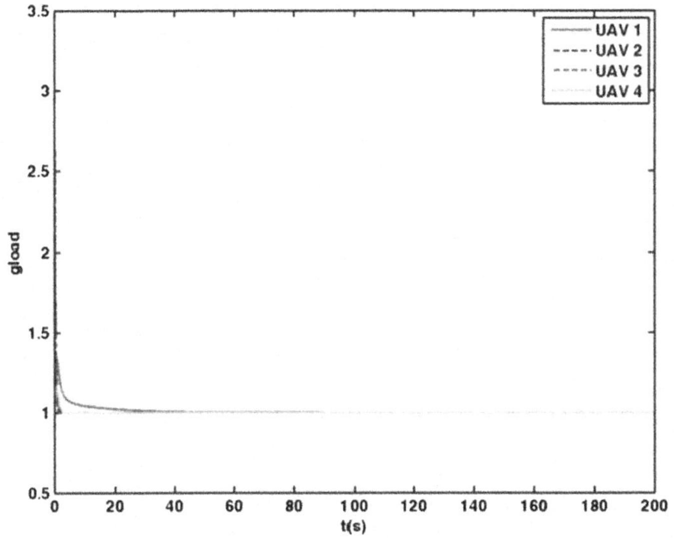

Figure 5.12 Time histories of the g-load n_{g_i}.

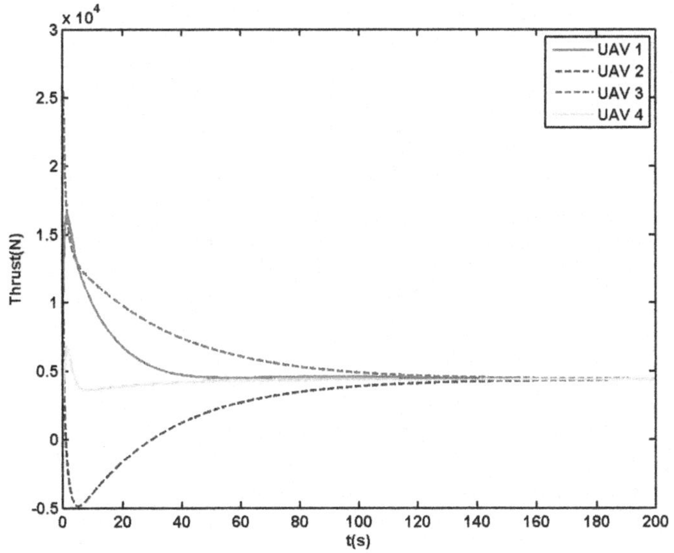

Figure 5.13 Time histories of the thrust T_{h_i}.

where the first term and the second term ensure that the desired formation and the coordination of velocities can be achieved synchronously. The tracking to the reference is guaranteed by the last term and thus the final velocity of the formation is coordinated to the velocity of the reference.

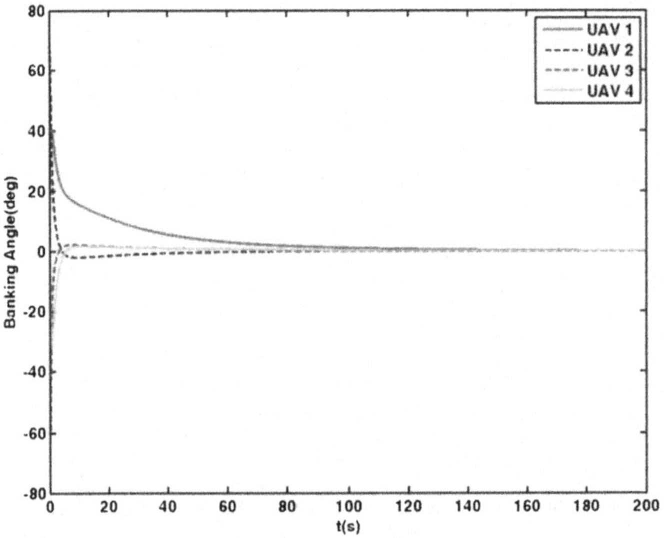

Figure 5.14 Time histories of the banking angle ϕ_{b_i}.

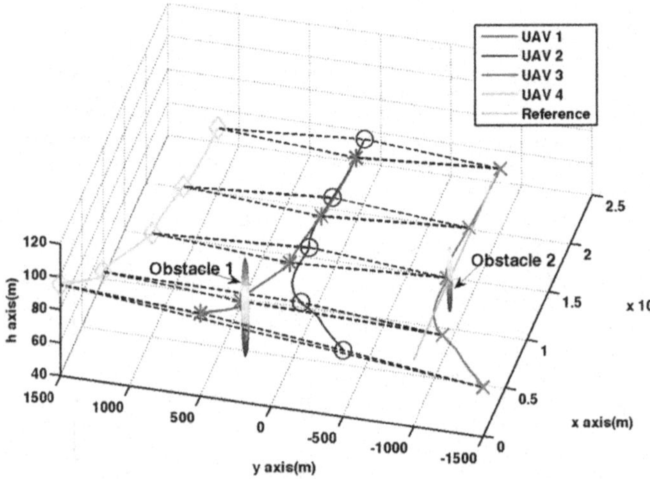

Figure 5.15 3D trajectories of UAVs' formation flying with two obstacles.

5.3.2 Formation control with two obstacles on the trajectories of UAVs

In this scenario, one obstacle is assumed to appear at (2015, 246, 85) m, which is on the trajectory of UAV 3. The radius of the obstacle and the detection region are set to $r_1 = 30$ m and $R_{d_1} = 500$ m respectively. The safe distance r_{safe} of obstacle 1 is set to 30 m. The other obstacle is assumed to appear at (8804, −987, 78) m, which is on the trajectory of UAV 1. The radius of the obstacle and the detection region are set to

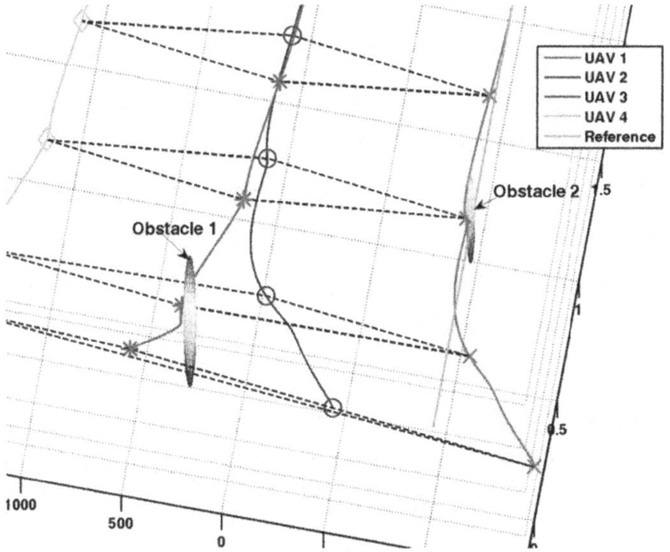

Figure 5.16 Zoom-in view of obstacle avoidance.

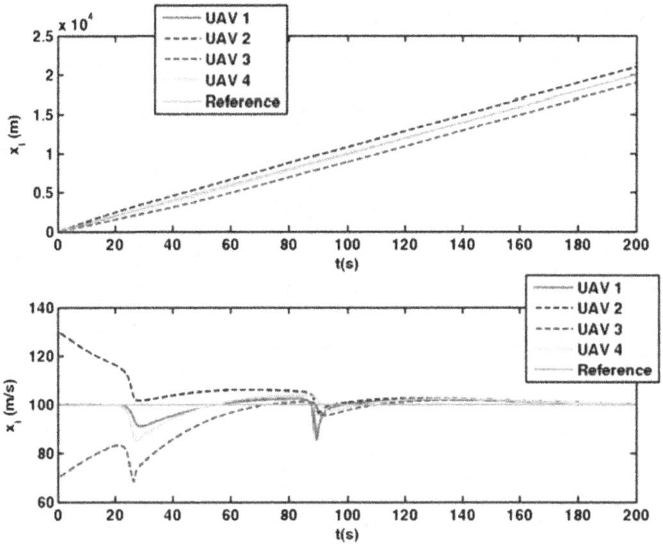

Figure 5.17 Time histories of the positions x_i and velocities $\dot{x}_i$.

$r_2 = 20$ m and $R_{d_2} = 500$ m, respectively. The safe distance r_{safe} of obstacle 2 is set to 20 m. The simulation results are given in Figs. 5.15–5.25.

Fig. 5.15 and Fig. 5.16 demonstrate the trajectories of the UAVs. As can be seen, the proposed optimal control law drives the UAV 3 and UAV 1 away from the obstacles and then all four UAVs reach the desired formation and the formation follows the

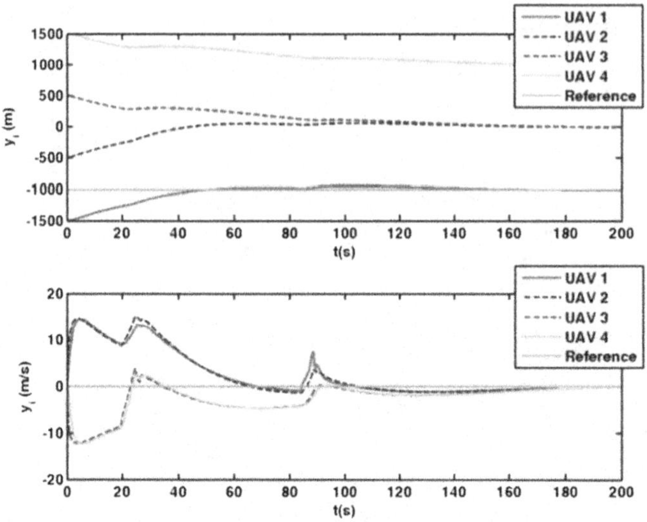

Figure 5.18 Time histories of the positions y_i and velocities $\dot{y}_i$.

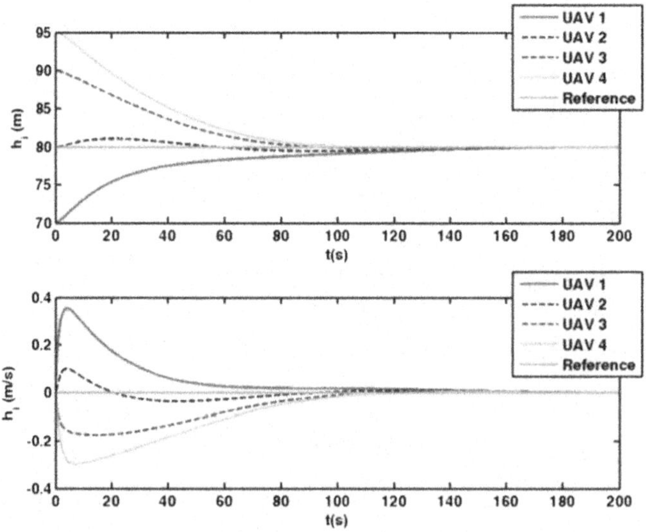

Figure 5.19 Time histories of the altitudes h_i and velocities $\dot{h}_i$.

reference trajectory. A zoom-in view of the trajectories is given in Fig. 5.16 to see the obstacle avoidance more clearly. The time histories of the UAVs' positions and velocities in Figs. 5.17–5.19 show that UAV 1 tracks the reference and all UAVs fly in formation synchronously with the final desired altitude of 80 m and the final desired velocity of $[100 \quad 0 \quad 0]^T$ m/s, which is the same as in the previous scenario.

The obstacle avoidance can be also seen in Figs. 5.17–5.22. When the UAV detects the obstacle, the obstacle avoidance law will regulate its velocity to v_{oa}, which is

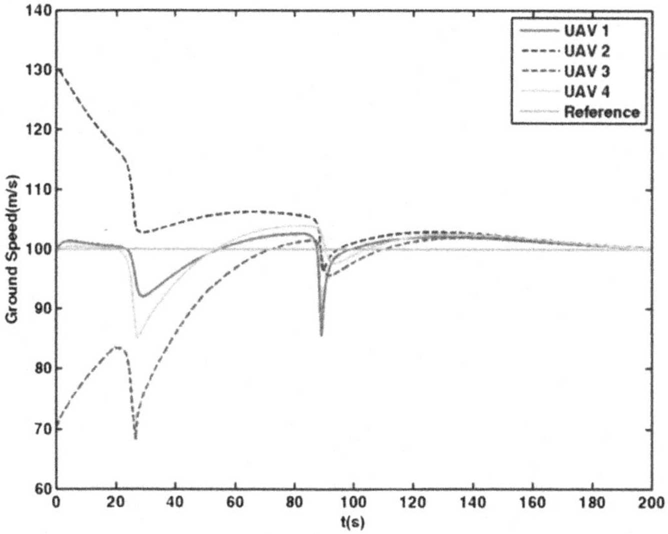

Figure 5.20 Time histories of the ground speed V_{g_i}.

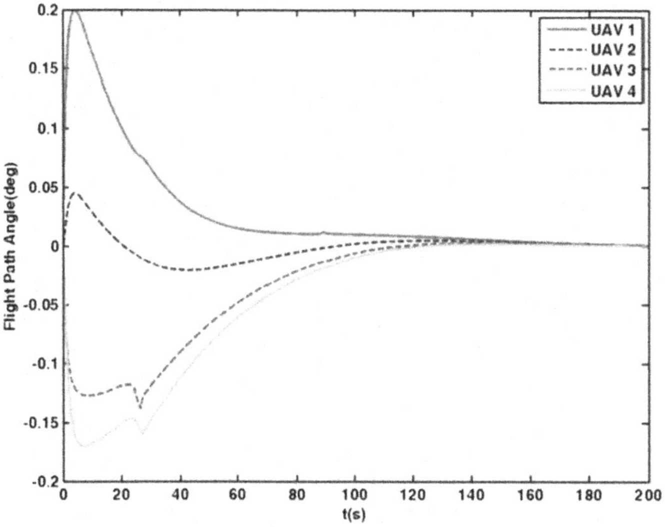

Figure 5.21 Time histories of the flight path angle γ_i.

shown in Fig. 5.4 and calculated by Eq. (5.27). In the time intervals of $[20, 30]$ s and $[80, 90]$ s, the obstacle avoidance takes effect and the UAVs succeed to avoid the obstacles. Figs. 5.23–5.25 illustrate the actual control responses and they are all within the prescribed constraints.

Note that the penalty function $g(\hat{X})$ in this simulation scenario contains both the tracking penalty function $g_{tr}(\hat{X})$ and the obstacle avoidance penalty function $g_{oa}(\hat{X})$. The proposed optimal control law is Eq. (5.28).

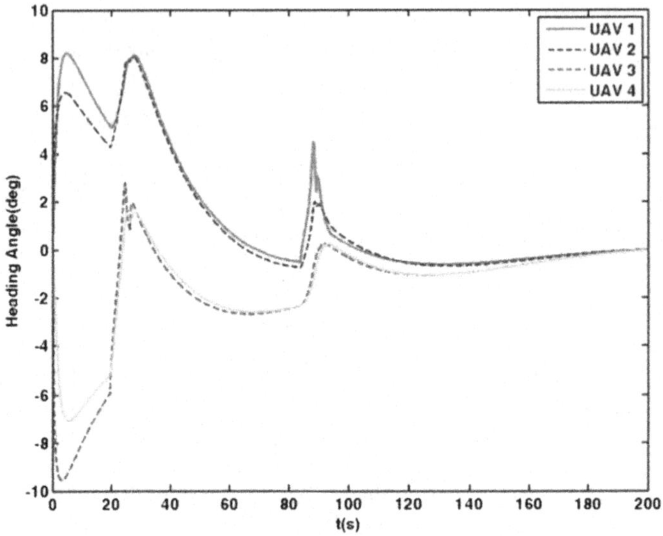

Figure 5.22 Time histories of the heading angle χ_i.

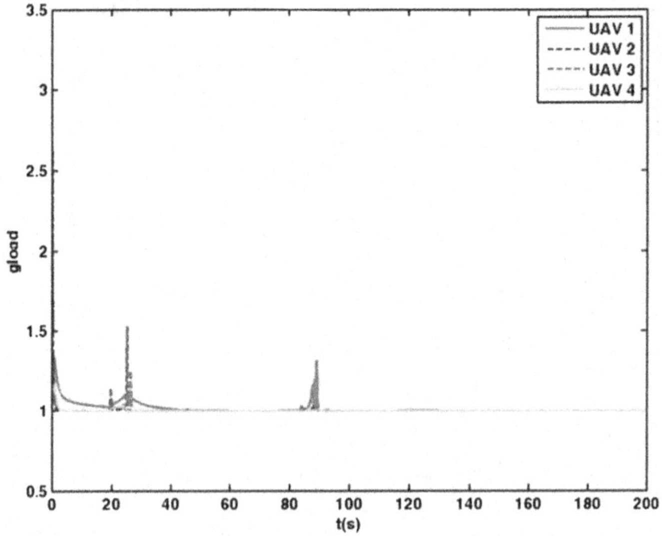

Figure 5.23 Time histories of the g-load n_{g_i}.

5.4 Conclusion remarks

In this chapter, we investigated the multi-UAV formation control problem with tracking and obstacle avoidance capabilities. Feedback linearization technology was utilized to obtain the reduced double-integrator model of UAVs. The integrated formation

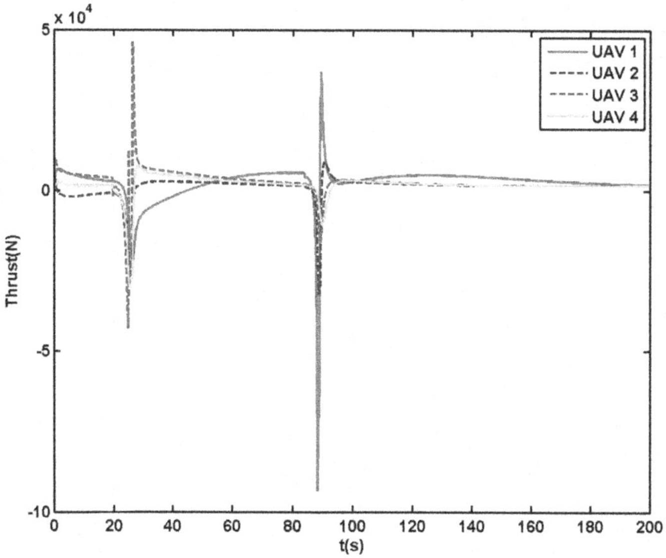

Figure 5.24 Time histories of the thrust T_{h_i}.

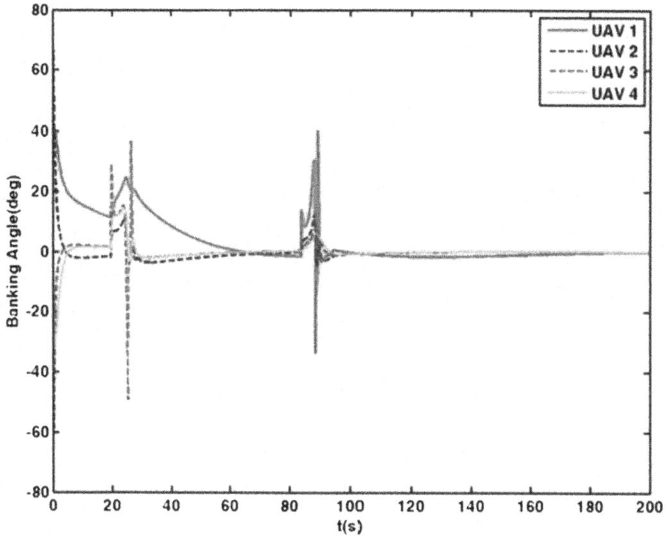

Figure 5.25 Time histories of the bank angle ϕ_{b_i}.

control law was designed using the inverse optimal control approach. The stability and optimality of the closed-loop system have been proved. Simulations were conducted to show that all the desired formation flying characteristics have been achieved using the proposed formation control law.

Optimal coverage control of multi-robot systems

6

6.1 Problem formulation

Consider a convex d-dimensional set $\mathcal{Q} \in \mathbb{R}^d$ to be covered by N robots which have integrator dynamics:

$$\dot{x}_i = u_i, \tag{6.1}$$

where $x_i \in \mathcal{Q}$ denotes the position of i-th robot, and u_i is the control input.

We define the density function, or referred to sensory function, as a measure of information importance or probability in the corresponding area with $\phi : \mathcal{Q} \to \mathbb{R}_+$ (note that ϕ is a strictly positive value). A point with a density value indicates the importance of quantity to measure at that point.

The problem considered in this chapter is to find the best locations of N robots in $\mathcal{Q}$ according to a given density function $\phi(q)$. The following assumption needs to be imposed.

Assumption 6.1. The graph $\mathcal{G}$ describing the information interaction among N robots is undirected and connected.

The following preliminaries are needed to derive the main results of this chapter.

6.1.1 *Voronoi tessellation and locational optimization*

In mathematics, a Voronoi tessellation is a partitioning of a plane into regions based on distance to points in a specific subset of the plane. The set of points (called seeds, sites, or generators) is specified beforehand, and for each seed there is a corresponding region consisting of all points closer to that seed than to any other. These regions are called Voronoi cells. Let $\mathcal{P} = \{\mathbf{x}_1, \mathbf{x}_2, \ldots, \mathbf{x}_N\}$ be the collection of positions of N robots, and $\mathcal{V} = \{\mathcal{V}_1, \mathcal{V}_2, \ldots, \mathcal{V}_N\}$ be the Voronoi partition of $\mathcal{Q}$,

$$\mathcal{V}_i(\mathcal{P}) = \{\mathbf{q} \in Q \,|\, \|\mathbf{q} - \mathbf{x}_i\| \leq \|\mathbf{q} - \mathbf{x}_j\|, \forall i \in [1, N], \forall j \neq i\},$$

where the positions of the robots are generator points and $\|.\|$ denotes 2-norm of vector.

Next, we define three notions of rigid body: mass, moment of inertia, and centroid of a Voronoi region as follows

$$M_{\mathcal{V}_i} = \int_{\mathcal{V}_i} \phi(\mathbf{q}) d\mathbf{q}, \;\; \mathcal{I}_{\mathcal{V}_i} = \int_{\mathcal{V}_i} \mathbf{q}\phi(\mathbf{q}) d\mathbf{q}, \;\; C_{\mathcal{V}_i} = \frac{\mathcal{I}_{\mathcal{V}_i}}{M_{\mathcal{V}_i}}.$$

Cooperative Control of Multi-Agent Systems. https://doi.org/10.1016/B978-0-12-820118-3.00016-8

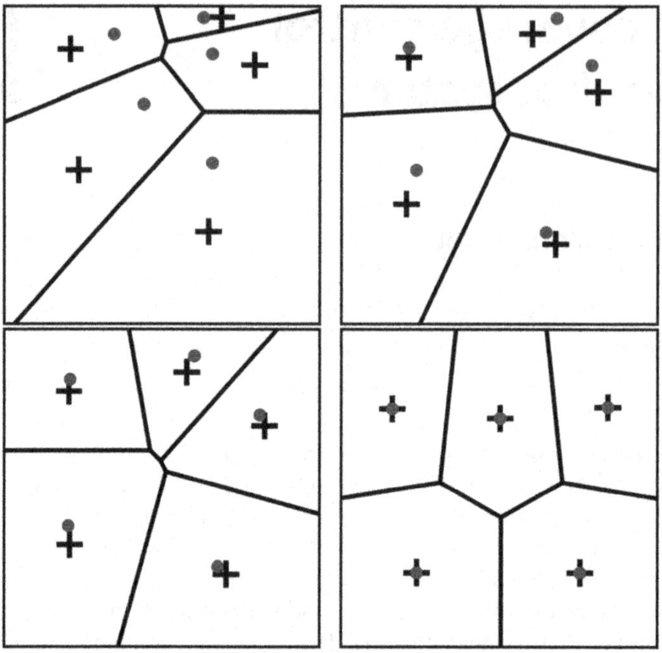

Figure 6.1 Illustration of the Voronoi iteration. The red spots (dark gray in print version) denote the current points of the robots and the plus signs denote the centroids of the Voronoi cells. In the last image, the points are very near the centroids of the Voronoi cells. A centroidal Voronoi tessellation has been found.

Illustration of the Voronoi iteration is shown in Fig. 6.1.

Remark 6.1. The Voronoi diagram of a set of points is dual to its Delaunay triangulation. The Voronoi partition of the robots are generated and the i-th robot only needs information from its Voronoi neighbor $\mathcal{N}_i$. In other words, the Delaunay triangulation is employed to generate the graph connection topology of network. By using the Delaunay triangulation, we always have a connected graph topology at all time, $\forall t > 0$.

The sensing capability at point $\mathbf{q} \in \mathcal{Q}$ sensed by $\mathbf{x}_i$, which is expressed as an isotropic, strictly increasing and convex function $f(\|\mathbf{q} - \mathbf{x}_i\|)$ with $f : \mathbb{R}_+ \to \mathbb{R}_+$, decreases proportionally with the distance $\|\mathbf{q} - \mathbf{x}_i\|$.

With the given definitions and assumptions, the cost function of this problem can be formulated as

$$\mathcal{H}(\mathcal{P}) = \sum_{i=1}^{N} \int_{\mathcal{V}_i} f(\|\mathbf{q} - \mathbf{x}_i\|)\phi(\mathbf{q})d\mathbf{q}. \tag{6.2}$$

Thus, the problem of covering the environment $\mathcal{Q}$ is now formulated as moving the robots to a configuration position to minimize $\mathcal{H}(\mathcal{P})$ with

$$\text{minimize}_{\mathbf{x}_i} \mathcal{H}(\mathcal{P}) = \sum_{i=1}^{N} \int_{\mathcal{V}_i} f(\|\mathbf{q} - \mathbf{x}_i\|)\phi(\mathbf{q})d\mathbf{q}.$$

Define the sensing incapability function as $f(\|\mathbf{q} - \mathbf{x}_i\|) = \|\mathbf{q} - \mathbf{x}_i\|^2$. For each robot, a standard result from locational optimization is that

$$\frac{\partial \mathcal{H}}{\partial x_i} = -\int_{\mathcal{V}_i} 2(\|\mathbf{q} - \mathbf{x}_i\|)\phi(\mathbf{q})d\mathbf{q} = -2M_{\mathcal{V}_i}\left(C_{\mathcal{V}_i} - x_i\right).$$

Definition 6.1 (Optimal coverage configuration [158]). a robot network is said to be in a (locally) optimal coverage configuration if every robot is positioned at the centroid of its Voronoi region, $x_i = C_{\mathcal{V}_i}$ for all i.

Next, we state some basic definitions of convex optimization.

Definition 6.2. A set $\Omega \in \mathcal{R}^n$ is called convex if for any two points in Ω, all points along the line segment joining them are also in Ω. Formally,

$$ax + (1-a)y \in \Omega, \quad \forall x, y \in \Omega \text{ and } \forall a \in [0, 1].$$

Definition 6.3. A function $f : \Omega \longmapsto \mathcal{R}$ is called convex if $f(ax + (1-a)y) \leq \alpha f(x) + (1-\alpha)f(y), \forall x, y \in \Omega$ and $\forall \alpha \in [0, 1]$.

6.1.2 Potential field method for collision avoidance

An object is considered to have a repulsive potential function expressed as

$$\mathcal{K} = \frac{1}{2}\epsilon\left(\frac{1}{\|\mathbf{x}_i - \mathbf{p}_j\|} - \frac{1}{d_0}\right)^2 \tag{6.3}$$

where $\epsilon \in \mathbb{R}_+$ is a positive gain, $\mathbf{x}_i$ denotes the position of i-th robot, $\forall i \in [1, N]$, $\mathbf{p}_j$ denotes the position of another object, e.g., another robot or obstacle, $\forall j \in [1, M]$, and M is the number of robots or obstacles, d_0 denotes the minimum distance between two objects, which is calculated from the center of associated objects.

From the potential function (6.3), the corresponding repulsive force produced to drive i-th robot away from hindering objects is expressed as

$$\dot{\mathbf{x}}_{ri} = \nabla_x \mathcal{K}.$$

6.2 Coverage controller design with known density function

6.2.1 Coverage controller design

First, we consider the case when the density function $\phi(q)$ is known by the robots in the network. Based on the locational optimization analysis given in the last section, we design the optimal coverage algorithm for each robot as

$$u_{ia} = k_0 M_{\mathcal{V}_i} \left(C_{\mathcal{V}_i} - x_i\right), \tag{6.4}$$

where $k_0 > 0$ is a constant value to adjust the coverage speed.

As mentioned, during the robots moving, the trajectory of i-th robot generated by the optimization solver has a possibility to meet the trajectory of j-th robot or cross an obstacle. Thus, a collision avoidance protocol is required due to the safety of robots' trajectories. Following the method in [178], we design the collision avoidance protocols for each agent as follows:

$$u_{ib} = \sum_{j=1}^{N} \mu_b b_{ij} a_{ij} \left(\frac{1}{\|\mathbf{x}_i - \mathbf{x}_j\|} - \frac{1}{d_b} \right) \frac{\mathbf{x}_i - \mathbf{x}_j}{\|\mathbf{x}_i - \mathbf{x}_j\|^2},$$

$$u_{ic} = \sum_{k=1}^{M} \mu_c c_{ik} \left(\frac{1}{\|\mathbf{x}_i - \mathbf{w}_k\|} - \frac{1}{d_c} \right) \frac{\mathbf{x}_i - \mathbf{w}_k}{\|\mathbf{x}_i - \mathbf{w}_k\|^2},$$

where $\mathbf{w}_k$, a_{ij} are the position of k-th obstacle and the adjacency matrix of the Delaunay-triangulation graph $\mathcal{G}$, respectively. Let d_b be the distance of the repulsive region between two robots to prevent collision among them, while d_c be the distance of the repulsive region between robot and obstacle calculated from the centers of two corresponding objects, then the parameters b_{ij} and c_{ik}, $\forall i, j \in [1, N]$ and $\forall k \in [1, M]$, are described as follows:

$$b_{ij} = \begin{cases} 1 & \forall\ \|\mathbf{x}_i - \mathbf{x}_j\| \leq d_b, \\ 0 & \text{otherwise,} \end{cases}$$

$$c_{ik} = \begin{cases} 1 & \forall\ \|\mathbf{x}_i - \mathbf{w}_k\| \leq d_c, \\ 0 & \text{otherwise.} \end{cases}$$

While μ_b and μ_c are the control gains of collision-avoidance term to be described later. In this work, obstacles are assumed to be circles in $\mathbb{R}^2$, or balls in $\mathbb{R}^3$, with radius $r < d_c$.

Finally, the controller for each robot is given as

$$u_i = u_{ia} + u_{ib} + u_{ic}. \tag{6.5}$$

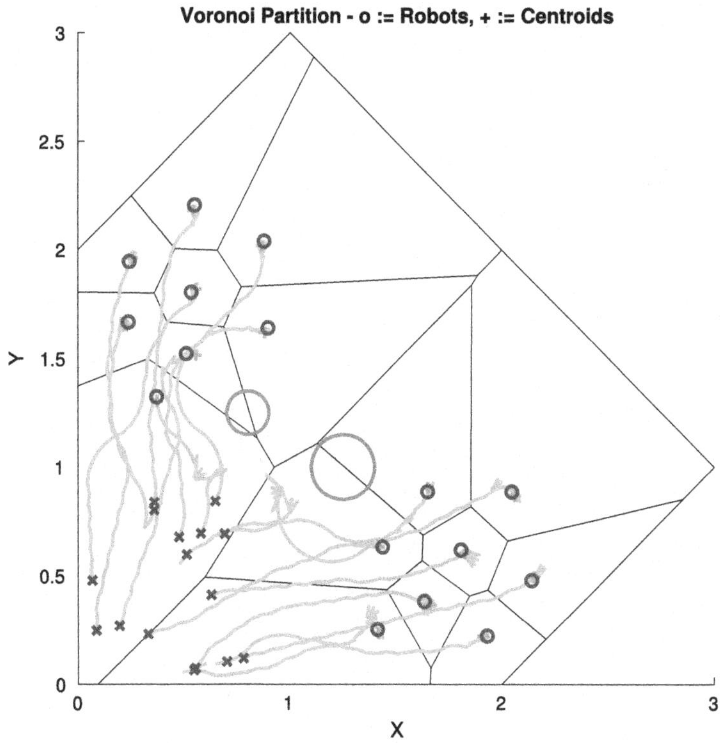

Figure 6.2 Trajectories and centroidal Voronoi regions of 16 robots in the case of a known density function.

6.2.2 Numerical examples

We deploy $N = 16$ robots in a farm area $\mathcal{Q}$, whose boundary points are $\{(0,0), (0,2), (1,3), (3,1), (2,0)\}$, with the known density function

$$\begin{aligned}\phi(x,y) = {} & \kappa \exp(-k_1(a_1(x-x_{c1})^2 + b_1(y-y_{c1})^2 + r_1)) \\ & + \kappa \exp(-k_2(a_2(x-x_{c2})^2 + b_2(y-y_{c2})^2 + r_2)),\end{aligned}$$

and two obstacles at (0.8,1.25) and (1.25,1) with $M = 2$. We choose $\kappa = 1000$, $k_1 = 2.5$, $x_{c1} = 1.8$, $y_{c1} = 0.5$, $a_1 = b_1 = r_1 = 1$, $k_2 = 2.6$, $x_{c2} = 0.5$, $y_{c2} = 1.8$, $a_2 = b_2 = r_2 = 1$, $d_b = 0.1$, $d_c = 0.4$. By using the proposed protocol (6.5), the trajectories of robots together with the Voronoi partition are depicted in Fig. 6.2, and 2-norm errors $\|\mathbf{x} - \mathbf{C}_{\mathcal{V}}\|$ are depicted in Fig. 6.3. Obstacles are illustrated by green circles (mid gray in print version).

The convergence results of the objective function (6.2) of this scenario are provided in Fig. 6.4. From those figures, it can be concluded that the robots can be driven to cover the area and the positions of all robots converge to the Voronoi centroids.

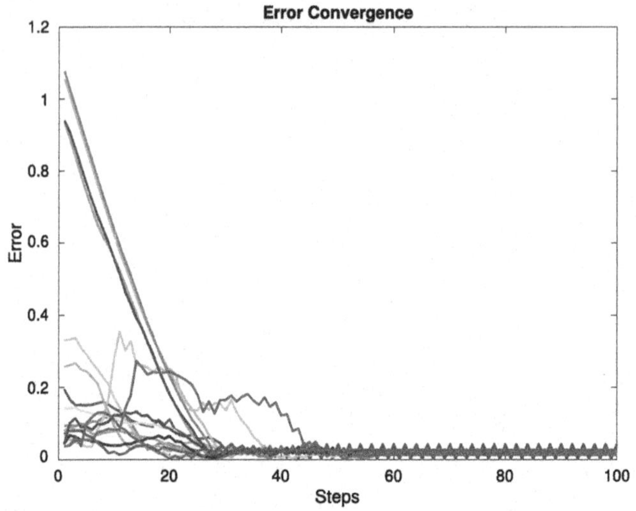

Figure 6.3 Euclidean 2-norm errors between 16 robots' positions and centroids.

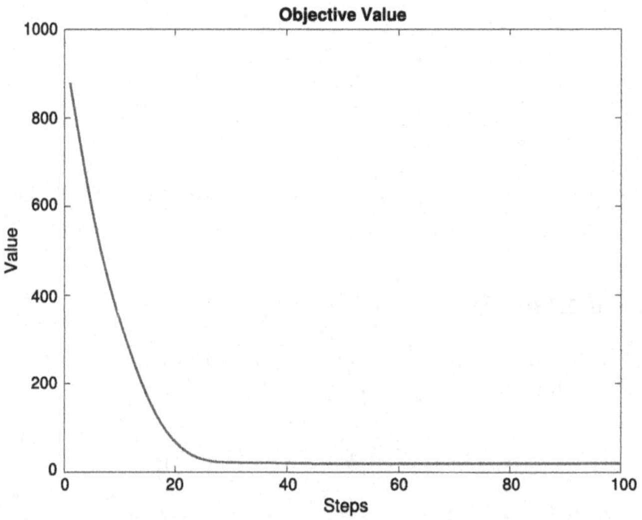

Figure 6.4 Total objective function of locational optimization with 16 robots in the case of a known density function.

6.3 Coverage controller design with density function estimation

6.3.1 Estimation based coverage controller design

In last section, it is assumed that density function $\phi(q)$ is known by the robots in the network. However, this is not practicable in some cases. In this subsection, we consider that the density function $\phi(q)$ is not known by the robots in the network, but the robots have sensors that can measure $\phi(p_i)$ at the i-th robot's position x_i. For our analysis, we require the following assumption from [158].

Assumption 6.2. There exists an ideal parameter vector $a \in \mathcal{R}^m$ such that

$$\phi(q) = \mathcal{K}(q)^T a,$$

and a is unknown to the robots. Furthermore,

$$a \geq \mathbf{1} a_{\min},$$

where $a_{\min}$ is a lower bound known by each robot.

Next, using $\hat{\phi}_i(\mathbf{q})$ we have the estimated centroid of the Voronoi region as

$$\hat{M}_{\mathcal{V}_i} = \int_{\mathcal{V}_i} \hat{\phi}_i(\mathbf{q}) d\mathbf{q}, \quad \hat{I}_{\mathcal{V}_i} = \int_{\mathcal{V}_i} \mathbf{q}\hat{\phi}_i(\mathbf{q}) d\mathbf{q}, \quad \hat{C}_{\mathcal{V}_i} = \frac{\hat{I}_{\mathcal{V}_i}}{\hat{M}_{\mathcal{V}_i}}, \tag{6.6}$$

and the parameter error

$$\tilde{a}(t) = \hat{a}_i(t) - a,$$

where the parameters, $\hat{a}_i, i = 1, 2, \ldots, N$, are used to calculate $\hat{C}_{\mathcal{V}_i}$. Before giving the expressions of $\hat{a}_i$, we define three quantities:

$$\begin{aligned}
\Lambda_i(t) &= \int_0^t w(t)\mathcal{K}_i(\tau)\mathcal{K}_i(\tau)^T \mathrm{d}\tau, \\
\lambda_i(t) &= \int_0^t w(t)\mathcal{K}_i(\tau)\phi_i(\tau)^T \mathrm{d}\tau, \\
F_i &= \frac{\int_{\mathcal{V}_i} \mathcal{K}_i(q)(\mathbf{q} - \mathbf{x}_i)^T d\mathbf{q} K \int_{\mathcal{V}_i} (\mathbf{q} - \mathbf{x}_i)\mathcal{K}_i(q)^T d\mathbf{q}}{\int_{\mathcal{V}_i} \hat{\phi}_i(\mathbf{q}) d\mathbf{q}},
\end{aligned}$$

where $w(t) \geq 0$ is a data collection weighting function.

Following the method in [158], the adaptation law for $\hat{a}_i$ is defined as follows:

$$\begin{aligned}
\dot{\hat{a}}_{\mathrm{pre}i} &= -F_i \hat{a}_i - \gamma(\Lambda_i \hat{a}_i - \lambda_i) - \rho \sum_{j=1}^{N} l_{ij} \left(\hat{a}_i - \hat{a}_j\right), \\
\dot{\hat{a}}_i &= \Gamma(\dot{\hat{a}}_{\mathrm{pre}i} - I_{\mathrm{preji}} \dot{\hat{a}}_{\mathrm{pre}i}),
\end{aligned}$$

where $\gamma > 0$ and $\rho > 0$ are learning rate gains, $\Gamma \in \mathcal{R}^{m \times m}$ is a diagonal, positive-definite adaption gain matrix, and diagonal matrix I_{preji} is defined as

$$I_{\mathrm{preji}} = \begin{cases} 0 & \text{for } \hat{a}_i(j) > a_{\min}, \\ 0 & \text{for } \hat{a}_i(j) = a_{\min} \text{ and } \dot{\hat{a}}_{\mathrm{pre}i}(j) \geq 0, \\ 1 & \text{otherwise.} \end{cases}$$

Finally, we design the optimal coverage algorithm for each robot as

$$u_i = u_{ia} + u_{ib} + u_{ic}, \tag{6.7}$$

with

$$\begin{aligned} u_{ia} =& k_0 \hat{M}_{\mathcal{V}_i} \left(\hat{C}_{\mathcal{V}_i} - x_i \right), \\ u_{ib} =& \sum_{j=1}^{N} \mu_b b_{ij} a_{ij} \left(\frac{1}{\|\mathbf{x}_i - \mathbf{x}_j\|} - \frac{1}{d_b} \right) \frac{\mathbf{x}_i - \mathbf{x}_j}{\|\mathbf{x}_i - \mathbf{x}_j\|^2}, \\ u_{ic} =& \sum_{k=1}^{M} \mu_c c_{ik} \left(\frac{1}{\|\mathbf{x}_i - \mathbf{w}_k\|} - \frac{1}{d_c} \right) \frac{\mathbf{x}_i - \mathbf{w}_k}{\|\mathbf{x}_i - \mathbf{w}_k\|^2}. \end{aligned}$$

6.3.2 Numerical examples

Similar to the scenario in the last section, we deploy $N = 16$ robots in the same farm area, but the density function is not known by all the robots. There is one obstacle at (1.0, 1.0) with $M = 1$. We assume that $\mathcal{K}(q)^T = [\mathcal{K}_1(q), \mathcal{K}_2(q), \cdots, \mathcal{K}_6(q)]$ with

$$\mathcal{K}_i(q)^T = \frac{1}{2\pi\sigma^2} \exp - \frac{(q - \mu_i)^2}{2\sigma^2},$$

where $\sigma = 0.18$ and μ_i is chosen so that one of the six Gaussians is centered at the middle of each grid triangular. By selecting $a = [100, a_{\min}, a_{\min}, a_{\min}, a_{\min}, 100]$, $a_{\min} = 0.1$, $K = 3I_2$, $\Gamma = I_6$, $\gamma = 1$, $\rho = 0.5$, $d_b = 0.1$, $d_c = 0.35$ and using the proposed control law (6.7), the trajectories of robots along with the Voronoi partition are illustrated in Fig. 6.5, and the errors $\|\mathbf{x} - \mathbf{C}_{\mathcal{V}}\|$ are depicted in Fig. 6.6. The convergence results of the objective function are provided in Fig. 6.7.

Based on those figures, we can see that the robots can be driven to converge to the Voronoi centroids with avoiding crash obstacles or other agents. The oscillations in Fig. 6.6 are caused by the configured steps of the robots during the simulation to shorten the simulation time. Furthermore, the collision-avoidance term may also cause the robots to move back and forth, to turn left or right to avoid other objects.

6.4 Conclusion remarks

In this chapter, distributed coverage optimization and control problem with collision avoidance and parameter estimation are studied. First, we consider the case when the density function $\phi(q)$ is known by all the robots in the network. By using the interaction term of the Voronoi neighbor, coverage optimization and control protocols are designed in a distributed way such that the best position of each robot can be determined and all the robots can move to the best positions without collision. Then, we

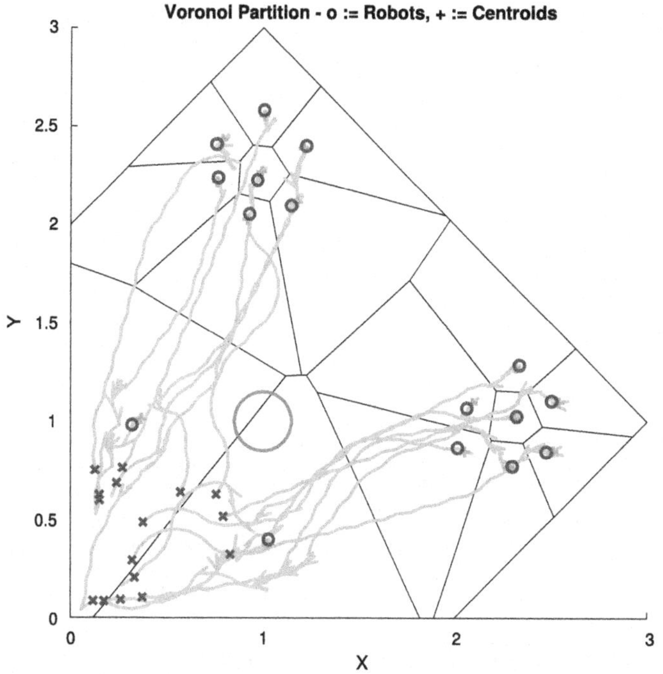

Figure 6.5 Trajectories and centroidal Voronoi regions of 16 robots in the case of unknown density function.

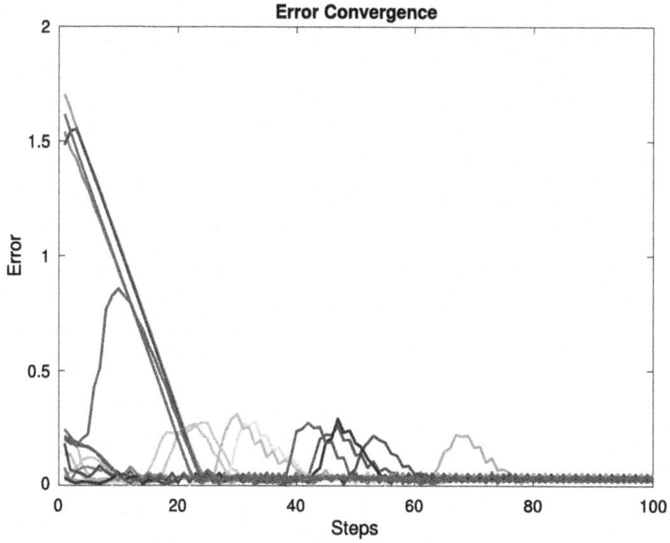

Figure 6.6 Euclidean 2-norm errors between 16 robots' positions and centroids in the case of unknown density function.

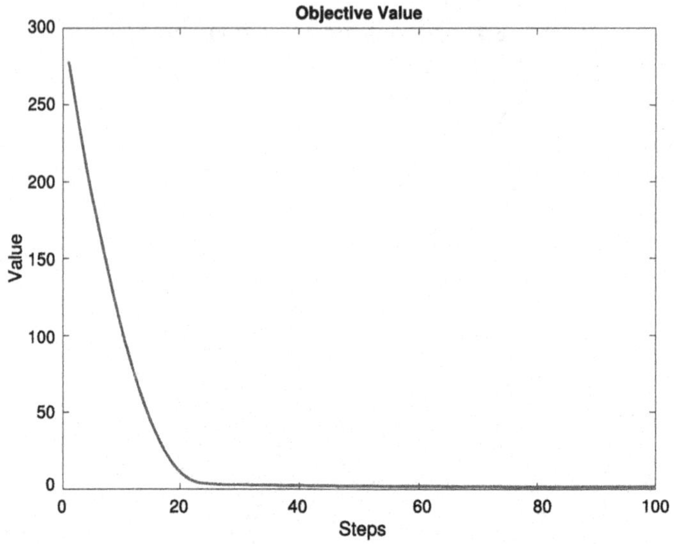

Figure 6.7 Total objective function of locational optimization with 16 robots in the case of unknown density function.

consider the case when the density function $\phi(q)$ is not known by the robots. By using the adaptive technique, the density function $\phi(q)$ can be estimated in a fast and distributed way. Finally, the proposed coverage control schemes are applied to remote sensing for precision farming and the effectiveness of the strategy is validated through some simulation results.

Part Three

Robust cooperative control

Jianan Wang, Chunyan Wang, Zhengtao Ding, Jiayuan Shan

Robust consensus control of multi-agent systems with input delay

7

7.1 Problem formulation

In this section, we consider control design for a group of N agents, each represented by a nonlinear subsystem that is subject to input delay and the Lipschitz nonlinearity,

$$\dot{x}_i(t) = Ax_i(t) + Bu_i(t-h) + \phi(x_i), \tag{7.1}$$

where for agent i, $i = 1, 2, \ldots, N$, $x_i \in \mathbb{R}^n$ is the state vector, $u_i \in \mathbb{R}^m$ is the control input vector, $A \in \mathbb{R}^{n\times n}$ and $B \in \mathbb{R}^{n\times m}$ are constant matrices with (A, B) being controllable, $h > 0$ is input delay, and the initial conditions $x_i(\theta)$, $\theta \in [-h, 0]$, are given and bounded, and $\phi : \mathbb{R}^n \to \mathbb{R}^n$, $\phi(0) = 0$, is a Lipschitz nonlinear function with a Lipschitz constant γ, i.e., for any two constant vectors $a, b \in \mathbb{R}^n$,

$$\|\phi(a) - \phi(b)\| \leq \gamma \|a - b\|.$$

Assumption 7.1. Zero is a simple eigenvalue of the Laplacian matrix $\mathcal{L}$.

Remark 7.1. This assumption implies that there is a directed spanning tree in the network. For undirected graph or strongly connected and balanced graph, the condition $\bar{x}^T\mathcal{L}\bar{x} \geq 0$, $\forall\, \bar{x} \in \mathbb{R}^N$, holds. However, for the general direction graph in Assumption 7.1, the Laplacian matrix $\mathcal{L}$ is asymmetric and $\bar{x}^T\mathcal{L}\bar{x}$ can be sign-indefinite [87]. The decomposition method developed in [88] for the undirected multi-agent systems cannot be applied here due to this unfavorable feature.

The consensus control problem considered in this section is to design a control strategy, using the relative state information, to ensure that all agents asymptotically converge to an identical trajectory.

Before moving to the main results, we will first introduce the basic idea of the predictor-based feedback design. Consider a linear input-delayed system

$$\dot{x}(t) = Ax(t) + Bu(t-h), \tag{7.2}$$

where $x \in \mathbb{R}^n$ denotes the state, $u \in \mathbb{R}^m$ denotes the control input, $A \in \mathbb{R}^{n\times n}$ and $B \in \mathbb{R}^{n\times m}$ are constant matrices, $h \in \mathbb{R}_+$ is input delay, which is known and constant.

If delay is absent, we only need to find a stabilizing gain vector K such that the matrix $A + BK$ is Hurwitz. Accordingly, for the input-delayed system (7.2), if we can have a control that achieves

$$u(t-h) = Kx(t), \tag{7.3}$$

Cooperative Control of Multi-Agent Systems. https://doi.org/10.1016/B978-0-12-820118-3.00018-1

the control problem of the input-delayed system (7.2) is solved with a stabilizing gain vector K.

The controller (7.3) can be alternatively written as

$$u(t) = Kx(t+h). \tag{7.4}$$

It is unrealistic since the controller requires future values of the state x at time $t+h$ which cannot be obtained with direct measurement. However, with the variation-of-constants formula, the vector $x(t+h)$ can be calculated as follows:

$$\begin{aligned} x(t+h) &= e^{A(t+h)}x(0) + \int_0^{t+h} \mathrm{e}^{A(t+h-\tau)}Bu(\tau-h)\mathrm{d}\tau \\ &= e^{Ah}x(t) + \int_{t-h}^{t} \mathrm{e}^{A(t-\tau)}Bu(\tau)\mathrm{d}\tau, \qquad \forall t \geq 0. \end{aligned} \tag{7.5}$$

Therefore we can express the controller as

$$u(t) = K\left[e^{Ah}x(t) + \int_{t-h}^{t} \mathrm{e}^{A(t-\tau)}Bu(\tau)\mathrm{d}\tau\right], \qquad \forall t \geq 0,$$

which is implementable, but infinite-dimensional since it contains the distributed delay term involving past controls [79], $\int_{t-h}^{t} \mathrm{e}^{A(t-\tau)}Bu(\tau)\mathrm{d}\tau$. The closed-loop system is fully delay-compensated,

$$\dot{x}(t) = (A+BK)x(t), \qquad t \geq h. \tag{7.6}$$

During the interval $t \in [0, h]$, the system state is governed by

$$x(t) = e^{Ah}x(0) + \int_0^{t} \mathrm{e}^{A(t-\tau)}Bu(\tau-h)\mathrm{d}\tau, \forall t \in [0, h]. \tag{7.7}$$

Based on the basic idea of the predictor feedback design, the model reduction method for linear system with input delay was first developed by Kwon and Pierson in [81]. The results were then extended to a time-varying system with distributed delays by Artstein in [3]. The outline of the method is as follows.

For the system (7.2), define a new state

$$z(t) = x(t) + \int_t^{t+h} \mathrm{e}^{A(t-\tau)}Bu(\tau-h)\mathrm{d}\tau. \tag{7.8}$$

Differentiating $z(t)$ against time yields

$$\begin{aligned} \dot{z}(t) &= Ax(t) + Bu(t-h) + \mathrm{e}^{-Ah}Bu(t) - Bu(t-h) \\ &\quad + A\int_t^{t+h} \mathrm{e}^{A(t-\tau)}Bu(\tau-h)\mathrm{d}\tau \\ &= Az(t) + Du(t), \end{aligned} \tag{7.9}$$

where $D = \mathrm{e}^{-Ah}B$. The general form of the Leibniz integral rule has been used for this derivation.

The controllability of (A, B) and $(A, e^{-Ah}B)$ are equivalent as proved in [9]. We consider a controller

$$u(t) = Kz(t). \tag{7.10}$$

From Eqs. (7.8) and (7.10), we have

$$\|x(t)\| \leq \|z(t)\| + h\left(\max_{-h\leq\theta\leq 0} \|e^{A\theta}\|\right)\|B\|\|K\|\|z_t(\theta)\|,$$

where $z_t(\theta) := z(t+\theta)$, $-h \leq \theta \leq 0$. Thus, $x(t) \to 0$ as $z(t) \to 0$. In other words, if the controller (7.10) stabilizes the transformed system (7.9), then the original system (7.2) is also stable with the same controller [81].

Remark 7.2. With any given bounded initial condition $u(\theta)$, $\theta \in [-h, 0]$, a stable feedback controller (7.10) implies that $u(t)$ in Eq. (7.8) is locally integrable, which allows for the model reduction as Eq. (7.9).

Remark 7.3. By introducing a state transformation, the input-delayed linear system is transformed to a delay-free one which is finite-dimensional, where the reduction in dimension is achieved.

7.2 Consensus of Lipschitz nonlinear systems with input delay: model reduction method

7.2.1 *Stability analysis for single nonlinear system*

In this section, we first consider the Artstein model reduction method for a single nonlinear system. Consider an input-delayed system

$$\dot{x}(t) = Ax(t) + Bu(t-h) + \phi(x(t)), \tag{7.11}$$

with $\phi : \mathbb{R}^n \to \mathbb{R}^n$, $\phi(0) = 0$, being a Lipschitz nonlinear function, and the initial conditions $x(\theta)$, $\theta \in [-h, 0]$, being bounded. Let

$$z(t) = x(t) + \int_t^{t+h} \mathrm{e}^{A(t-\tau)}Bu(\tau - h)\mathrm{d}\tau. \tag{7.12}$$

Differentiating $z(t)$ in time yields

$$\begin{aligned}\dot{z}(t) &= Ax(t) + \phi(x(t)) + \mathrm{e}^{-Ah}Bu(t) + A\int_t^{t+h} \mathrm{e}^{A(t-\tau)}Bu(\tau - h)\mathrm{d}\tau \\ &= Az(t) + Du(t) + \phi(x(t)),\end{aligned} \tag{7.13}$$

where $D = \mathrm{e}^{-Ah}B$.

We consider a controller

$$u(t) = Kz(t). \tag{7.14}$$

From Eqs. (7.12) and (7.14), we have

$$\|x(t)\| \leq \|z(t)\| + h\left(\max_{-h\leq\theta\leq 0} \|e^{A\theta}\|\right)\|B\|\|K\|\|z_t(\theta)\|,$$

where $z_t(\theta) := z(t+\theta)$, $-h \leq \theta \leq 0$. Thus, $x(t) \to 0$ as $z(t) \to 0$. In other words, if the controller (7.14) stabilizes the transformed system (7.13), then the original system (7.11) is also stable with the same controller [81].

7.2.2 Consensus analysis

For the multi-agent systems (7.1), we use Eq. (7.12) to transform the agent dynamics to

$$\dot{z}_i(t) = Az_i(t) + Du_i(t) + \phi(x_i(t)), \tag{7.15}$$

where $D = \mathrm{e}^{-Ah}B$.

We propose a control design using the relative state information. The control input takes the structure

$$u_i(t) = -K\sum_{j=1}^{N} l_{ij}z_j(t) = -K\sum_{j=1}^{N} a_{ij}\left(z_i(t) - z_j(t)\right) \tag{7.16}$$

where $K \in \mathbb{R}^{m\times n}$ is a constant control gain matrix to be designed later, a_{ij} and l_{ij} being the elements of the graph adjacency matrix $\mathcal{A}$ and the Laplacian matrix $\mathcal{L}$, respectively.

Remark 7.4. From Eqs. (7.12) and (7.16), we have

$$u_i(t) = -K\sum_{j=1}^{N} l_{ij}\left(x_j(t) + \int_t^{t+h} \mathrm{e}^{A(t-\tau)}Bu_j(\tau-h)\mathrm{d}\tau\right).$$

It is observed that certain compensation is added in the controller design based on the model reduction method. With the state transformation (7.12), the original input-delayed multi-agent systems (7.11) are reduced to delay-free systems (7.15). In this way, conventional finite-dimensional techniques can be used for the consensus analysis and controller design.

The closed-loop system is then described by

$$\dot{z}(t) = (I_N \otimes A - \mathcal{L} \otimes DK)z(t) + \Phi(x),$$

where

$$z(t)=\begin{bmatrix} z_1(t) \\ z_2(t) \\ \vdots \\ z_N(t) \end{bmatrix}, \quad \Phi(x)=\begin{bmatrix} \phi(x_1) \\ \phi(x_2) \\ \vdots \\ \phi(x_N) \end{bmatrix}.$$

With Lemmas 2.6 and 2.7, we can define $r^T \in \mathbb{R}^N$ as the left eigenvector of $\mathcal{L}$ corresponding to the eigenvalue at 0, that is, $r^T \mathcal{L} = 0$. Furthermore, let r be scaled such that $r^T \mathbf{1} = 1$ and let the first row of T^{-1} be $(T^{-1})_1 = r^T$, where T is the transformation matrix defined in Lemma 2.7.

Based on the vector r, we introduce a state transformation

$$\overline{\xi}_i(t) = z_i(t) - \sum_{j=1}^{N} r_j z_j(t), \tag{7.17}$$

for $i = 1, 2 \ldots, N$. Let

$$\overline{\xi} = [\overline{\xi}_1^T, \overline{\xi}_2^T, \cdots, \overline{\xi}_N^T]^T.$$

We have

$$\overline{\xi} = z - ((\mathbf{1}r^T) \otimes I_n)z = (M \otimes I_n)z,$$

where $M = I_N - \mathbf{1}r^T$. Since $r^T \mathbf{1} = 1$, it can be shown that $M\mathbf{1} = 0$. Therefore the consensus of the system (7.15) is achieved when $\lim_{t\to\infty} \overline{\xi}(t) = 0$, as $\overline{\xi} = 0$ implies $z_1 = z_2 = \cdots = z_N$, due to the fact that the null space of M is span($\mathbf{1}$). The dynamics of $\overline{\xi}$ can then be obtained as

$$\begin{aligned} \dot{\overline{\xi}} &= \dot{z} - (\mathbf{1}r^T \otimes I_N)\,\dot{z} = (I_N \otimes A - \mathcal{L} \otimes DK)z \\ &\quad - (\mathbf{1}r^T \otimes I_N)[I_N \otimes A - \mathcal{L} \otimes DK]z + (M \otimes I_n)\Phi(x) \\ &= (I_N \otimes A - \mathcal{L} \otimes DK)\overline{\xi} + (M \otimes I_n)\Phi(x). \end{aligned}$$

To explore the structure of $\mathcal{L}$, let us introduce another state transformation

$$\overline{\eta} = (T^{-1} \otimes I_n)\overline{\xi}. \tag{7.18}$$

Then with Lemma 2.7, we have

$$\dot{\overline{\eta}} = (I_N \otimes A - J \otimes DK)\overline{\eta} + \Psi(x), \tag{7.19}$$

where $\Psi(x) = (T^{-1}M \otimes I_n)\Phi(x)$, and

$$\overline{\eta}=\begin{bmatrix} \overline{\eta}_1 \\ \overline{\eta}_2 \\ \vdots \\ \overline{\eta}_N \end{bmatrix}, \qquad \Psi(x)=\begin{bmatrix} \psi_1(x) \\ \psi_2(x) \\ \vdots \\ \psi_N(x) \end{bmatrix},$$

with $\overline{\eta}_i \in \mathbb{R}^n$ and $\psi_i : \mathbb{R}^{n\times N} \to \mathbb{R}^n$ for $i = 1, 2, \ldots, N$.

Then from Eqs. (7.17) and (7.18), we have

$$\overline{\eta}_1 = \left(r^T \otimes I_n\right)\overline{\xi} = \left((r^T M) \otimes I_n\right) z \equiv 0.$$

The nonlinear term $\Psi(x)$ in the transformed system dynamic model (7.19) is expressed as a function of the state x. For the stability analysis, first we need to establish a bound of this nonlinear function in terms of the transformed state $\overline{\eta}$. The following lemma gives a bound of $\Psi(x)$.

Lemma 7.1. *For the nonlinear term $\Psi(x)$ in the transformed system dynamics (7.19), a bound can be established in terms of the state $\overline{\eta}$ as*

$$\|\Psi\|^2 \leq \gamma_0^2(\|\overline{\eta}\|^2 + 4\lambda_\sigma^2(\mathcal{A})\,\|\delta\|^2),$$

with

$$\gamma_0 = 2\sqrt{2N}\gamma\,\|r\|\,\|T\|_F\,\lambda_\sigma(T^{-1}),$$
$$\delta = -\int_t^{t+h} e^{A(t-\tau)}BK\overline{\eta}(\tau - h)d\tau.$$

Proof. Based on the state transformations (7.17) and (7.18), we have

$$\Psi(x) = \left(T^{-1} \otimes I_n\right)(M \otimes I_n)\,\Phi(x) = \left(T^{-1} \otimes I_n\right)\mu,$$

where $\mu = (M \otimes I_n)\Phi(x)$. Then, we have

$$\|\Psi(x)\| \leq \lambda_\sigma(T^{-1})\,\|\mu\|,$$

where $\mu = [\mu_1, \mu_2, \ldots, \mu_N]^T$.

Recalling that $M = I_N - \mathbf{1}r^T$, we have

$$\mu_i = \phi(x_i) - \sum_{k=1}^N r_k\phi(x_k) = \sum_{k=1}^N r_k\left(\phi(x_i) - \phi(x_k)\right).$$

It then follows that

$$\|\mu_i\| \leq \gamma \sum_{k=1}^N |r_k|\,\|x_i - x_k\|.$$

From the state transformation (7.12), we have

$$x_i - x_k = (z_i - \sigma_i) - (z_k - \sigma_k) = (z_i - z_k) - (\sigma_i - \sigma_k),$$

where

$$\sigma_i = \int_t^{t+h} e^{A(t-\tau)}Bu_i(\tau - h)d\tau.$$

Then, we have

$$\|\mu_i\| \le \gamma \sum_{k=1}^{N} |r_k| \left(\|z_i - z_k\| + \|\sigma_i - \sigma_k\|\right). \tag{7.20}$$

From $\overline{\eta} = (T^{-1} \otimes I_n)\overline{\xi}$, we obtain $\overline{\xi} = (T \otimes I_n)\overline{\eta}$, and from the state transformations (7.17), we have

$$z_i - z_k = \overline{\xi}_i - \overline{\xi}_k = ((t_i - t_k) \otimes I_n)\,\overline{\eta} = \sum_{j=1}^{N} (t_{ij} - t_{kj})\overline{\eta}_j,$$

where t_k denotes the k-th row of T. Then, we obtain

$$\|z_i - z_k\| \le \left(\|t_i\| + \|t_k\|\right) \|\overline{\eta}\|. \tag{7.21}$$

Here we used the inequality

$$\sum_{i=1}^{N} (a_i b_i) \le \|a\| \, \|b\| = \sqrt{\sum_{i=1}^{N} a_i^2 \sum_{i=1}^{N} b_i^2}.$$

We next deal with the derived terms σ_i and σ_k. We have

$$\begin{aligned} \sum_{k=1}^{N} |r_k| \, \|\sigma_i - \sigma_k\| &\le \sum_{k=1}^{N} |r_k| \, \|\sigma_i\| + \sum_{k=1}^{N} |r_k| \, \|\sigma_k\| \\ &\le \|r\| \sqrt{N} \, \|\sigma_i\| + \|r\| \, \|\sigma\|, \end{aligned} \tag{7.22}$$

where $\sigma = [\sigma_1^T, \sigma_2^T, \cdots, \sigma_N^T]^T$, and we used the inequality

$$\sum_{i=1}^{N} |a_i| \le \sqrt{N} \, \|a\|.$$

Then, from Eqs. (7.20), (7.21) and (7.22), we can obtain that

$$\begin{aligned} \|\mu_i\| &\le \gamma \sum_{k=1}^{N} |r_k| \left(\|t_i\| + \|t_k\|\right) \|\overline{\eta}\| + \gamma \sqrt{N} \, \|r\| \, \|\sigma_i\| + \gamma \, \|r\| \, \|\sigma\| \\ &\le \gamma \left(\|r\| \sqrt{N} \, \|t_i\| + \|r\| \, \|T\|_F\right) \|\overline{\eta}\| + \gamma \sqrt{N} \, \|r\| \, \|\sigma_i\| + \gamma \, \|r\| \, \|\sigma\| \\ &= \gamma \, \|r\| \left((\sqrt{N} \, \|t_i\| + \|T\|_F) \, \|\overline{\eta}\| + \sqrt{N} \, \|\sigma_i\| + \|\sigma\| \right). \end{aligned}$$

It then follows that

$$\|\mu\|^2 = \sum_{i=1}^{N} (\|\mu_i\|)^2 \le \left(\sum_{i=1}^{N} \|\mu_i\| \right)^2$$

$$
\begin{aligned}
&\leq 4\gamma^2 \|r\|^2 \sum_{i=1}^{N} \left(N \|t_i\|^2 + \|T\|_F^2 \right) \|\overline{\eta}\|^2 \\
&\quad + 4\gamma^2 \|r\|^2 \sum_{i=1}^{N} \left(N \|\sigma_i\|^2 + \|\sigma\|^2 \right) \\
&= 4\gamma^2 \|r\|^2 N \left(2 \|T\|_F^2 \|\overline{\eta}\|^2 + 2 \|\sigma\|^2 \right) \\
&= 8\gamma^2 \|r\|^2 N \left(\|T\|_F^2 \|\overline{\eta}\|^2 + \|\sigma\|^2 \right),
\end{aligned}
\tag{7.23}
$$

where we have used

$$
\sum_{k=1}^{N} \|t_k\|^2 = \|T\|_F^2,
$$

and the inequality

$$
(a+b+c+d)^2 \leq 4(a^2+b^2+c^2+d^2).
$$

Next we need to deal with $\|\sigma\|^2$. From Eq. (7.16), we can get

$$
\begin{aligned}
\sigma_i &= \int_t^{t+h} \mathrm{e}^{A(t-\tau)} B u_i(\tau - h) \mathrm{d}\tau \\
&= -\int_t^{t+h} \mathrm{e}^{A(t-\tau)} B K \sum_{j=1}^{N} l_{ij} z_j(\tau - h) \mathrm{d}\tau.
\end{aligned}
$$

From the relationship between $\mathcal{A}$ and $\mathcal{L}$, we have

$$
\begin{aligned}
\sum_{j=1}^{N} l_{ij} z_j &= \sum_{j=1}^{N} a_{ij} (z_i - z_j) = \sum_{j=1}^{N} a_{ij} \left((t_i - t_j) \otimes I_n \right) \overline{\eta} \\
&= \sum_{j=1}^{N} a_{ij} \sum_{l=1}^{N} (t_{il} - t_{jl}) \overline{\eta}_l.
\end{aligned}
$$

Here we define δ_l

$$
\delta_l = -\int_t^{t+h} \mathrm{e}^{A(t-\tau)} B K \overline{\eta}_l(\tau - h) \mathrm{d}\tau. \tag{7.24}
$$

Then we can obtain that

$$
\sigma_i = \sum_{j=1}^{N} a_{ij} \sum_{l=1}^{N} (t_{il} - t_{jl}) \delta_l.
$$

It then follows that

$$\|\sigma_i\| \le \sum_{j=1}^{N} a_{ij} \left(\|t_i\| + \left\|t_j\right\|\right) \|\delta\|, \tag{7.25}$$

where $\delta = [\delta_1^T, \delta_2^T, \cdots, \delta_N^T]^T$. With Eq. (7.25), the sum of the $\|\sigma_i\|$ can be obtained

$$\begin{aligned}
\sum_{i=1}^{N} \|\sigma_i\| &\le \|\delta\| \sum_{i=1}^{N}\sum_{j=1}^{N} a_{ij}(\|t_i\| + \left\|t_j\right\|) \\
&= \|\delta\| \sum_{i=1}^{N}\sum_{j=1}^{N} a_{ij} \|t_i\| + \|\delta\| \sum_{i=1}^{N}\sum_{j=1}^{N} a_{ij} \left\|t_j\right\| \\
&\le \lambda_\sigma(\mathcal{A}) \|T\|_F \|\delta\| + \lambda_\sigma(\mathcal{A}^T) \|T\|_F \|\delta\| \\
&= \left(\lambda_\sigma(\mathcal{A}) + \lambda_\sigma(\mathcal{A}^T)\right) \|T\|_F \|\delta\| \\
&= 2\lambda_\sigma(\mathcal{A}) \|T\|_F \|\delta\|,
\end{aligned} \tag{7.26}$$

with $\lambda_\sigma(\mathcal{A}) = \lambda_\sigma(\mathcal{A}^T)$. In Eq. (7.26), we have used the following inequalities:

$$\sum_{i=1}^{N}\sum_{j=1}^{N} a_{ij} \|t_i\| = \begin{bmatrix} a_{11} & \cdots & a_{N1} \\ \vdots & \ddots & \vdots \\ a_{1N} & \cdots & a_{NN} \end{bmatrix} \begin{bmatrix} \|t_1\| \\ \vdots \\ \|t_N\| \end{bmatrix} \le \lambda_\sigma\left(\mathcal{A}^T\right) \|T\|_F,$$

$$\sum_{i=1}^{N}\sum_{j=1}^{N} a_{ij} \left\|t_j\right\| = \begin{bmatrix} a_{11} & \cdots & a_{1N} \\ \vdots & \ddots & \vdots \\ a_{N1} & \cdots & a_{NN} \end{bmatrix} \begin{bmatrix} \|t_1\| \\ \vdots \\ \|t_N\| \end{bmatrix} \le \lambda_\sigma(\mathcal{A}) \|T\|_F.$$

Therefore we have

$$\|\sigma\|^2 = \sum_{i=1}^{N} (\|\sigma_i\|)^2 \le \left(\sum_{i=1}^{N} \|\sigma_i\|\right)^2 \le 4\lambda_\sigma^2(\mathcal{A}) \|T\|_F^2 \|\delta\|^2. \tag{7.27}$$

Hence, together with Eqs. (7.23) and (7.27), we get

$$\|\mu\|^2 \le 8\gamma^2 \|r\|^2 N \|T\|_F^2 \left(\|\overline{\eta}\|^2 + 4\lambda_\sigma^2(\mathcal{A}) \|\delta\|^2\right).$$

Finally, we obtain the bound for Ψ as

$$\|\Psi\|^2 \le \lambda_\sigma^2(T^{-1}) \|\mu\|^2 \le \gamma_0^2 \left(\|\overline{\eta}\|^2 + 4\lambda_\sigma^2(\mathcal{A}) \|\delta\|^2\right),$$

with

$$\gamma_0 = 2\sqrt{2N}\gamma \|r\| \|T\|_F \lambda_\sigma(T^{-1}),$$

$$\delta = -\int_t^{t+h} e^{A(t-\tau)} BK\overline{\eta}(\tau - h)d\tau.$$

This completes the proof. □

With the control law shown in Eq. (7.16), the control gain matrix K is chosen as

$$K = D^T P, \tag{7.28}$$

where P is a positive definite matrix. In the remaining part of this section, we will use a Lyapunov-function-based analysis to identify a condition for P to ensure that consensus is achieved by using the control algorithm (7.16) with the control gain K in Eq. (7.28).

The stability analysis will be carried out in terms of $\overline{\eta}$. As discussed earlier, the consensus control can be guaranteed by showing that $\overline{\eta}$ converges to zero, what is implied by showing that $\overline{\eta}_i$ converges to zero for $i = 2, 3, \dots, N$, since we have shown that $\overline{\eta}_1 = 0$.

From the structure of the Laplacian matrix shown in Lemma 2.7, we can see that

$$N_k = 1 + \sum_{j=1}^{k} n_j,$$

for $k = 1, 2, \dots, q$. Note that $N_q = N$.

The agent state variables $\overline{\eta}_i$ from $i = 2$ to N_p are the state variables which are associated with the Jordan blocks of real eigenvalues, and $\overline{\eta}_i$ for $i = N_p + 1$ to N are with Jordan blocks of complex eigenvalues.

For the state variables associated with the Jordan blocks J_k of real eigenvalues, i.e., for $k \le p$, we have the dynamics given by

$$\dot{\overline{\eta}}_i = (A - \lambda_k DD^T P)\overline{\eta}_i - DD^T P\overline{\eta}_{i+1} + \psi_i(x),$$

for $i = N_{k-1} + 1,\ N_{k-1} + 2, \cdots, N_k - 1$, and

$$\dot{\overline{\eta}}_i = (A - \lambda_k DD^T P)\overline{\eta}_i + \psi_i(x),$$

for $i = N_k$.

For the state variables associated with the Jordan blocks J_k, i.e., for $k > p$, corresponding to complex eigenvalues, we consider the dynamics of the state variables in pairs. For notational convenience, let

$$i_1(j) = N_{k-1} + 2j - 1,$$
$$i_2(j) = N_{k-1} + 2j,$$

for $j = 1, 2, \dots, n_k/2$. The dynamics of $\overline{\eta}_{i_1}$ and $\overline{\eta}_{i_2}$ for $j = 1, 2, \dots, n_k/2 - 1$ are expressed by

$$\dot{\overline{\eta}}_{i_1} = (A - \alpha_k DD^T P)\overline{\eta}_{i_1} - \beta_k DD^T P\overline{\eta}_{i_2} - DD^T P\overline{\eta}_{i_1+2} + \psi_{i_1},$$

$$\dot{\overline{\eta}}_{i_2} = (A - \alpha_k DD^T P)\overline{\eta}_{i_2} + \beta_k DD^T P\overline{\eta}_{i_1} - DD^T P\overline{\eta}_{i_2+2} + \psi_{i_2}.$$

For $j = n_k/2$, we have

$$\dot{\overline{\eta}}_{i_1} = (A - \alpha_k DD^T P)\overline{\eta}_{i_1} - \beta_k DD^T P\overline{\eta}_{i_2} + \psi_{i_1},$$
$$\dot{\overline{\eta}}_{i_2} = (A - \alpha_k DD^T P)\overline{\eta}_{i_2} + \beta_k DD^T P\overline{\eta}_{i_1} + \psi_{i_2}.$$

Let

$$V_i = \overline{\eta}_i^T P\overline{\eta}_i,$$

for $i = 2, 3\ldots, N$. Let

$$V_0 = \sum_{i=2}^{N} \overline{\eta}_i^T P\overline{\eta}_i.$$

For the convenience of presentation, we borrow the following results for V_0 from [33].

Lemma 7.2. *For a network-connected dynamic system (7.1) with the transformed state $\overline{\eta}$, $\dot{V}_0$ has following bounds specified in one of the following two cases:*

1) If the eigenvalues of the Laplacian matrix $\mathcal{L}$ are distinct, i.e., $n_k = 1$ for $k = 1, 2, \ldots, q$, $\dot{V}_0$ satisfies

$$\dot{V}_0 \leq \sum_{i=2}^{N} \overline{\eta}_i^T \left(A^T P + PA - 2\alpha PDD^T P + \kappa PP\right)\overline{\eta}_i + \frac{1}{\kappa}\|\Psi\|^2,$$

with κ being any positive real number and

$$\alpha = \min\{\lambda_1, \lambda_2, \ldots, \lambda_p, \alpha_{p+1}, \alpha_{p+2}, \ldots, \alpha_q\}.$$

2) If the Laplacian matrix $\mathcal{L}$ has multiple eigenvalues, i.e., $n_k > 1$ for any $k \in \{1, 2, \cdots, q\}$, $\dot{V}_0$ satisfies

$$\dot{V}_0 \leq \sum_{i=2}^{N} \overline{\eta}_i^T \left(A^T P + PA - 2(\alpha - 1)PDD^T P + \kappa PP\right)\overline{\eta}_i + \frac{1}{\kappa}\|\Psi\|^2,$$

with κ being any positive real number.

Using Lemmas 7.1 and 7.2, we easily obtain

$$\dot{V}_0 \leq \sum_{i=2}^{N} \overline{\eta}_i^T \left(A^T P + PA - 2\alpha PDD^T P + \kappa PP + \frac{\gamma_0^2}{\kappa} I_n\right)\overline{\eta}_i + \frac{4\gamma_0^2}{\kappa}\lambda_\sigma^2(\mathcal{A})\widetilde{\Delta}, \tag{7.29}$$

for Case 1) with $\widetilde{\Delta} = \delta^T \delta$, and

$$\dot{V}_0 \le \sum_{i=2}^{N} \overline{\eta}_i^T \left(A^T P + PA - 2(\alpha - 1) PDD^T P + \kappa PP + \frac{\gamma_0^2}{\kappa} I_n \right) \overline{\eta}_i + \frac{4\gamma_0^2}{\kappa} \lambda_\sigma^2(\mathcal{A}) \widetilde{\Delta}, \tag{7.30}$$

for Case 2). Here we have used $\|\overline{\eta}\|^2 = \sum_{i=2}^{N} \|\overline{\eta}_i\|^2$.

The remaining analysis is to explore the bound of $\widetilde{\Delta}$. With δ_l in Eq. (7.24) and Lemma 2.14, we have

$$\begin{aligned}\widetilde{\Delta}_i &= \int_t^{t+h} \overline{\eta}_i^T(\tau - h) K^T B^T \mathrm{e}^{A^T(t-\tau)} \mathrm{d}\tau \int_t^{t+h} \mathrm{e}^{A(t-\tau)} BK \overline{\eta}_i(\tau - h) \mathrm{d}\tau \\ &\le h \int_t^{t+h} \overline{\eta}_i^T(\tau - h) PDD^T \mathrm{e}^{A^T h} \mathrm{e}^{A^T(t-\tau)} \mathrm{e}^{A(t-\tau)} \mathrm{e}^{Ah} DD^T P \overline{\eta}_i(\tau - h) \mathrm{d}\tau.\end{aligned}$$

In view of Lemma 2.15, with $P = I_n$, provided that

$$\bar{R} = -A^T - A + \omega_1 I_n > 0, \tag{7.31}$$

we have

$$\mathrm{e}^{A^T t} \mathrm{e}^{At} < \mathrm{e}^{\omega_1 t} I_n,$$

and

$$\begin{aligned}\widetilde{\Delta}_i &\le h \int_t^{t+h} \mathrm{e}^{\omega_1(t-\tau)} \overline{\eta}_i^T(\tau - h) PDD^T e^{A^T h} \mathrm{e}^{Ah} DD^T P \overline{\eta}_i(\tau - h) \mathrm{d}\tau \\ &\le h \mathrm{e}^{\omega_1 h} \int_t^{t+h} \mathrm{e}^{\omega_1(t-\tau)} \overline{\eta}_i^T(\tau - h) PDD^T DD^T P \overline{\eta}_i(\tau - h) \mathrm{d}\tau \\ &\le \rho^2 h \mathrm{e}^{\omega_1 h} \int_t^{t+h} \mathrm{e}^{\omega_1(t-\tau)} \overline{\eta}_i^T(\tau - h) \overline{\eta}_i(\tau - h) \mathrm{d}\tau \\ &\le \rho^2 h \mathrm{e}^{2\omega_1 h} \int_t^{t+h} \overline{\eta}_i^T(\tau - h) \overline{\eta}_i(\tau - h) \mathrm{d}\tau,\end{aligned}$$

where ρ is a positive real number satisfying

$$\rho^2 I_n \ge PDD^T DD^T P. \tag{7.32}$$

Then the summation of $\widetilde{\Delta}_i$ can be obtained as

$$\widetilde{\Delta} = \sum_{i=2}^{N} \widetilde{\Delta}_i \le \sum_{i=2}^{N} \rho^2 h \mathrm{e}^{2\omega_1 h} \int_t^{t+h} \overline{\eta}_i^T(\tau - h) \overline{\eta}_i(\tau - h) \mathrm{d}\tau. \tag{7.33}$$

For the integral term $\widetilde{\Delta}$ shown in Eq. (7.33), we consider the following Krasovskii functional

$$\widetilde{W}_i = \int_t^{t+h} e^{\tau-t}\overline{\eta}_i^T(\tau-h)\overline{\eta}_i(\tau-h)d\tau + \int_t^{t+h} \overline{\eta}_i^T(\tau-2h)\overline{\eta}_i(\tau-2h)d\tau.$$

A direct evaluation gives that

$$\begin{aligned}\dot{\widetilde{W}}_i &= -\int_t^{t+h} e^{\tau-t}\overline{\eta}_i^T(\tau-h)\overline{\eta}_i(\tau-h)d\tau \\ &\quad - \overline{\eta}_i(t-2h)^T\overline{\eta}_i(t-2h) + e^h\overline{\eta}_i^T(t)\overline{\eta}_i(t) \\ &\leq -\int_t^{t+h} \overline{\eta}_i^T(\tau-h)\overline{\eta}_i(\tau-h)d\tau + e^h\overline{\eta}_i^T(t)\overline{\eta}_i(t).\end{aligned}$$

With $\widetilde{W}_0 = \sum_{i=2}^N \widetilde{W}_i$, we have

$$\dot{\widetilde{W}}_0 = \sum_{i=2}^N \dot{\widetilde{W}}_i \leq -\sum_{i=2}^N \int_t^{t+h} \overline{\eta}_i^T(\tau-h)\overline{\eta}_i(\tau-h)d\tau + \sum_{i=2}^N e^h\overline{\eta}_i^T(t)\overline{\eta}_i(t). \tag{7.34}$$

Let

$$V = V_0 + \rho^2 h e^{2\omega_1 h}\frac{4\gamma_0^2}{\kappa}\lambda_\sigma^2(\mathcal{A})\widetilde{W}_0.$$

From Eqs. (7.29), (7.30), (7.33) and (7.34), we obtain that

$$\dot{V} \leq \overline{\eta}^T(t)\,(I_N \otimes H_1)\,\overline{\eta}(t),$$

where

$$H_1 := A^TP + PA - 2\alpha PDD^TP + \kappa PP + \frac{\gamma_0^2}{\kappa}\left(1 + \lambda_\sigma^2(\mathcal{A})\rho^2 h e^{(2\omega_1+1)h}\right) I_n, \tag{7.35}$$

for Case 1), and

$$\begin{aligned}H_1 &:= A^TP + PA - 2(\alpha-1)PDD^TP + \kappa PP \\ &\quad + \frac{\gamma_0^2}{\kappa}\left(1 + \lambda_\sigma^2(\mathcal{A})\rho^2 h e^{(2\omega_1+1)h}\right) I_n,\end{aligned} \tag{7.36}$$

for Case 2).

7.2.3 Controller design

The above expressions can be used for consensus analysis of network-connected systems with Lipschitz nonlinearity and input delay. The following theorem summarizes the results.

Theorem 7.1. *For an input-delayed multi-agent system (7.1) with the associated Laplacian matrix that satisfies Assumption 7.1, the consensus control problem can be solved by the control algorithm (7.16) with the control gain $K = D^T P$ specified in one of the following two cases:*

1) If the eigenvalues of the Laplacian matrix $\mathcal{L}$ are distinct, the consensus is achieved when the following conditions are satisfied for $W = P^{-1}$ and $\rho > 0$, $\omega_1 \geq 0$,

$$(A - \frac{1}{2}\omega_1 I_n)^T + (A - \frac{1}{2}\omega_1 I_n) < 0, \tag{7.37}$$

$$\rho W \geq DD^T, \tag{7.38}$$

$$\begin{bmatrix} WA^T + AW - 2\alpha DD^T + \kappa I_n & W \\ W & \frac{-\kappa I_n}{\gamma_0^2(1 + 4h_0\rho^2)} \end{bmatrix} < 0, \tag{7.39}$$

where κ is any positive real number and $h_0 = \lambda_\sigma^2(\mathcal{A})he^{(2\omega_1+1)h}$.

2) If the Laplacian matrix $\mathcal{L}$ has multiple eigenvalues, the consensus is achieved when the conditions (7.37), (7.38) and the following condition are satisfied for $W = P^{-1}$ and $\rho > 0$, $\omega_1 \geq 0$,

$$\begin{bmatrix} WA^T + AW - 2(\alpha - 1)DD^T + \kappa I_n & W \\ W & \frac{-\kappa I_n}{\gamma_0^2(1 + 4h_0\rho^2)} \end{bmatrix} < 0, \tag{7.40}$$

where κ is any positive real number and $h_0 = \lambda_\sigma^2(\mathcal{A})he^{(2\omega_1+1)h}$.

Proof. When the eigenvalues are distinct, from the analysis in this section, we know that the feedback law (7.16) will stabilize $\overline{\eta}$ if the conditions (7.31), (7.32) and $H_1 < 0$ in Eq. (7.35) are satisfied. Indeed, it is easy to see that the conditions (7.31) and (7.32) are equivalent to the conditions specified in Eqs. (7.37) and (7.38). From Eq. (7.35), it can be obtained that $H_1 < 0$ is equivalent to

$$P^{-1}A^T + AP^{-1} - 2\alpha DD^T + \kappa I_n + \frac{\gamma_0^2}{\kappa}(1 + 4h_0\rho^2)P^{-1}P^{-1} < 0,$$

which is further equivalent to Eq. (7.39). Hence we conclude that $\overline{\eta}$ converges to zero asymptotically.

When the Laplacian matrix has multiple eigenvalues, the feedback law (7.16) will stabilize $\overline{\eta}$ if the conditions (7.31), (7.31) and $H_1 < 0$ in Eq. (7.36) are satisfied. Following the similar procedure as in Case 1), we can show that, under the conditions (7.37), (7.38) and (7.40), $\overline{\eta}$ converges to zero asymptotically. The proof is completed. □

Remark 7.5. The conditions shown in Eqs. (7.37) to (7.40) can be checked by standard LMI routines for a set of fixed values ρ and ω_1. The iterative methods developed in [202] for single linear system may also be applied here.

7.3 Consensus of Lipschitz nonlinear systems with input delay: truncated predictor feedback method

7.3.1 Finite-dimensional consensus controller design

The consensus controller (7.16) designed in last section

$$u_i(t) = -K\sum_{j=1}^{N} l_{ij}z_j(t),$$

alternatively can be written as

$$u_i(t) = -K\sum_{j=1}^{N} l_{ij}\left(x_j(t) + \int_t^{t+h} \mathrm{e}^{A(t-\tau)}Bu_j(\tau-h)\mathrm{d}\tau\right).$$

It is obvious that the controller for each agent requires the relative input signals among the agents. It may be unreachable sometimes due to the sensor restriction. To overcome this problem, in this section, we will propose another alternative dynamic consensus protocol based on the TPF method. The control input takes the structure

$$u_i(t) = Ke^{Ah}\sum_{j=1}^{N} a_{ij}\left(x_i(t) - x_j(t)\right) = Ke^{Ah}\sum_{j=1}^{N} l_{ij}x_j(t), \tag{7.41}$$

where $K \in \mathbb{R}^{m\times n}$ is a constant control gain matrix to be designed later, a_{ij} and l_{ij} being the elements of the graph adjacency matrix $\mathcal{A}$ and the Laplacian matrix $\mathcal{L}$, respectively.

Remark 7.6. It is worth noting from Eq. (7.41) that the proposed control only uses the relative state information of the agents via network connections. The controller for each agent is finite-dimensional and easy to implement since the integral of the relative input information is not needed.

7.3.2 Overview of truncated predictor feedback approach

Consider a linear input-delayed system

$$\dot{x}(t) = Ax(t) + Bu(t-h),$$

where $x \in \mathbb{R}^n$ denotes the state, $u \in \mathbb{R}^m$ denotes the control input, $A \in \mathbb{R}^{n\times n}$ and $B \in \mathbb{R}^{n\times m}$ are constant matrices, $h > 0$ is input delay.

As introduced before, the main idea of the predictor feedback is to design the controller

$$u(t-h) = Kx(t), \quad \forall t \geq h. \tag{7.42}$$

The resultant closed-loop system is then given by

$$\dot{x}(t) = (A + BK)x(t),$$

where the control gain matrix K is chosen so that $A + BK$ is Hurwitz. The input (7.42) can be rewritten as

$$u(t) = Kx(t+h) = Ku_1(t) + Ku_2(t), \tag{7.43}$$

where

$$u_1(t) = Ke^{Ah}x(t),$$
$$u_2(t) = K\int_{t-h}^{t} \mathrm{e}^{A(t-\tau)}Bu(\tau)\mathrm{d}\tau.$$

As pointed out in [102], no matter how large the value of the input delay is, the infinite-dimensional predictor term $u_2(t)$ in Eq. (7.43) is dominated by the finite-dimensional predictor term $u_1(t)$ and thus might be safely neglected in $u(t)$ under certain conditions. As a result, the predictor feedback law (7.43) can be truncated as

$$u(t) = Ku_1(t) = Ke^{Ah}x(t), \tag{7.44}$$

what is referred to as the TPF control method. The resultant closed-loop system is then given by

$$\dot{x}(t) = (A + BK)x(t) + \widetilde{d}(t), \tag{7.45}$$

where

$$\widetilde{d}(t) = -BK\int_{t-h}^{t} \mathrm{e}^{A(t-\tau)}Bu(\tau)\mathrm{d}\tau.$$

Now the control problem is to find a proper gain matrix K to stabilize the resultant closed-loop system (7.45).

7.3.3 Consensus analysis

For the multi-agent systems (7.1), we have

$$x_i(t) = e^{Ah}x_i(t-h) + \int_{t-h}^{t} e^{A(t-\tau)}\left(Bu_i(\tau - h) + \phi\left(x_i(\tau)\right)\right)\mathrm{d}\tau.$$

Under the control algorithm (7.41), the multi-agent systems (7.1) can be written as

$$\begin{aligned}\dot{x}_i &= Ax_i + BK\sum_{j=1}^{N} l_{ij}x_j + \phi(x_i)\\ &\quad - BK\sum_{j=1}^{N} l_{ij}\int_{t-h}^{t} e^{A(t-\tau)}\left(Bu_j(\tau - h) + \phi(x_j)\right)\mathrm{d}\tau.\end{aligned}$$

The closed-loop system is then described by

$$\dot{x} = (I_N \otimes A + \mathcal{L} \otimes BK)x + (\mathcal{L} \otimes BK)(d_1 + d_2) + \Phi(x), \tag{7.46}$$

where

$$d_1 = -\int_{t-h}^{t} e^{A(t-\tau)} Bu(\tau - h)\mathrm{d}\tau,$$
$$d_2 = -\int_{t-h}^{t} e^{A(t-\tau)} \Phi(x)\mathrm{d}\tau,$$

with

$$x(t) = \left[x_1^T(t), x_2^T(t), \cdots, x_N^T(t)\right]^T,$$
$$u(t) = \left[u_1^T(t), u_2^T(t), \cdots, u_N^T(t)\right]^T,$$
$$\Phi(x) = \left[\phi^T(x_1), \phi^T(x_2), \cdots, \phi^T(x_N)\right]^T.$$

Based on the vector r in Lemma 2.15, we introduce a state transformation

$$\xi_i = x_i - \sum_{j=1}^{N} r_j x_j, \tag{7.47}$$

for $i = 1, 2, \ldots, N$. Let $\xi = \left[\xi_1^T, \xi_2^T, \cdots, \xi_N^T\right]^T$. Then we have

$$\xi = x - \left(\left(\mathbf{1}r^T\right) \otimes I_n\right) x = (M \otimes I_n)x,$$

where $M = I_N - \mathbf{1}r^T$. Since $r^T\mathbf{1} = 1$, it can be shown that $M\mathbf{1} = 0$. Therefore the consensus of system (7.46) is achieved when $\lim_{t\to\infty} \xi(t) = 0$, as $\xi = 0$ implies that $x_1 = x_2 = \cdots = x_N$, due to the fact that the null space of M is span$\{\mathbf{1}\}$. The dynamics of ξ can then be derived as

$$\begin{aligned}\dot{\xi} &= (I_N \otimes A + \mathcal{L} \otimes BK)x - \left(\mathbf{1}r^T \otimes I_n\right)[I_N \otimes A + \mathcal{L} \otimes BK]x \\ &\quad + (M \otimes I_n)(\mathcal{L} \otimes BK)(d_1 + d_2) + (M \otimes I_n)\Phi(x) \\ &= (I_N \otimes A + \mathcal{L} \otimes BK)\xi + (M \otimes I_n)\Phi(x) + (\mathcal{L} \otimes BK)(d_1 + d_2),\end{aligned}$$

where we have used $r^T\mathcal{L} = 0$.

To explore the structure of $\mathcal{L}$, we propose another state transformation

$$\eta = \left(T^{-1} \otimes I_n\right)\xi, \tag{7.48}$$

with $\eta = \left[\eta_1^T, \eta_2^T, \cdots, \eta_N^T\right]^T$. Then, based on Lemma 2.15, we have

$$\dot{\eta} = (I_N \otimes A + J \otimes BK)\eta + \Pi(x) + \Delta(x) + \Psi(x), \tag{7.49}$$

where

$$\Pi(x) = \left(T^{-1}\mathcal{L} \otimes BK\right) d_1,$$
$$\Delta(x) = \left(T^{-1}\mathcal{L} \otimes BK\right) d_2,$$
$$\Psi(x) = \left(T^{-1}M \otimes I_n\right) \Phi(x).$$

For the notational convenience, let

$$\Pi = \begin{bmatrix} \pi_1 \\ \pi_2 \\ \vdots \\ \pi_N \end{bmatrix}, \quad \Delta = \begin{bmatrix} \delta_1 \\ \delta_2 \\ \vdots \\ \delta_N \end{bmatrix}, \quad \Psi = \begin{bmatrix} \psi_1 \\ \psi_2 \\ \vdots \\ \psi_N \end{bmatrix},$$

with $\pi_i, \delta_i, \psi_i, : \mathbb{R}^{n\times N} \to \mathbb{R}^n$ for $i = 1, 2, \ldots, N$.

From the state transformations (7.47) and (7.48), we have

$$\eta_1 = \left(r^T \otimes I_n\right) \xi = \left((r^T M) \otimes I_n\right) x \equiv 0.$$

With the control law shown in Eq. (7.41), the control gain matrix K is chosen as

$$K = -B^T P,$$

where P is a positive definite matrix.

The consensus analysis will be carried out in terms of η. By Eq. (7.48), the consensus is achieved if η converges to zero, or equivalently if η_i converges to zero for $i = 2, 3, \ldots, N$, since it has been shown that $\eta_1 = 0$. Let

$$V_i = \eta_i^T P \eta_i,$$

for $i = 2, 3, \cdots, N$. By employing the similar Lyapunov functions developed in last section, we have the following results.

Lemma 7.3. *For the multi-agent systems (7.1) with the transformed state η, $\dot{V}_0$ has the following bounds specified in one of the following two cases:*

1) If the eigenvalues of the Laplacian matrix $\mathcal{L}$ are distinct, i.e., $n_k = 1$ for $k = 1, 2, \ldots, q$, $\dot{V}_0$ satisfies

$$\dot{V}_0 \leq \eta^T \left[I_N \otimes \left(A^T P + PA - 2\alpha PBB^T P + \sum_{\iota=1}^{3} \kappa_\iota PP \right) \right] \eta + \frac{1}{\kappa_1}\|\Pi\|^2 + \frac{1}{\kappa_2}\|\Delta\|^2 + \frac{1}{\kappa_3}\|\Psi\|^2,$$

with $\kappa_1, \kappa_2, \kappa_3$ being any positive real numbers and α is the real part of the smallest nonzero eigenvalue of the Laplacian matrix $\mathcal{L}$,

$$\alpha = \min\{\lambda_1, \lambda_2, \ldots, \lambda_p, \alpha_{p+1}, \alpha_{p+2}, \ldots, \alpha_q\}.$$

2) If the Laplacian matrix $\mathcal{L}$ has multiple eigenvalues, i.e., $n_k > 1$ for any $k \in \{1, 2, \cdots, q\}$, $\dot{V}_0$ satisfies

$$\dot{V}_0 \leq \eta^T \left[I_N \otimes \left(A^T P + PA - 2(\alpha - 1) PBB^T P + \sum_{\iota=1}^{3} \kappa_\iota PP \right) \right] \eta + \frac{1}{\kappa_1} \|\Pi\|^2 + \frac{1}{\kappa_2} \|\Delta\|^2 + \frac{1}{\kappa_3} \|\Psi\|^2, \tag{7.50}$$

with κ being any positive real number.

The following lemmas give the bounds of $\|\Pi\|^2$, $\|\Delta\|^2$ and $\|\Psi\|^2$.

Lemma 7.4. *For the integral term $\|\Pi\|^2$ shown in the transformed system dynamics (7.49), the bounds can be established as*

$$\|\Pi\|^2 \leq \gamma_0 \int_{t-h}^{t} \eta^T (\tau - h) \eta (\tau - h) \mathrm{d}\tau, \tag{7.51}$$

with

$$\gamma_0 = 4h\rho^4 e^{2\omega_1 h} \lambda_\sigma^2 \left(T^{-1} \right) \|\mathcal{L}\|_F^2 \|\mathcal{A}\|_F^2 \|T\|_F^2 ,$$

where $\mathcal{A}$ is the adjacency matrix, $\mathcal{L}$ is the Laplacian matrix, T is the nonsingular matrix, ρ and ω_1 are positive numbers such that

$$\rho^2 I \geq PBB^T BB^T P, \tag{7.52}$$

$$\omega_1 I > A^T + A. \tag{7.53}$$

Proof. By the definition of $\Pi(x)$ in Eq. (7.49), we have

$$\|\Pi\| = \left\| \left(T^{-1} \otimes I_n \right) (\mathcal{L} \otimes BK) d_1 \right\| \leq \lambda_\sigma \left(T^{-1} \right) \|\mu\| , \tag{7.54}$$

where $\mu = (\mathcal{L} \otimes BK) d_1$. For the notational convenience, let $\mu = \left[\mu_1^T, \mu_2^T, \ldots, \mu_N^T \right]^T$. Then from Eqs. (7.41) and (7.46), we have

$$\begin{aligned} \mu_i &= - BK \sum_{j=1}^{N} l_{ij} \int_{t-h}^{t} e^{A(t-\tau)} B u_j(\tau - h) \mathrm{d}\tau \\ &= BB^T P \sum_{j=1}^{N} l_{ij} \int_{t-h}^{t} e^{A(t-\tau)} BB^T P e^{Ah} \sum_{k=1}^{N} a_{jk} \left(x_k(\tau - h) - x_j(\tau - h) \right) \mathrm{d}\tau. \end{aligned} \tag{7.55}$$

From $\eta = \left(T^{-1} \otimes I_n\right)\xi$, we obtain $\xi = (T \otimes I_n)\,\eta$, and from the state transformations (7.47) and (7.48), we have

$$x_k(t) - x_j(t) = \xi_k(t) - \xi_j(t) = \left(\left(T_k - T_j\right) \otimes I_n\right)\eta(t) = \sum_{l=1}^{N}\left(T_{kl} - T_{jl}\right)\eta_l(t), \tag{7.56}$$

where T_k denotes the k-th row of T. We define

$$\sigma_l = BB^T P \int_{t-h}^{t} \mathrm{e}^{A(t-\tau)} BB^T P e^{Ah} \eta_l(\tau - h)\mathrm{d}\tau.$$

Then, from Eqs. (7.55) and (7.56), we can obtain that

$$\mu_i = \sum_{j=1}^{N} l_{ij} \sum_{k=1}^{N} a_{jk} \sum_{l=1}^{N}\left(T_{kl} - T_{jl}\right)\sigma_l.$$

For the notational convenience, let $\sigma = \left[\sigma_1^T, \sigma_2^T, \ldots, \sigma_N^T\right]^T$. It then follows that

$$\begin{aligned}
\|\mu_i\| &\le \sum_{j=1}^{N} \left|l_{ij}\right| \sum_{k=1}^{N} \left|a_{jk}\right| \|T_k\| \,\|\sigma\| + \sum_{k=1}^{N}\sum_{j=1}^{N} \left|l_{ij}\right| \left|a_{jk}\right| \left\|T_j\right\| \|\sigma\| \\
&\le \sum_{j=1}^{N} \left|l_{ij}\right| \left\|a_j\right\| \|T\|_F \|\sigma\| + \sum_{k=1}^{N}\sum_{j=1}^{N} \left|l_{ij}\right| \|a_k\| \,\|T\|_F \|\sigma\| \\
&\le \|l_i\| \,\|\mathcal{A}\|_F \|T\|_F \|\sigma\| + \|l_i\| \,\|\mathcal{A}\|_F \|T\|_F \|\sigma\| \\
&= 2\,\|l_i\| \,\|\mathcal{A}\|_F \|T\|_F \|\sigma\|,
\end{aligned}$$

where l_i and a_i denote the i-th row of $\mathcal{L}$ and $\mathcal{A}$, respectively. Therefore, we have

$$\begin{aligned}
\|\mu\|^2 = \sum_{i=1}^{N} \|\mu_i\|^2 &\le 4 \sum_{i=1}^{N} \|l_i\|^2 \|\mathcal{A}\|_F^2 \|T\|_F^2 \|\sigma\|^2 \\
&= 4\,\|L\|_F^2 \|\mathcal{A}\|_F^2 \|T\|_F^2 \|\sigma\|^2 .
\end{aligned} \tag{7.57}$$

Next we need to deal with $\|\sigma\|^2$. With the Jensen Inequality in Lemma 2.14 and the condition (7.46), we have

$$\begin{aligned}
\|\sigma_i\|^2 \le h &\int_{t-h}^{t} \eta_i^T(\tau - h) e^{A^T h} PBB^T e^{A^T(t-\tau)} PBB^T \\
&\times BB^T P e^{A(t-\tau)} BB^T P e^{Ah} \eta_i(\tau - h)\mathrm{d}\tau \\
\le h\rho^2 &\int_{t-h}^{t} \eta_i^T(\tau - h) e^{A^T h} PBB^T e^{A^T(t-\tau)} e^{A(t-\tau)} \\
&\times BB^T P e^{Ah} \eta_i(\tau - h)\mathrm{d}\tau.
\end{aligned}$$

In view of Lemma 2.15, with the condition (7.47), we have

$$\begin{aligned}\|\sigma_i\|^2 &\le h\rho^2 \int_{t-h}^{t} e^{\omega_1(t-\tau)}\eta_i^T(\tau-h)e^{A^T h}PBB^TBB^TPe^{Ah}\eta_i(\tau-h)\mathrm{d}\tau \\ &\le h\rho^4 e^{\omega_1 h}\int_{t-h}^{t}\eta_i^T(\tau-h)e^{A^T h}e^{Ah}\eta_i(\tau-h)\mathrm{d}\tau \\ &\le h\rho^4 e^{2\omega_1 h}\int_{t-h}^{t}\eta_i^T(\tau-h)\eta_i(\tau-h)\mathrm{d}\tau.\end{aligned}$$

Then, $\|\sigma\|^2$ can be bounded as

$$\|\sigma\|^2 = \sum_{i=1}^{N}\|\sigma_i\|^2 \le h\rho^4 e^{2\omega_1 h}\int_{t-h}^{t}\eta^T(\tau-h)\eta(\tau-h)\mathrm{d}\tau. \tag{7.58}$$

Therefore, from Eqs. (7.54), (7.57) and (7.58), we have

$$\|\Pi\|^2 \le \gamma_0 \int_{t-h}^{t}\eta^T(\tau-h)\eta(\tau-h)\mathrm{d}\tau.$$

This completes the proof. □

Lemma 7.5. *For the integral term $\Delta(x)$ in the transformed system dynamics (7.49), a bound can be established as*

$$\|\Delta\|^2 \le \gamma_1 \int_{t-h}^{t}\eta^T(\tau)\eta(\tau)\mathrm{d}\tau, \tag{7.59}$$

where

$$\gamma_1 = 4\rho^2 h e^{\omega_1 h}\gamma^2\lambda_\sigma^2\left(T^{-1}\right)\lambda_\sigma^2(\mathcal{A})\,\|T\|_F^2,$$

with ρ and ω_1 defined in Eqs. (7.52) and (7.53).

Proof. In a way similar to Lemma 7.4, we have

$$\|\Delta(x)\| = \left\|\left(T^{-1}\otimes I_n\right)(\mathcal{L}\otimes BK)d_2\right\| \le \lambda_\sigma\left(T^{-1}\right)\|\bar{\delta}\|, \tag{7.60}$$

where $\bar{\delta} = (\mathcal{L}\otimes BK)d_2$. Let $\bar{\delta} = \left[\bar{\delta}_1^T, \bar{\delta}_2^T, \ldots, \bar{\delta}_N^T\right]^T$. Then, from Eqs. (7.44) and (7.46), we have

$$\bar{\delta}_i = \sum_{j=1}^{N} a_{ij}BB^TP\int_{t-h}^{t} e^{A(t-\tau)}\left[\phi(x_i)-\phi(x_j)\right]\mathrm{d}\tau.$$

It follows that

$$\|\bar{\delta}_i\|^2 = \sum_{j=1}^{N} a_{ij}^2 \int_{t-h}^{t} [\phi(x_i) - \phi(x_j)]^T e^{A^T(t-\tau)} \mathrm{d}\tau \\ \times PBB^T BB^T P \int_{t-h}^{t} e^{A(t-\tau)} [\phi(x_i) - \phi(x_j)] \mathrm{d}\tau.$$

With Jensen's Inequality in Lemma 2.14 and the condition (7.52), we have

$$\begin{aligned} \|\bar{\delta}_i\|^2 &\le h \sum_{j=1}^{N} a_{ij}^2 \int_{t-h}^{t} [\phi(x_i) - \phi(x_j)]^T e^{A^T(t-\tau)} PBB^T \\ &\quad \times BB^T P e^{A(t-\tau)} [\phi(x_i) - \phi(x_j)] \mathrm{d}\tau \\ &\le \rho^2 h \sum_{j=1}^{N} a_{ij}^2 \int_{t-h}^{t} [\phi(x_i) - \phi(x_j)]^T e^{A^T(t-\tau)} e^{A(t-\tau)} [\phi(x_i) - \phi(x_j)] \mathrm{d}\tau. \end{aligned}$$

In view of Lemma 2.15, with the condition (7.53), we have

$$\begin{aligned} \|\bar{\delta}_i\|^2 &\le \rho^2 h \sum_{j=1}^{N} a_{ij}^2 \int_{t-h}^{t} e^{\omega_1(t-h)} \|\phi(x_i) - \phi(x_j)\|^2 \mathrm{d}\tau \\ &\le \rho^2 h e^{\omega_1 h} \gamma^2 \sum_{j=1}^{N} a_{ij}^2 \int_{t-h}^{t} \|x_i(\tau) - x_j(\tau)\|^2 \mathrm{d}\tau. \end{aligned}$$

From the state transformations, we have

$$x_i(t) - x_j(t) = \xi_i(t) - \xi_j(t) = \left((t_i - t_j) \otimes I_n\right) \eta(t) = \sum_{l=1}^{N} \left(t_{il} - t_{jl}\right) \eta_l(t).$$

Let us define $\overline{\sigma}_l = \int_{t-h}^{t} \eta_l(\tau)\mathrm{d}\tau$ for $l = 1, 2, \cdots, N$. Then,

$$\begin{aligned} \|\bar{\delta}_i\|^2 &\le \rho^2 h e^{\omega_1 h} \gamma^2 \sum_{j=1}^{N} a_{ij}^2 \sum_{l=1}^{N} |t_{il} - t_{jl}|^2 \|\overline{\sigma}_l\|^2 \\ &\le 2\rho^2 h e^{\omega_1 h} \gamma^2 \sum_{j=1}^{N} a_{ij}^2 \sum_{l=1}^{N} \left(|t_{il}|^2 + |t_{jl}|^2\right) \|\overline{\sigma}_l\|^2 \\ &\le 2\rho^2 h e^{\omega_1 h} \gamma^2 \sum_{j=1}^{N} a_{ij}^2 \left(\|t_i\|^2 + \|t_j\|^2\right) \|\overline{\sigma}\|^2, \end{aligned}$$

where $\overline{\sigma} = \left[\overline{\sigma}_1^T, \overline{\sigma}_2^T, \ldots, \overline{\sigma}_N^T\right]^T$. Consequently,

$$\|\bar{\delta}\|^2 = \sum_{i=1}^{N} \|\bar{\delta}_i\|^2 \le 2\rho^2 h e^{\omega_1 h} \gamma^2 \|\overline{\sigma}\|^2 \sum_{i=1}^{N} \sum_{j=1}^{N} a_{ij}^2 \left(\|t_i\|^2 + \|t_j\|^2\right)$$

$$\leq 2\rho^2 h e^{\omega_1 h}\gamma^2\lambda_\sigma^2(\mathcal{A})\,\|T\|_F^2\,\|\overline{\sigma}\|^2 + 2\rho^2 h e^{\omega_1 h}\gamma^2\lambda_\sigma^2(\mathcal{A})\,\|T\|_F^2\,\|\overline{\sigma}\|^2$$
$$= 4\rho^2 h e^{\omega_1 h}\gamma^2\lambda_\sigma^2(\mathcal{A})\,\|T\|_F^2\,\|\overline{\sigma}\|^2. \tag{7.61}$$

Putting Eqs. (7.60) and (7.61) together, we have

$$\|\Delta\|^2 \leq \gamma_1 \int_{t-h}^{t} \eta^T(\tau)\eta(\tau)\mathrm{d}\tau$$

with

$$\gamma_1 = 4\rho^2 h N e^{\omega_1 h}\gamma^2\lambda_\sigma^2(T^{-1})\lambda_\sigma^2(\mathcal{A})\,\|T\|_F^2.$$

□

Lemma 7.6. *For the nonlinear term $\Psi(x)$ in the transformed system dynamics (7.49), a bound can be established as*

$$\|\Psi\|^2 \leq \gamma_2\,\|\eta\|^2, \tag{7.62}$$

where

$$\gamma_2 = 4N\gamma^2\,\|r\|^2\,\lambda_\sigma^2\left(T^{-1}\right)\|T\|_F^2.$$

Proof. By the definition of $\Psi(x)$ in Eq. (7.49), we have

$$\|\Psi\| = \left\|\left(T^{-1}\otimes I_n\right)(M\otimes I_n)\,\Phi(x)\right\| \leq \lambda_\sigma\left(T^{-1}\right)\|\bar{z}\|,$$

where $\bar{z} = (M\otimes I_n)\,\Phi(x)$. For the notational convenience, let $\bar{z} = \left[\bar{z}_1^T, \bar{z}_2^T, \ldots, \bar{z}_N^T\right]^T$. Then from Eq. (7.47), we have

$$\bar{z}_i = \phi(x_i) - \sum_{k=1}^{N} r_k\phi(x_k) = \sum_{k=1}^{N} r_k(\phi(x_i) - \phi(x_k)).$$

It then follows that

$$\|\bar{z}_i\| \leq \sum_{k=1}^{N} |r_k|\,\|(\phi(x_i) - \phi(x_k))\| \leq \gamma\sum_{k=1}^{N} |r_k|\,\|x_i - x_k\|.$$

In light of Eq. (7.56), we have

$$\|\bar{z}_i\| \leq \gamma\sum_{k=1}^{N} |r_k|\,(\|T_i\| + \|T_k\|)\,\|\eta\| \leq \gamma\,\|\eta\|\left(\sum_{k=1}^{N} |r_k|\,\|T_i\| + \|r\|\,\|T\|_F\right).$$

Therefore we have

$$\|\bar{z}\|^2 = \sum_{i=1}^{N} \|\bar{z}_i\|^2 \leq 2\gamma^2\,\|\eta\|^2\sum_{i=1}^{N}\left(\|T_i\|^2\left(\sum_{k=1}^{N} |r_k|\right)^2 + \|r\|^2\,\|T\|_F^2\right)$$

$$\leq 2\gamma^2 \|\eta\|^2 \sum_{i=1}^{N} \left(\|T_i\|^2 N \|r\|^2 + \|r\|^2 \|T\|_F^2 \right)$$
$$= 4N\gamma^2 \|r\|^2 \|T\|_F^2 \|\eta\|^2,$$

and

$$\|\Psi\|^2 \leq \gamma_2 \|\eta\|^2.$$

This completes the proof. □

Using Eqs. (7.50), (7.51), (7.59) and (7.62), we can obtain

$$\begin{aligned}\dot{V}_0 \leq{}& \eta^T \left[I_N \otimes \left(A^T P + PA - 2\hat{\alpha} PBB^T P + \sum_{\iota=1}^{3} \kappa_\iota PP + \frac{\gamma_2}{\kappa_3} I_n \right) \right] \eta \\ &+ \frac{\gamma_0}{\kappa_1} \int_{t-h}^{t} \eta^T(\tau - h)\eta(\tau - h)\mathrm{d}\tau + \frac{\gamma_1}{\kappa_2} \int_{t-h}^{t} \eta^T(\tau)\eta(\tau)\mathrm{d}\tau, \end{aligned} \tag{7.63}$$

where $\hat{\alpha} = \alpha$ for Case 1) and $\hat{\alpha} = \alpha - 1$ for Case 2) in Lemma 7.3.

For the first integral term shown in Eq. (7.63), we consider the following Krasovskii functional

$$W_3 = e^h \int_{t-h}^{t} e^{\tau - t} \eta^T(\tau - h)\eta(\tau - h)\mathrm{d}\tau + e^h \int_{t-h}^{t} \eta^T(\tau)\eta(\tau)\mathrm{d}\tau.$$

A direct evaluation gives that

$$\begin{aligned}\dot{W}_3 ={}& -e^h \int_{t-h}^{t} e^{\tau - t} \eta^T(\tau - h)\eta(\tau - h)\mathrm{d}\tau \\ &- \eta^T(t - 2h)\eta(t - 2h) + e^h \eta^T(t)\eta(t) \\ \leq{}& -\int_{t-h}^{t} \eta^T(\tau - h)\eta(\tau - h)\mathrm{d}\tau + e^h \eta(t)^T \eta(t). \end{aligned} \tag{7.64}$$

For the second integral term shown in Eq. (7.63), we consider the following Krasovskii functional

$$W_4 = e^h \int_{t-h}^{t} e^{\tau - t} \eta^T(\tau)\eta(\tau)\mathrm{d}\tau.$$

A direct evaluation gives that

$$\begin{aligned}\dot{W}_4 ={}& -e^h \int_{t-h}^{t} e^{\tau - t} \eta^T(\tau)\eta(\tau)\mathrm{d}\tau + e^h \eta^T(t)\eta(t) - \eta^T(t - h)\eta(t - h) \\ \leq{}& -\int_{t-h}^{t} \eta^T(\tau)\eta(\tau)\mathrm{d}\tau + e^h \eta^T(t)\eta(t). \end{aligned} \tag{7.65}$$

Let

$$V = V_0 + \frac{\gamma_0}{\kappa_1} W_3 + \frac{\gamma_1}{\kappa_2} W_4.$$

From Eqs. (7.63), (7.64), and (7.65), we obtain that

$$\dot{V} \leq \eta^T(t) (I_N \otimes H_3) \eta(t), \tag{7.66}$$

where

$$H_3 := A^T P + PA - 2\alpha PBB^T P + \sum_{\iota=1}^{3} \kappa_\iota PP + \left(\frac{\gamma_0}{\kappa_1} e^h + \frac{\gamma_1}{\kappa_2} e^h + \frac{\gamma_2}{\kappa_3} \right) I_n \tag{7.67}$$

for Case 1) and

$$\begin{aligned} H_3 :=& A^T P + PA - 2(\alpha - 1) PBB^T P \\ &+ \sum_{\iota=1}^{3} \kappa_\iota PP + \left(\frac{\gamma_0}{\kappa_1} e^h + \frac{\gamma_1}{\kappa_2} e^h + \frac{\gamma_2}{\kappa_3} \right) I_n \end{aligned} \tag{7.68}$$

for Case 2).

Based on the analysis above, the following theorem presents sufficient conditions to ensure that the consensus problem is solved by using the control algorithm (7.41).

Theorem 7.2. *For the Lipschitz nonlinear multi-agent systems (1) with input delay:*

1) If the eigenvalues of the Laplacian matrix $\mathcal{L}$ are distinct, the consensus control problem can be solved by the control algorithm (7.41) with $K = -B^T P$ when there exists a positive definite matrix P and constants $\omega_1 \geq 0$, $\rho, \kappa_1, \kappa_2, \kappa_3 > 0$ such that

$$\rho W \geq BB^T, \tag{7.69}$$

$$\left(A - \frac{1}{2}\omega_1 I_n \right)^T + \left(A - \frac{1}{2}\omega_1 I_n \right) < 0, \tag{7.70}$$

$$\begin{bmatrix} WA^T + AW - 2\alpha BB^T + (\kappa_1 + \kappa_2 + \kappa_3) I_n & W \\ W & -\frac{I_n}{\Gamma} \end{bmatrix} < 0, \tag{7.71}$$

are satisfied with $W = P^{-1}$ and

$$\Gamma = \frac{\gamma_0}{\kappa_1} e^h + \frac{\gamma_1}{\kappa_2} e^h + \frac{\gamma_2}{\kappa_3}.$$

2) If the Laplacian matrix $\mathcal{L}$ has multiple eigenvalues, the consensus control problem can be solved by the control algorithm (7.41) with $K = -B^T P$ when there exists *a positive definite matrix P and constants $\omega_1 \geq 0$, $\rho, \kappa_1, \kappa_2, \kappa_3 > 0$ such that the conditions (7.69), (7.70) and*

$$\begin{bmatrix} WA^T + AW - 2(\alpha - 1)BB^T + (\kappa_1 + \kappa_2 + \kappa_3)I_n & W \\ W & -\frac{I_n}{\Gamma} \end{bmatrix} < 0,$$

are satisfied.

Proof. From the analysis in this section, we know that the feedback law (7.41) will stabilize η if the conditions (7.52), (7.53) and $H_3 < 0$ in Eq. (7.66) are satisfied. Indeed, it is easy to see the conditions (7.52) and (7.53) are equivalent to the conditions specified in Eqs. (7.69) and (7.70). From Eq. (7.67), it can be obtained that $H_3 < 0$ is equivalent to

$$WA^T + AW - 2\hat{\alpha}BB^T + (\kappa_1 + \kappa_2 + \kappa_3)I_n + \left(\frac{\gamma_0}{\kappa_1}e^h + \frac{\gamma_1}{\kappa_2}e^h + \frac{\gamma_2}{\kappa_3}\right)WW < 0,$$

which is further equivalent to condition (7.71). It implies that η converges to zero asymptotically. Hence, the consensus control is achieved. □

It is observed that condition (7.71) is more likely to be satisfied if the values of $\rho, \omega_1, \kappa_1, \kappa_2, \kappa_3$ are small. Therefore, the algorithm for finding a feasible solution of the conditions shown in (7.69) to (7.71) can be designed by following the iterative methods developed in [202] for an individual linear system. In particular, we suggest the following step by step algorithm.

1) Set $\omega_1 = \lambda_{\max}(A + A^T)$ if $\lambda_{\max}(A + A^T) > 0$; otherwise set $\omega_1 = 0$.

2) Fix the value of $\rho, \omega_1, \kappa_1, \kappa_2, \kappa_3$ to some constants $\tilde{\omega}_1 > \omega_1$ and $\tilde{\rho}, \tilde{\kappa}_1, \tilde{\kappa}_2, \tilde{\kappa}_3 > 0$; make an initial guess for the values of $\tilde{\rho}, \tilde{\omega}_1, \tilde{\kappa}_1, \tilde{\kappa}_2, \tilde{\kappa}_3$.

3) Solve the LMI equation (7.71) for W with the fixed values; if a feasible value of W cannot be found, return to Step 2) and reset the values of $\tilde{\rho}, \tilde{\omega}_1, \tilde{\kappa}_1, \tilde{\kappa}_2, \tilde{\kappa}_3$.

4) Solve the LMI equation (7.69) for ρ with the feasible value of W obtained in Step 3) and make sure that the value of ρ is minimized.

5) If the condition $\tilde{\rho} \geq \rho$ is satisfied, then $(\tilde{\rho}, \tilde{\omega}_1, \tilde{\kappa}_1, \tilde{\kappa}_2, \tilde{\kappa}_3, W)$ is a feasible solution for Theorem 7.2; otherwise, set $\tilde{\rho} = \rho$ and return to Step 3).

Remark 7.7. Given the input delay h and the Lipschitz constant γ, it is concluded that the existence of a feasible solution is related to the matrices (A, B) and the Laplacian matrix $\mathcal{L}$. Additionally, since the values of h and γ are fixed and they are not the decision variables of the LMIs, a feasible solution may not exist if the values of h and γ are too large. Therefore, a trigger should be added in the algorithm to stop the iteration procedure if the values of $\tilde{\rho}, \tilde{\omega}_1, \tilde{\kappa}_1, \tilde{\kappa}_2, \tilde{\kappa}_3$ are out of the preset range.

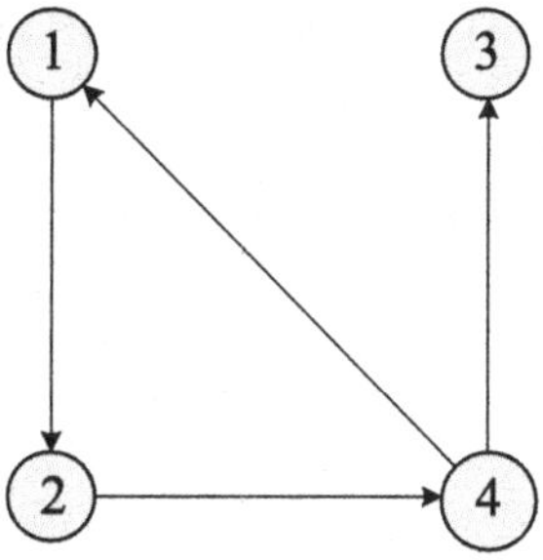

Figure 7.1 Communication topology.

7.4 Numerical examples

7.4.1 A circuit example

In this section, we will illustrate in some details the proposed consensus control design through a circuit example. The system under consideration is a connection of four agents (i.e. $N = 4$) as shown in Fig. 7.1, each of which is described by a second-order dynamic model as

$$\begin{cases} \dot{p}_i(t) = v_i(t), \\ \dot{v}_i(t) = f(v_i) + u_i(t-h), \end{cases} \tag{7.72}$$

where $p_i = [p_{ix}, p_{iy}, p_{iz}]^T \in \mathbb{R}^3$ denotes the position vector of agent i, $v_i = [v_{ix}, v_{iy}, v_{iz}]^T \in \mathbb{R}^3$ the velocity vector, $f(v_i) \in \mathbb{R}^3$ the intrinsic dynamics of agent i, governed by the chaotic Chua circuit [185],

$$f(v_i) = \begin{bmatrix} -0.59v_{ix} + v_{iy} - 0.17(|v_{ix}+1| - |v_{ix}-1|) \\ v_{ix} - v_{iy} + v_{iz} \\ -v_{iy} - 5v_{iz} \end{bmatrix}.$$

Let $x_i = [p_i^T, v_i^T]^T \in \mathbb{R}^6$. The dynamic equation (7.72) of each agent can be rearranged as the state space model (7.1) with

$$A = \begin{bmatrix} 0 & 0 & 0 & 1 & 0 & 0 \\ 0 & 0 & 0 & 0 & 1 & 0 \\ 0 & 0 & 0 & 0 & 0 & 1 \\ 0 & 0 & 0 & -0.59 & 1 & 0 \\ 0 & 0 & 0 & 1 & -1 & 1 \\ 0 & 0 & 0 & 0 & -1 & -5 \end{bmatrix}, \quad B = \begin{bmatrix} 0 & 0 & 0 \\ 0 & 0 & 0 \\ 0 & 0 & 0 \\ 1 & 0 & 0 \\ 0 & 1 & 0 \\ 0 & 0 & 1 \end{bmatrix},$$

and $\phi(x_i) = [0, 0, 0, -0.17(|v_{ix} + 1| - |v_{ix} - 1|), 0, 0]^T$. The adjacency matrix is given by

$$\mathcal{A} = \begin{bmatrix} 0 & 0 & 0 & 1 \\ 1 & 0 & 0 & 0 \\ 0 & 0 & 0 & 1 \\ 0 & 1 & 0 & 0 \end{bmatrix},$$

and the resultant Laplacian matrix is obtained as

$$\mathcal{L} = \begin{bmatrix} 1 & 0 & 0 & -1 \\ -1 & 1 & 0 & 0 \\ 0 & 0 & 1 & -1 \\ 0 & -1 & 0 & 1 \end{bmatrix}.$$

The eigenvalues of $\mathcal{L}$ are $\left\{0, 1, 3/2 \pm j\sqrt{3}/2\right\}$, and therefore Assumption 7.1 is satisfied. Furthermore, the eigenvalues are distinct. We obtain that

$$J = \begin{bmatrix} 0 & 0 & 0 & 0 \\ 0 & 1 & 0 & 0 \\ 0 & 0 & \frac{3}{2} & \frac{\sqrt{3}}{2} \\ 0 & 0 & -\frac{\sqrt{3}}{2} & \frac{3}{2} \end{bmatrix},$$

with the matrix

$$T = \begin{bmatrix} 1 & 0 & \frac{1}{2} & \frac{\sqrt{3}}{2} \\ 1 & 0 & -1 & 0 \\ 1 & -2 & \frac{1}{2} & \frac{\sqrt{3}}{2} \\ 1 & 0 & \frac{1}{2} & -\frac{\sqrt{3}}{2} \end{bmatrix},$$

and $r^T = [1/3, 1/3, 0, 1/3]^T$.

The nonlinear function $\phi(x_i)$ in each agent dynamics is globally Lipschitz with a Lipschitz constant $\gamma = 0.34$, which gives $\gamma_0 = 3.7391$. Based on the Laplacian matrix $\mathcal{L}$, we have $\alpha = 1$. In simulation, the input delay is set as $h = 0.03$ s. A positive definite matrix P can be obtained with $\kappa = 0.01$, $\omega_1 = 1.5$ and $\rho = 2$, as

$$P = \begin{bmatrix} 5.03 & -0.53 & 0.18 & 2.58 & 0.29 & 0.08 \\ -0.53 & 5.37 & 0.43 & 0.28 & 2.39 & 0.47 \\ 0.18 & 0.43 & 7.75 & -0.08 & -0.38 & 1.58 \\ 2.58 & 0.28 & -0.08 & 2.65 & 0.93 & 0.17 \\ 0.29 & 2.39 & -0.38 & 0.93 & 2.17 & 0.25 \\ 0.08 & 0.47 & 1.58 & 0.17 & 0.25 & 0.92 \end{bmatrix},$$

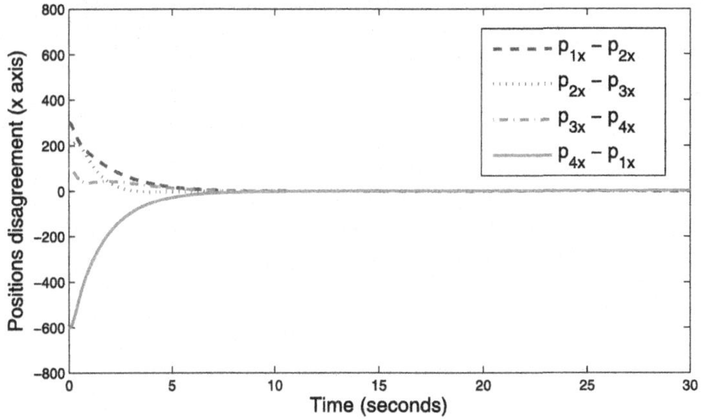

Figure 7.2 The positions disagreement of 4 agents in x axis: $h = 0.03$ s.

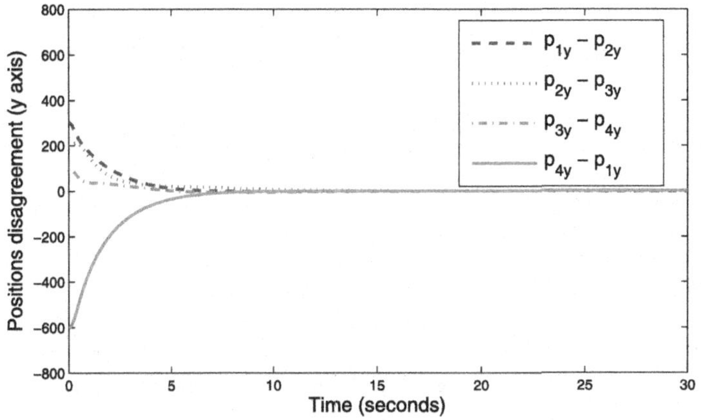

Figure 7.3 The positions disagreement of 4 agents in y axis: $h = 0.03$ s.

to satisfy the conditions of Theorem 7.1. Consequently, the control gain is obtained as

$$K = \begin{bmatrix} -2.19 & -0.12 & -0.01 & -2.46 & -0.74 & -0.15 \\ -0.13 & -2.10 & 0.30 & -0.75 & -2.08 & -0.32 \\ -0.09 & -0.43 & -1.64 & -0.18 & -0.18 & -1.27 \end{bmatrix}.$$

Simulation study has been carried out with the results shown in Figs. 7.2–7.4 for the positions state disagreement of each agent. Clearly, the conditions specified in Theorem 7.1 are sufficient for the control gain to achieve consensus control for the multi-agent systems. The same control gain has also been used for different values of input delay. The results shown in Figs. 7.5–7.7 indicate that the conditions could be conservative in the control gain design for a given input delay.

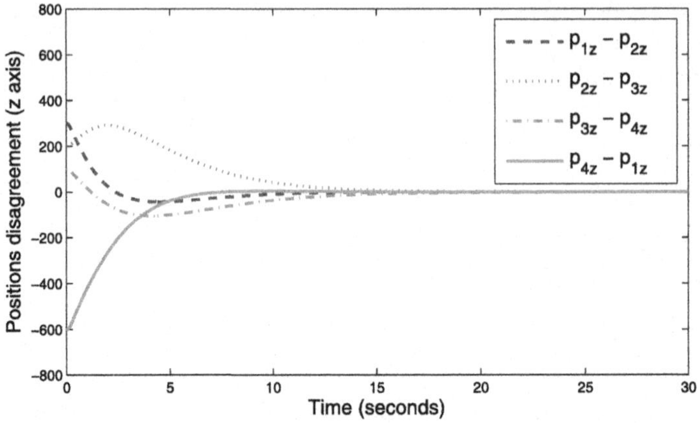

Figure 7.4 The positions disagreement of 4 agents in z axis: $h = 0.03$ s.

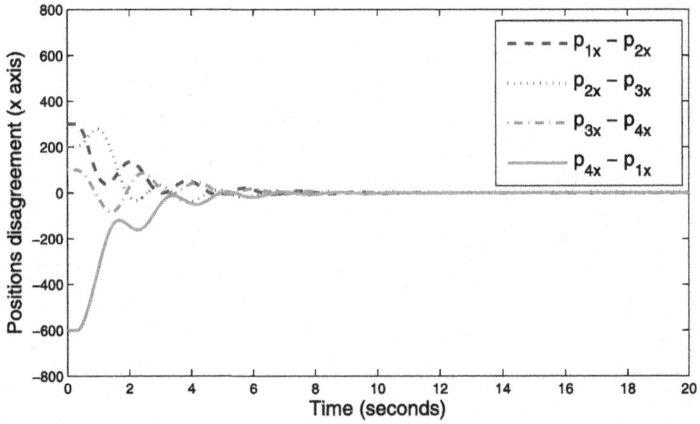

Figure 7.5 The positions disagreement of 4 agents in x axis: $h = 0.3$ s.

7.4.2 Numerical examples

In this section, a simulation study is carried out to demonstrate the effectiveness of the proposed TPF controller design. Consider a connection of four agents as shown in Fig. 7.1. The dynamics of the i-th agent is described by a second-order model as

$$\dot{x}_i(t) = \begin{bmatrix} -0.09 & 1 \\ -1 & -0.09 \end{bmatrix} x_i(t) + g \begin{bmatrix} \sin(x_{i1}(t)) \\ 0 \end{bmatrix} + \begin{bmatrix} 0 \\ 1 \end{bmatrix} u(t - 0.1).$$

The linear part of the system represents a decayed oscillator. The time delay of the system is 0.1 seconds, and the Lipschitz constant $\gamma = g$. The eigenvalues of $\mathcal{L}$ are $\left\{0, 1, 3/2 \pm j\sqrt{3}/2\right\}$ and $r^T = \left[\frac{1}{3}, \frac{1}{3}, 0, \frac{1}{3}\right]$.

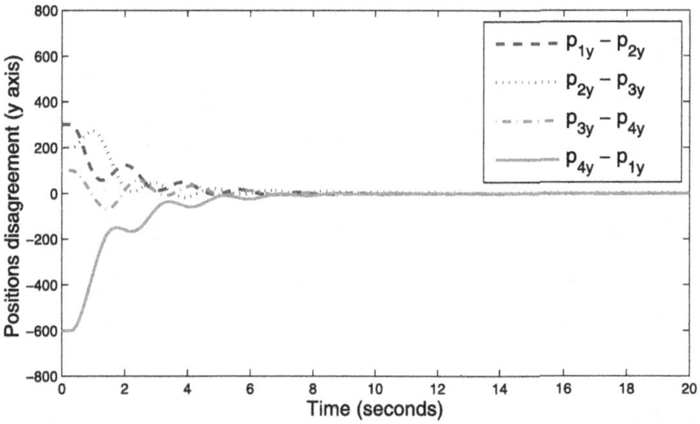

Figure 7.6 The positions disagreement of 4 agents in y axis: $h = 0.3$ s.

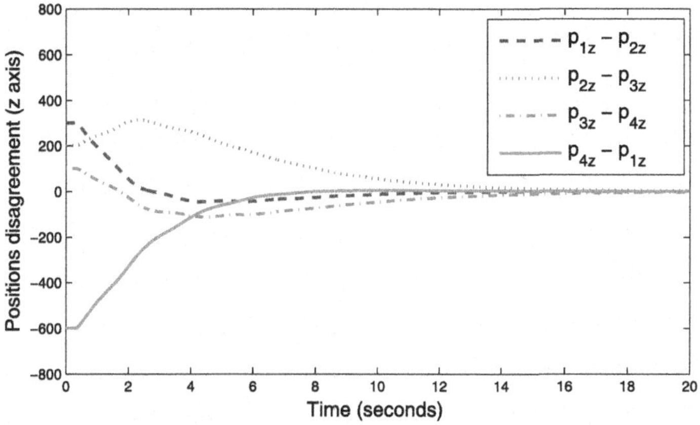

Figure 7.7 The positions disagreement of 4 agents in z axis: $h = 0.3$ s.

In this case, we choose $\gamma = g = 0.03$, and the initial conditions for the agents as $x_1(\theta) = [1,\ 1]^T$, $x_2(\theta) = [0,\ 0]^T$, $x_3(\theta) = [0.3,\ 0.5]^T$, $x_4(\theta) = [0.5,\ 0.3]^T$, $u(\theta) = [0,\ 0,\ 0,\ 0]^T$, for $\theta \in [-h, 0]$. With the values of $\omega_1 = 0$, $\rho = 0.3$, $\kappa_1 = \kappa_2 = 0.05$, and $\kappa_3 = 0.5$, a feasible solution of the feedback gain K is found to be

$$K = \begin{bmatrix} -0.1480 & -0.6359 \end{bmatrix}.$$

Figs. 7.8 and 7.9 show the simulation results for the state disagreement of each agent. Clearly the conditions specified in Theorem 7.2 are sufficient for the control gain to achieve consensus control. Without retuning the control gain, the consensus control is still achieved for the multi-agent systems with a larger delay of 0.5 of second and a bigger Lipschitz constant of $g = 0.15$, as shown in Figs. 7.10 and 7.11.

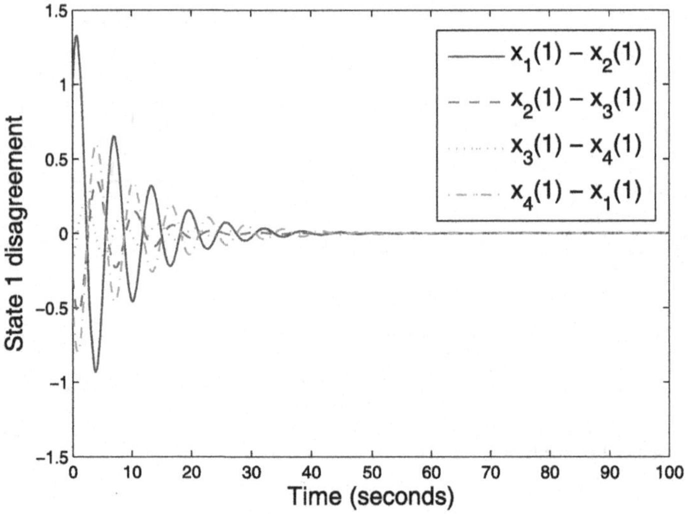

Figure 7.8 The state 1 disagreement of agents with $h = 0.1$ and $g = 0.03$.

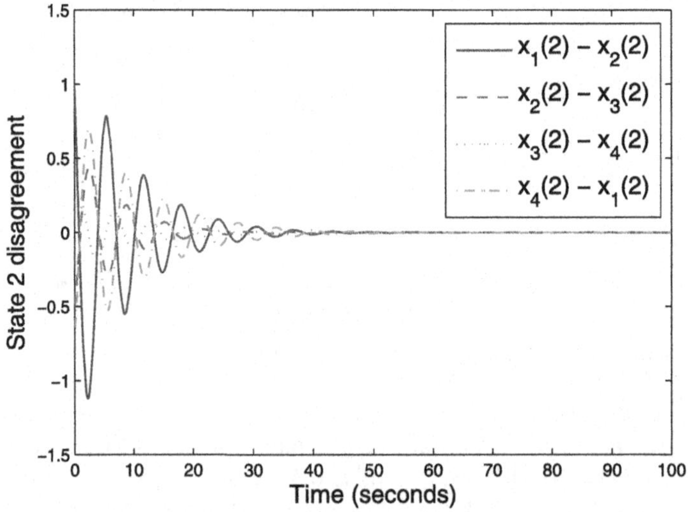

Figure 7.9 The state 2 disagreement of agents with $h = 0.1$ and $g = 0.03$.

7.5 Conclusion remarks

This chapter has investigated the impacts of nonlinearity and input delay in robust consensus control. This input delay may represent some delays in the network communication. Sufficient conditions are derived for the multi-agent systems to guarantee the global consensus using Lyapunov–Krasovskii method in the time domain.

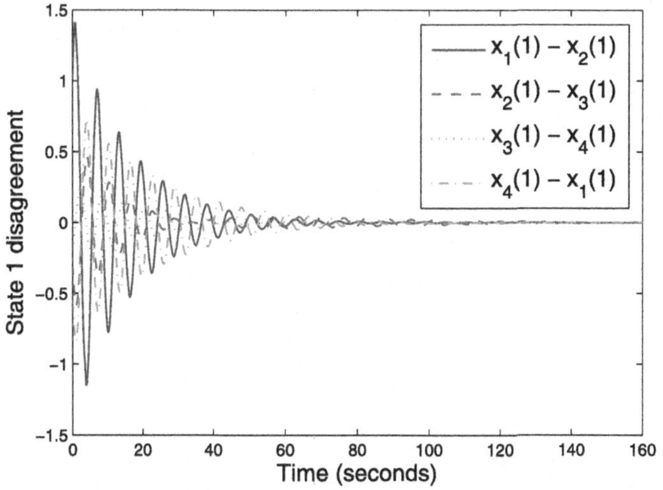

Figure 7.10 The state 1 disagreement of agents with $h = 0.5$ and $g = 0.15$.

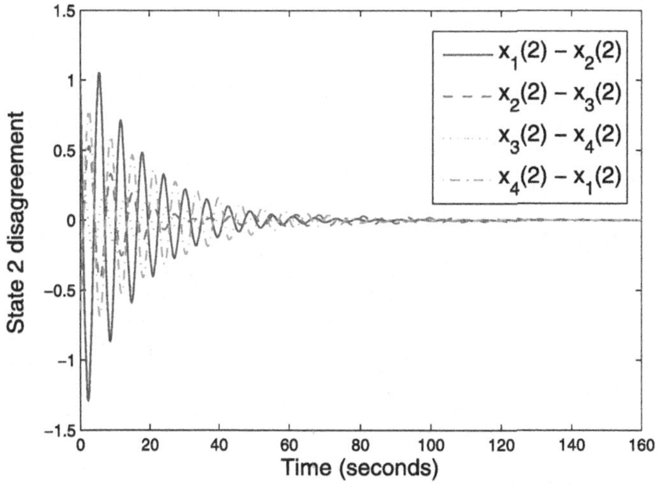

Figure 7.11 The state 2 disagreement of agents with $h = 0.5$ and $g = 0.15$.

The significance of this research is to provide feasible methods to deal with consensus control of a class of Lipschitz nonlinear multi-agent systems with input delay which includes some common circuits such as Chua circuits. In particular, consensus can be achieved under the same control gain for multi-agent systems with different values of time delay and the Lipschitz nonlinear constant.

Robust consensus control of multi-agent systems with disturbance rejection

8

8.1 Problem formulation

In this chapter, we consider a set of N linear subsystems, of which the subsystem dynamics are described by

$$\dot{x}_i = Ax_i + Bu_i + D\omega_i, \tag{8.1}$$

for $i = 1, \ldots, N$, where $A \in R^{n\times n}$, $B \in R^{n\times m}$, and $D \in R^{n\times s}$ are constant matrices, $x_i \in R^n$ is the state vector, $u_i \in R^m$ is the input, $\omega_i \in R^s$ is a disturbance that is generated by an exosystem, i.e.,

$$\dot{\omega}_i = S\omega_i, \tag{8.2}$$

with $S \in R^{s\times s}$ being a known constant matrix.

The consensus disturbance rejection problem is to design control inputs using the relative state information collected from the network to ensure that all the subsystem states converge to a common trajectory when the subsystems are subject to different disturbances. The control inputs are based on the consensus disturbance observers that estimate the part of disturbances in the subsystems that cause differences from a common trajectory. We make three assumptions about the dynamics of the subsystems and the connections between the subsystems.

Assumption 8.1. The disturbance is matched, i.e., there exists a matrix $F \in R^{s\times s}$ such that $D = BF$.

Assumption 8.2. The eigenvalues of the exosystem matrix S are distinct, and on the imaginary axis, and the pair (S, D) is observable.

Assumption 8.3. The Laplacian matrix of the connection graph has a single eigenvalue at 0.

From Assumption 8.3, we know that the Laplacian matrix $\mathcal{L}$ only has one eigenvalue at 0, and all other eigenvalues are with positive real part. In this case, the least positive real part is also known as the connectivity of the network, which is denoted by α.

Remark 8.1. Assumption 8.1 on the matched disturbance could be relaxed in some circumstances because unmatched disturbances under uncertain conditions may be

Cooperative Control of Multi-Agent Systems. https://doi.org/10.1016/B978-0-12-820118-3.00019-3

converted to matched ones based on the output regulation theory [69]. The assumption on the eigenvalues of S is commonly used for disturbance rejection and output regulation. It allows the disturbances to be sinusoidal functions and constants. Other functions may be approximated by sinusoidal functions with a bias. It is also common in consensus and cooperative control to assume the common exosystem dynamics, to reflect common dynamics and environment of the subsystems [168]. From an engineering point of view, if the exosystem dynamics reflect certain unmodeled dynamics of the subsystems, it would be natural to assume that they are common in consensus control as one would expect that the subsystem dynamics are common. The assumption on the observability of (S, D) is reasonable, as any unobservable components would have no impact on system states.

Remark 8.2. The condition specified in Assumption 8.3 is satisfied if the connection graph contains a directed spanning tree.

Next follows some preliminary results. When the disturbance ω is known, the disturbance rejection is fairly straightforward by adding the term $-F\omega_i$ in the control input u_i. The key issue is to estimate ω_i using the relative state information. At least, we would expect that the disturbance state ω_i is observable from the system state measurement x_i. For this, we have the following lemma.

Lemma 8.1. *If the pair (S, D) is observable, the pair (A_H, H) is observable, with $A_H = \begin{bmatrix} A & D \\ 0 & S \end{bmatrix}$ and $H = [I \quad 0]$.*

Proof. Let us establish the result by seeking a contradiction. Suppose that (A_H, H) is not observable, for any eigenvalue of A_H, i.e., λ_j, the matrix

$$\begin{bmatrix} \lambda_j I - A & -D \\ 0 & \lambda_j I - S \\ I & 0 \end{bmatrix}$$

is rank deficient, i.e., there exists a nonzero vector $v = [v_1^T, v_2^T]^T \in R^{n+s}$ such that

$$\begin{bmatrix} \lambda_j I - A & -D \\ 0 & \lambda_j I - S \\ I & 0 \end{bmatrix} \begin{bmatrix} v_1 \\ v_2 \end{bmatrix} = 0.$$

This implies that

$$v_1 = 0, \quad \begin{bmatrix} -D \\ \lambda_j I - S \end{bmatrix} v_2 = 0. \tag{8.3}$$

Since $v_1 = 0$, we have $v_2 \neq 0$, which implies, together with Eq. (8.3), that (S, D) is not observable. This is a contradiction; therefore, (A_H, H) must be observable. Since (A_H, H) is observable, there exists a positive definite matrix Q that satisfies

$$QA_H + A_H^T Q - 2H^T H < 0.$$

□

8.2 Disturbance rejection for a directed graph

8.2.1 Consensus controller design with disturbance rejection

In this section, we consider consensus disturbance rejection with directed graphs. Further to the condition specified in Assumption 8.3, we state the following condition.

Assumption 8.4. The minimum value of the network connectivity α is known.

Note that the connectivity can be easily calculated from the Laplacian matrix of a network connection. By doing so, it does require all the elements of a_{ij}, which are not restricted to the neighborhood of a subsystem. In the succeeding section, the assumption on the network connectivity is relaxed by a fully distributed consensus disturbance rejection scheme.

The disturbance rejection design is based on disturbance estimation and rejection. The estimation is based on the relative state information. Define, for $i = 1, \ldots, N$,

$$\zeta_i = \sum_{i=1}^{N} a_{ij}(x_i - x_j). \tag{8.4}$$

From the definition of l_{ij}, it is easy to see that $\zeta_i = \sum_{i=1}^{N} l_{ij} x_j$. The disturbance estimation and rejection method proposed in this chapter will be based on signals ζ_i.

The control input for disturbance rejection is designed as follows:

$$u_i = -K\chi_i - Fz_i, \tag{8.5}$$

where $\chi_i \in R^n$ and $z_i \in R^s$ are generated by

$$\dot{\chi}_i = (A - BK)\chi_i + cG_x(\zeta_i - \sum_{i=1}^{N} l_{ij}\chi_j), \tag{8.6}$$

and

$$\dot{z}_i = Sz_i + cG_\omega(\zeta_i - \sum_{i=1}^{N} l_{ij}\chi_j), \tag{8.7}$$

where $c \geq 1/\alpha$ is a positive real constant, K is a constant gain matrix with a proper dimension such that $(A - BK)$ is Hurwitz, and $[G_x^T, G_\omega^T]^T := G = Q^{-1}H^T$. Note that Eq. (8.6) is designed as a state observer as only the relative state information is available, and Eq. (8.7) is designed as a consensus disturbance observer. Let

$$e_i = \begin{bmatrix} x_i - \chi_i \\ \omega_i - z_i \end{bmatrix}.$$

From Eqs. (8.1), (8.2), (8.6) and (8.7), we can establish that

$$\dot{e}_i = A_h e_i + cG \sum_{j=1}^{N} l_{ij} H e_j.$$

Let $e = [e_1^T, \ldots, e_N^T]^T$. The dynamics of e can be obtained as

$$\dot{e} = (I_N \otimes A_H - c\mathcal{L} \otimes GH)e. \tag{8.8}$$

Let us define $r \in R^N$ as the left eigenvector of $\mathcal{L}$ corresponding to the eigenvalue at 0, i.e., $r^T\mathcal{L} = 0$; furthermore, we set $r^T\mathbf{1} = 1$ with $\mathbf{1} = [1, \ldots, 1]^T \in R^N$. Based on the vector r, we can introduce a state transformation $\varepsilon_i = e_i - \sum_{j=1}^{N} r_j e_j$ for $i = 1, \ldots, N$. With $\varepsilon = [\varepsilon_1^T, \ldots, \varepsilon_N^T]^T$, the state transformation is given by

$$\begin{aligned} \varepsilon &= e - ((\mathbf{1}r)^T \otimes I_{n+s})e \\ &= (M \otimes I_{n+s})e, \end{aligned} \tag{8.9}$$

where $M = I_N - \mathbf{1}r^T$.

It can be shown that the rank of M is $N-1$ with one eigenvalue at 0 and others at 1. The eigenvector corresponding to the eigenvalue at 0 is 1, and then, $\varepsilon = 0$ implies that $e_1 = e_2 = \ldots = e_N$.

For each subsystem, we have

$$\begin{aligned} \dot{x}_i &= Ax_i - BK\chi_i + BF(\omega_i - z_i) \\ &= (A - BK)x_i + B[K, F]e_i. \end{aligned}$$

Hence, $\lim_{t\to\infty} \varepsilon(t) = 0$ ensures

$$\lim_{t\to\infty} x_1(t) = \lim_{t\to\infty} x_2(t) = \cdots = \lim_{t\to\infty} x_N(t),$$

i.e., the solution of the consensus disturbance rejection problem.

8.2.2 Consensus analysis

Theorem 8.1. *For the network-connected systems with the subsystem dynamics (8.1), the control inputs (8.5) with the state and disturbance observers (8.6) and (8.7) solve the consensus disturbance rejection problem under Assumption 8.1.*

Proof. The dynamics of ε can then be obtained as

$$\dot{\varepsilon} = (I_N \otimes A_H - c\mathcal{L} \otimes GH)\varepsilon.$$

To exploit the structure of $\mathcal{L}$, let us introduce another state transformation

$$\eta = (T^{-1} \otimes I_{n+s})\varepsilon \tag{8.10}$$

where T is the transformation such that $T^{-1}\mathcal{L}T = J$ with J being in the real Jordan canonical form, as shown in [33]. Its dynamics is obtained as

$$\dot{\eta} = (I_N \otimes A_H - cJ \otimes GH)\eta. \tag{8.11}$$

Hence, the consensus disturbance rejection problem is solved by showing that η converges to zero. Moreover, note that the transformations shown in Eqs. (8.10) and (8.9) also imply that

$$\eta_1 = (r^T \otimes I_{n+s})e \equiv 0.$$

The matrix J is in the real Jordan canonical form with the first diagonal element being equal to 0. Assume that λ_i is a single real eigenvalue of $\mathcal{L}$. From Eq. (8.11), the dynamics of η_i is obtained as

$$\dot{\eta}_i = (A_H - c\lambda_i GH)\eta_i = (A_H - c\lambda_i Q^{-1}H^T H)\eta_i.$$

Let $V_i = \eta_i^T Q\eta_i$. From Eq. (8.11), we have

$$\begin{aligned} \dot{V}_i &= \eta_i^T(QA_H + A_H Q - 2c\lambda_i H^T H)\eta_i \\ &\leq \eta_i^T(QA_H + A_H Q - 2H^T H)\eta_i \\ &\leq -\epsilon\|\eta_i\|^2 \end{aligned} \tag{8.12}$$

for some positive real constant ϵ because of $c\lambda_i \geq c\alpha \geq 1$.

Following the procedures in the proof of [33] we can establish results similar to Eq. (8.12) for complex eigenvalues, multiple eigenvalues, and multiple complex eigenvalues. Let

$$V_\eta = \sum_{i=2}^{N} V_i,$$

and it can then be obtained that

$$\dot{V} \leq -\epsilon\|\eta\|^2.$$

Therefore, η converges to 0 exponentially. □

8.3 Fully distributed consensus disturbance rejection

8.3.1 Local information based disturbance rejection controller

The control in the previous section takes the relative state information from the subsystems in the neighborhood and needs the knowledge of the connectivity. This is a common assumption in most of the consensus control design. However, the connectivity is defined as the minimum positive real part of the eigenvalues of the Laplacian matrices. The evaluation of this value need the Laplacian matrix or the adjacency matrix, i.e., the entire a_{ij} for $i, j = 1, \ldots, N$. Hence, the connectivity is a global information of the network connection. The proposed disturbance rejection algorithm in the previous section is not fully distributed in this sense.

In this section, we shall present a design that only uses the local connection information, a_{ij} values for the i-th subsystem, and the relative state information within the neighborhood. This fully distributed consensus disturbance rejection is for connections with undirected graph. Thus, we have the following assumption.

Assumption 8.5. The network connection graph is undirected and connected.

Adaptive control technique is used to deal with unknown connectivity. The control input is designed in the same way as shown in Eq. (8.5)

$$u_i = -K\chi_i - Fz_i. \tag{8.13}$$

However, the estimated state χ_i and disturbance z_i are now generated by

$$\dot{\chi}_i = (A - BK)\chi_i + G_x \sum_{j=1}^{N} c_{ij} a_{ij}(\rho_i - \rho_j), \tag{8.14}$$

and

$$\dot{z}_i = Sz_i + G_\omega \sum_{j=1}^{N} c_{ij} a_{ij}(\rho_i - \rho_j), \tag{8.15}$$

where $\rho_i := x_i - \chi_i$, and c_{ij} are adaptive parameters generated by

$$\dot{c}_{ij} = (\rho_j - \rho_i)^T(\rho_j - \rho_i),$$

with $c_{ij}(0) = c_{ji}(0)$.

Note that the only difference in the control design, compared with the one shown in the previous section, is that the observer gain parameter c in the previous design is replaced by adaptive parameters c_{ij}. The setting $c_{ij}(0) = c_{ji}(0)$ for the initial values is to keep the symmetry of the adaptive parameters.

The adaptive parameters also make a difference in stability analysis. The dynamics of e_i are now obtained as

$$\dot{e}_i = A_H e_i + G \sum_{j=1}^{N} c_{ij} a_{ij}(e_j - e_i),$$

where e_i follows the same definition as in the previous section. Let us redefine ε_i as $\varepsilon_i = e_i - (1/N)\sum_{j=1}^{N} e_j$. It can be obtained that

$$\dot{\varepsilon}_i = A_H \varepsilon_i + G \sum_{j=1}^{N} c_{ij} a_{ij} H(\varepsilon_j - \varepsilon_i).$$

To obtain the given expression, we have observed that $(e_j - e_i) = (\varepsilon_j - \varepsilon_i)$ and $(1/N)G\sum_{j=1}^{N}\sum_{k=1}^{N} c_{jk}a_{jk}H(e_j - e_k) = 0$. The consensus disturbance rejection problem is solved if we can show that ε_i converges to 0, for $i = 1, \dots, N$, in the same way as discussed in the previous section.

8.3.2 Consensus analysis

Theorem 8.2. *For the network-connected system with the subsystem dynamics (8.1), the control inputs (8.13) with the state and disturbance observers (8.14) and (8.15) solve the distributed consensus disturbance rejection problem under Assumptions 8.1 and 8.5.*

Proof. Using the notation ε_i, the adaptive laws can be written as

$$\dot{c}_{ij} = (\varepsilon_j - \varepsilon_i)^T H^T H(\varepsilon_j - \varepsilon_i).$$

Let

$$V_i = \varepsilon_i^T Q \varepsilon_i.$$

From the dynamics of ε_i and c_{ij} shown earlier, we have

$$\begin{aligned} \dot{V}_i &= \varepsilon_i^T (QA_H + A_H^T Q)\varepsilon_i + 2\varepsilon_i^T QG \sum_{j=1}^{N} c_{ij} a_{ij} H(\varepsilon_j - \varepsilon_i) \qquad (8.16) \\ &= \varepsilon_i^T (QA_H + A_H^T Q - 2H^T H)\varepsilon_i + 2\varepsilon_i^T H^T H \varepsilon_i \\ &\quad + 2\varepsilon_i^T H H^T \sum_{j=1}^{N} c_{ij} a_{ij} (\varepsilon_j - \varepsilon_i) \\ &\le -\epsilon \|\varepsilon_i\|^2 + 2\varepsilon_i^T H^T H \varepsilon_i + 2\varepsilon_i^T H H^T \sum_{j=1}^{N} c_{ij} a_{ij} (\varepsilon_j - \varepsilon_i), \end{aligned}$$

where ϵ is a positive real constant. Let

$$V_c = \frac{1}{2}\sum_{i=1}^{N}\sum_{j=1}^{N} a_{ij}(c_{ij} - \frac{1}{\alpha})^2.$$

Its derivative is obtained as

$$\dot{V}_c = \sum_{i=1}^{N}\sum_{j=1}^{N} a_{ij}(c_{ij} - \frac{1}{\alpha})(\varepsilon_j - \varepsilon_i)^T H^T H(\varepsilon_j - \varepsilon_i) \tag{8.17}$$
$$= -2\sum_{i=1}^{N}\sum_{j=1}^{N} a_{ij}(c_{ij} - \frac{1}{\alpha})\varepsilon_i^T H^T H(\varepsilon_j - \varepsilon_i),$$

where the last equation is obtained by using $c_{ij} = c_{ji}$. We can further expand the terms in Eq. (8.17) to obtain that

$$\dot{V}_c = -2\sum_{i=1}^{N}\sum_{j=1}^{N} a_{ij}c_{ij}\varepsilon_i^T H^T H(\varepsilon_j - \varepsilon_i) - 2\frac{1}{\alpha}\sum_{i=1}^{N}\sum_{j=1}^{N} l_{ij}\varepsilon_i^T H^T H\varepsilon_i \tag{8.18}$$
$$\leq -2\sum_{i=1}^{N}\sum_{j=1}^{N} a_{ij}c_{ij}\varepsilon_i^T H^T H(\varepsilon_j - \varepsilon_i) - 2\sum_{i=1}^{N}\varepsilon_i^T H^T H\varepsilon_i,$$

where the inequality is obtained by viewing $H\varepsilon_i$ as x_i.

Let

$$V = \sum_{i=1}^{N} V_i + V_c.$$

From the results shown in Eqs. (8.16) and (8.18), we have

$$\dot{V} \leq \sum_{i=1}^{N} -\epsilon\|\varepsilon_i\|^2.$$

Hence, we can conclude that all the variables in the entire system are bounded, and $\varepsilon_i \in L_2 \cap L_\infty$ for $i = 1, \ldots, N$. Since the derivatives of ε_i are bounded, we conclude that ε_i for $i = 1, \ldots, N$ converge to zero by Babalat's lemma [37]. This completes the proof. □

8.4 Disturbance rejection in leader-follower format

8.4.1 Leader-follower disturbance rejection controller

In the previous sections, control schemes are proposed for disturbance rejection in consensus control without a leader, i.e., the final common value depends on all the subsystems involved in the network. Another form of consensus control is that one subsystem is identified as the leader and other subsystems are controlled to follow this leader. In this section, we propose a control design for consensus disturbance rejection with a leader. Without loss of generality, we assume the subsystem at node 1 in the connection graph is the leader. Based on this, we have an assumption about the network connection.

Assumption 8.6. There exists a spanning tree in the connection graph with the subsystem at node 1 as the root.

When a network contains a spanning tree, the corresponding Laplacian matrix will only have one eigenvalue at 0. Therefore, Assumption 8.6 supersedes Assumption 8.3. With the subsystem 1 as the leader, the control objective is to ensure the state variables of other subsystems to follow the state of the leader x_1 under the disturbances. Let

$$\bar{x}_i = x_{i+1} - x_1$$

for $i = 1, \ldots, N-1$. As a common practice in leader-follower consensus control, the leader does not take any control from consensus control design, i.e., we have $u_1 = 0$. The dynamics of $\bar{x}_i$, for $i = 1, \ldots, N-1$, are described by

$$\dot{\bar{x}}_i = A\bar{x}_i + B\bar{u}_i + D\bar{\omega}_i,$$

$$\dot{\bar{\omega}}_i = S\bar{\omega}_i,$$

where $\bar{\omega}_i = \omega_{i+1} - \omega_1$ and $\bar{u}_i = u_{i+1}$.

We still use the signal ζ_i defined in Eq. (8.4) of Section 8.3 for the disturbance observer and control design. To express ζ_i in $\bar{x}_i$, we have, for $i = 2, \ldots, N$,

$$\begin{aligned}\zeta_i &= \sum_{j=2}^{N} l_{ij}x_j + l_{i1}x_1 = \sum_{j=2}^{N} l_{ij}x_j + \sum_{j=2}^{N} l_{ij}x_1 \\ &= \sum_{j=2}^{N} l_{ij}\bar{x}_{j-1} = \sum_{j=1}^{N-1} \bar{l}_{(i-1)j}\bar{x}_j.\end{aligned}$$

For the notational convenience, we let $\bar{\zeta}_i = \zeta_{i+1}$. Hence, we have

$$\bar{\zeta}_i = \sum_{j=1}^{N-1} \bar{l}_{ij}\bar{x}_j.$$

The control input for disturbance rejection in the leader-follower setup is designed as follows, for $i = 1, \ldots, N-1$:

$$\bar{u}_i = -K\chi_i - Fz_i, \tag{8.19}$$

where χ_i and z_i are generated by

$$\dot{\chi}_i = (A - BK)\chi_i + cG_x(\bar{\zeta}_i - \sum_{j=1}^{N-1} \bar{l}_{ij}\chi_j), \tag{8.20}$$

and

$$\dot{z}_i = Sz_i + cG_\omega(\bar{\chi}_i - \sum_{j=1}^{N-1} \bar{l}_{ij}\chi_j), \tag{8.21}$$

where $c \geq 2p_{max}/r_0$ is a positive real constant with $p_{max} = max\{p_1, \ldots, p_{N-1}\}$, and we still use $[G_x^T, G_\omega^T]^T = Q^{-1}H^T$.

Let

$$\bar{e}_i = \begin{bmatrix} \bar{x}_i - \chi_i \\ \bar{\omega}_i - z_i \end{bmatrix},$$

for $i = 1, \ldots, N-1$. The dynamics of $\bar{e}_i$ can be obtained as

$$\dot{\bar{e}}_i = A_h\bar{e}_i + cG\sum_{j=1}^{N-1} \bar{l}_{ij}H\bar{e}_j.$$

Let $\bar{e} = [\bar{e}_1^T, \ldots, \bar{e}_{N-1}^T]^T$. The dynamics of $\bar{e}$ can be obtained as

$$\dot{\bar{e}} = (I_{N-1} \otimes A_H - c\bar{\mathcal{L}} \otimes GH)\bar{e}. \tag{8.22}$$

It can be seen that the dynamics of (8.22) are in the same format as the dynamics shown in Eq. (8.8). However, in this case, $\bar{\mathcal{L}}$ is a nonsingular matrix, and the stability analysis will be different. In this leader-follower case, we will establish that all the $\bar{e}_i$ for $i = 1, \ldots, N-1$ converge to 0, instead of showing they converge to the same values as in the leaderless consensus control case.

8.4.2 Consensus analysis

Theorem 8.3. *For the network-connected systems with the subsystem dynamics (8.1) the control inputs (8.19) with the disturbance observers (8.20) and (8.21) solve the consensus disturbance rejection problem in the leader-follower case under Assumptions 8.1, 8.2, and 8.6.*

Proof. Let

$$V_e = \bar{e}^T(P \otimes Q)\bar{e}.$$

By the result in Lemma 7.3, its derivative can be obtained as

$$\begin{aligned}
\dot{V}_e &= \bar{e}^T[P \otimes (QA_H + A_H^TQ) - c(P\bar{\mathcal{L}} + \bar{\mathcal{L}}^TP) \otimes H^TH]\bar{e} \\
&\leq \bar{e}^T[P \otimes (QA_H + A_H^TQ) - cr_0I_{N-1} \otimes H^TH]\bar{e} \\
&\leq \bar{e}^T[P \otimes (QA_H + A_H^TQ - 2H^TH)]\bar{e} \\
&\leq -\epsilon\bar{e}^T\bar{e},
\end{aligned}$$

where ϵ is a positive real number. Therefore, we can conclude that $\bar{e}$ converges to zero exponentially. As the closed-loop dynamics of $\bar{x}_i$ can be written as

$$\dot{\bar{x}}_i = (A - BK)\bar{x}_i + [BK, D]\bar{e}_i,$$

we further conclude that $\bar{x}_i$ converges to zero exponentially for $i = 1, \ldots, N-1$. □

8.5 Numerical examples

In this section, an example is used to demonstrate the potential applications of the proposed consensus disturbance rejection methods. We consider an example of consensus disturbance rejection of a possible vibration in a control surface for formation control of unmanned aerial vehicles (UAVs). The formation control of YF-22 research UAVs was presented in [60] where the longitudinal dynamics is described by

$$\begin{bmatrix} \dot{V} \\ \dot{\alpha} \\ \dot{q} \\ \dot{\theta} \end{bmatrix} = \begin{bmatrix} -0.284 & -23.096 & 0 & -0.171 \\ 0 & -4.117 & 0.778 & 0 \\ 0 & -33.884 & -3.573 & 0 \\ 0 & 0 & 1 & 0 \end{bmatrix} \cdot \begin{bmatrix} V \\ \alpha \\ q \\ \theta \end{bmatrix} + \begin{bmatrix} 20.168 \\ 0.544 \\ -39.085 \\ 0 \end{bmatrix} i_H$$

where V, α, q, and θ are the speed in meters per second, angle of attack in degrees, the pitch rate in degrees per second, and pitch in degrees, respectively, and i_H is the stabilizer incidence angle in degrees. The existing control law [60] is set as $i_H = 0.12q + 0.500$.

As shown in [60], the formation control has been introduced, and our task in this example is to reject harmonic disturbances in the input channel, which can represent an undesired vibration. Since the system model is linear in this operating point, we can add the control input designed by the proposed consensus disturbance rejection method to the existing control. For i-th system, we use x_i to denote the state (V, α, q, θ), u_i for the additional control input generated for consensus disturbance rejection, and $[1, 0]\omega_i$ as the input disturbance, of which w_i is generated by $\dot{\omega}_i = \begin{bmatrix} 0 & \omega \\ -\omega & 0 \end{bmatrix} \omega_i$, with ω as the vibration frequency in radians per second. Hence, the input disturbance problem is formulated as the consensus disturbance rejection problem with conditions (8.1) and (8.2) presented in Section 8.2, and we can identify the system matrices as

$$A = \begin{bmatrix} -0.284 & -23.096 & 2.420 & 9.913 \\ 0 & -4.117 & 0.843 & 0.272 \\ 0 & -33.884 & -8.263 & -19.543 \\ 0 & 0 & 1 & 0 \end{bmatrix}, \quad B = \begin{bmatrix} 20.168 \\ 0.544 \\ -39.085 \\ 0 \end{bmatrix}, \quad D = B\begin{bmatrix} 1 & 0 \end{bmatrix}.$$

Note that the system matrix A has considered the exiting feedback control. It can be checked that Assumptions 8.1 and 8.2 are satisfied, and the F matrix is given by $F = [1, 0]$.

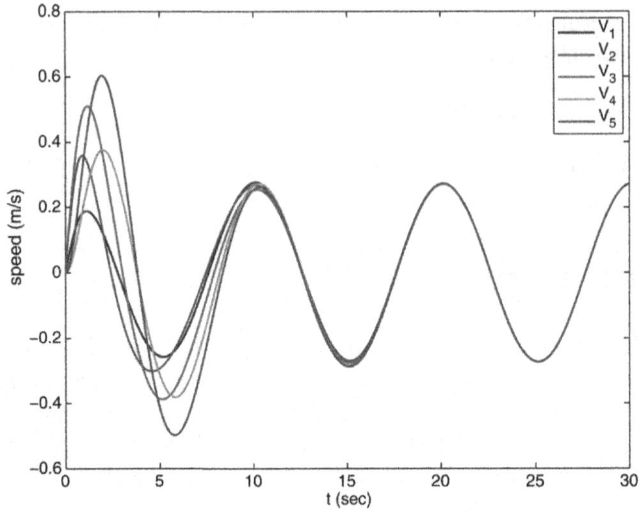

Figure 8.1 Speed variations of UAVs.

In this example, we demonstrate the consensus disturbance rejection method under the leader-follower setup of five UAVs, whose connection graph is specified by the adjacency matrix

$$\mathcal{A} = \begin{bmatrix} 0 & 0 & 0 & 0 & 0 \\ 1 & 0 & 1 & 0 & 0 \\ 0 & 0 & 0 & 0 & 1 \\ 0 & 1 & 0 & 0 & 0 \\ 0 & 0 & 0 & 1 & 0 \end{bmatrix}.$$

Note that the first row of $\mathcal{A}$ are all zeros, as the subsystem at node 1 is taken as the leader. Following the result shown in Lemma 7.3, we obtain that $P = diag\{1/4 \ 1/7 \ 1/5 \ 1/6\}$ and $r_0 = 0.0537$. With $p_{\max} = 1/4$ and $2p_{\max}/r_0 = 9.3110$, we set $c = 10$ in the disturbance observer (8.21) and the relative state observer (8.20). The observer gains G_w and G_x follow the standard design procedures shown in Section 8.4. Since A is stable in this example, we set $K = 0$ for the control input. Simulation study has been carried out with different disturbances for subsystems. The speeds of all the subsystems are shown in Fig. 8.1, the all four states in Fig. 8.2, and the control inputs in Fig. 8.3.

From the results shown in the figures, it can be seen that all the five UAVs converge to the same set of states, although they are under different disturbances. Therefore, the formation is preserved. Moreover, note that the consensus disturbance rejection only ensures the relative positions for the formation, not the complete rejection of the input disturbance in general, because the relative information is used. In this example, the disturbance of the leading UAV is not rejected at all. Such a complete rejection can be introduced when the information of the actual state is available, instead of the relative

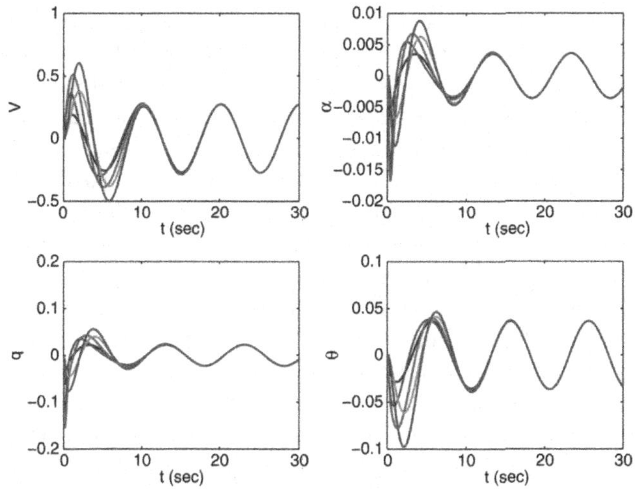

Figure 8.2 Four states of UAV longitudinal control.

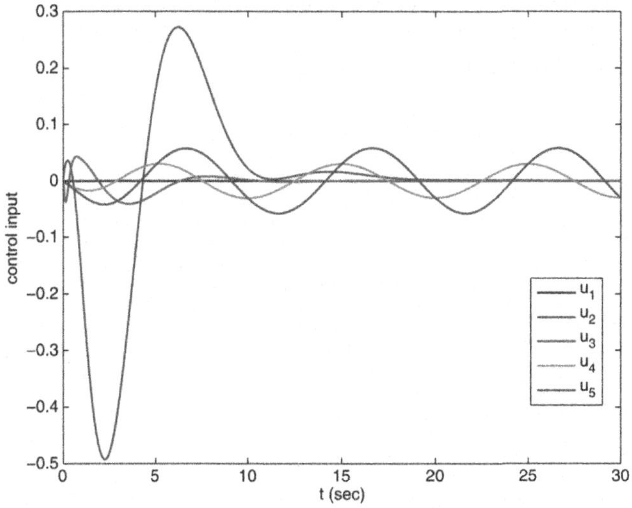

Figure 8.3 Control inputs for consensus disturbance rejection.

ones. In the case that there is no disturbance in the leading UAV, all the disturbances in the followers will be completely rejected by the proposed scheme.

8.6 Conclusion remarks

In this chapter, a number of fundamental design issues for consensus disturbance rejection have been addressed. Systematic designs for consensus disturbance rejection have been proposed and analyzed for network-connected multiple-input linear systems using relative state information of the subsystems in the neighborhood in leaderless and leader-follower consensus setups under some common assumptions on the network connections. Furthermore, in the leaderless consensus case, an adaptive scheme has been proposed to remove the requirement of the network connectivity in the disturbance observer design so that the proposed scheme is truly distributed in the sense that only the local information in the neighborhood is used for the control design. Consensus disturbance rejection ensures the states of the subsystems converge to a consensus trajectory among all the subsystems, whereas this common trajectory may be influenced by the disturbances, as the use of relative state information restricts the rejection only to the part of the disturbances that causes differences among the subsystem trajectories. This is a difference from the disturbance rejection of individual systems due to the restriction of using relative state information. The partial rejection of disturbances for consensus control would consume less energy compared with the complete rejection of the disturbances for certain applications such as formation control. The design and analysis methods for consensus disturbance rejection are also different from the corresponding methods for standard consensus state feedback control, due to the fact that the states for the disturbances are not controllable. The good simulation results suggest that the proposed methods have significant potentials in various engineering applications.

Robust consensus control of nonlinear odd power integrator systems

9

9.1 Problem formulation

In [199], the stabilization problem of the following power integrator system was investigated:

$$\begin{cases} \dot{x}_1 = x_2^p \\ \quad \vdots \\ \dot{x}_{n-1} = x_n^p \\ \dot{x}_n = u \end{cases}, \tag{9.1}$$

where $p \geq 1$ is an odd positive integer. $u \in \mathbb{R}$ and $\boldsymbol{x} = \begin{bmatrix} x_1 & , \cdots, & x_n \end{bmatrix}^T \in \mathbb{R}^n$ are the system input and state, respectively. For a physical model of this power integrator system one can refer to [139], where the odd power integer was derived from an irregular spring with cubic force-deformation relationship. The difficulty of controlling this system is that its linearized model around its origin is uncontrollable and thus conventional methods are inapplicable. Despite the aforementioned efforts in [199], many important problems remain open and unsolved, one of which is whether a network of such systems can be synchronized to a fixed value or a time-varying reference by using the neighborhood information through the communication network. It can be noticed that when $p = 1$, the system turns out to be a standard linear high-order integrator system, which has been well investigated in [149] and [72]. However when $p > 1$, the conventional methodologies, such as eigenvalues or linearization, are no longer applicable. To the authors' best knowledge, only [142] proposed a 'diagonally quasi-linear function of positive gain' condition ensuring system (9.1) with $n = 2$ to reach consensus to a unknown constant value. However, the controller becomes tremendously complicated when the system's order n increases.

In this chapter, we consider a network of N followers with the dynamics of agent i being

$$\begin{cases} \dot{x}_{i,1} = x_{i,2}^p + \varphi_{i,1}(x_{i,1}, x_{i,2}, t) \\ \dot{x}_{i,2} = x_{i,3}^p + \varphi_{i,2}(x_{i,1}, x_{i,2}, x_{i,3}, t) \\ \quad \vdots \\ \dot{x}_{i,n} = u_i + \varphi_{i,n}(x_{i,1}, \cdots, x_{i,n}, t) + \Delta_i(t) \\ y_i = x_{i,1} \end{cases}, \tag{9.2}$$

Cooperative Control of Multi-Agent Systems. https://doi.org/10.1016/B978-0-12-820118-3.00020-X

where $\boldsymbol{x}_i = [x_{i,1}, \cdots, x_{i,n}]^T \in \mathbb{R}^n$, $u_i \in \mathbb{R}$ and $y_i \in \mathbb{R}$ are the state, control and output of agent i, respectively. $p \geq 1$ is an odd positive integer. $\Delta_i(t)$ is the bounded disturbance in the input channel. $\varphi_{i,1}(x_{i,1}, x_{i,2,}, t), \cdots, \varphi_{i,n}(x_{i,1}, \cdots, x_{i,n}, t)$ is a set of model uncertainties in the triangular form.

Remark 9.1. Our recent research has indicated that there is a type of irregular springs whose dynamics can be modeled as high-order integrator system in Eq. (9.2) with $p = 3$ and $n = 1$. The control of such springs, even for the cooperative work of them are of importance for the whole system's stability. In this chapter, we first conceptualize and then formulate this robust cooperative output tracking problem, which can be solved by integrating the graph theory with the conventional control theory.

Assumption 9.1. The unknown disturbance $\Delta_i(t)$ in the input channel for agent i is bounded by $\bar{\Delta}$, i.e., $\forall i = 1, \cdots, N$, $|\Delta_i(t)| \leq \bar{\Delta}$.

Assumption 9.2. The model uncertainties $\varphi_{i,m}(x_{i,1}, \cdots, x_{i,m+1}, t)$ are bounded by constants ϕ_m, i.e., $\forall i = 1, \cdots, N$ and $\forall m = 1, \cdots, n$, $|\varphi_{i,m}(x_{i,1}, \cdots, x_{i,m+1}, t)| \leq \phi_m$.

In this chapter, the reference trajectory is denoted by $y_d(t)$ and the following assumption is commonly imposed in the tracking problem.

Assumption 9.3. For a desired $\mathbb{C}^1$ trajectory $y_d(t)$, its derivative $\dot{y}_d(t)$ is bounded by v_d, i.e., $|\dot{y}_d| \leq v_d$.

The following definitions are given.

Definition 9.1. The disagreement vector $\boldsymbol{\delta}(t) = \boldsymbol{y}(t) - y_d(t)\mathbf{1}_N \in \mathbb{R}^N$ where $\boldsymbol{y}(t) = [y_1, \cdots, y_N]^T$, is uniformly ultimately bounded (UUB) if there exists a compact set $\Omega \in \mathbb{R}^N$ such that $\forall \boldsymbol{\delta}(t_0) \in \Omega$ there exist a bound B and $t_f(B, \boldsymbol{\delta}(t_0))$, both independent of $t_0 \geq 0$ such that $||\boldsymbol{\delta}(t)|| \leq B$, $\forall t \geq t_0 + t_f$.

Definition 9.2. The neighboring tracking error of agent i is

$$\xi_{i,1} = \sum_{j \in \mathcal{N}_i} a_{ij}(y_i - y_j) + b_i(y_i - y_d), \tag{9.3}$$

where a_{ij} and b_i are derived from the graph $\mathcal{G}$ as defined in Section 9.2.

Then, we have the relation between the neighboring tracking error and disagreement vector,

$$\boldsymbol{\xi}_1 = (\mathcal{L} + \mathcal{B})\boldsymbol{\delta}(t), \tag{9.4}$$

where $\boldsymbol{\xi}_1 = [\xi_{1,1}, \cdots, \xi_{N,1}]^T \in \mathbb{R}^N$, $\mathcal{L}$ is the Laplacian matrix associated with the graph $\mathcal{G}$ and the diagonal matrix $\mathcal{B}$ contains the connection information between agents and reference. The control objective of this chapter is to design a distributed control law $u_i(\boldsymbol{x}_i, \xi_{i,1})$ for all the agents in the network denoted by $\mathcal{G}$ such that the disagreement vector $\boldsymbol{\delta}(t)$ is uniformly ultimately bounded (UUB) even under external disturbances as well as model uncertainties.

Remark 9.2. The stabilization or tracking problem of Eq. (9.2) has been well investigated in the early works, such as [199] and [140]. In [139], an underactuated physical model in the form of power integrator system has been proposed, in which the odd power integer is derived from an irregular spring with a cubic force-deformation relationship. But to our best knowledge, the robust cooperative output tracking of multiple such systems in the presence of disturbance and uncertainty has not been studied yet. To distinguish from the tracking control of a single system, the controller $u_i(\boldsymbol{x}_i, \xi_{i,1})$ needs the neighborhood information and the reference trajectory y_d if it is connected to the reference. Followers are transmitting their output information throughout the communication network to achieve the control objective. Another difference of the proposed control from the conventional control technique is to guarantee all the outputs simultaneously synchronized to the desired trajectory.

9.2 Distributed controller for nonlinear odd power integrator systems

In this section, we recursively design a distributed controller for each agent to achieve the control objective described in Section 9.1. Before we present the main result of this chapter, the following lemma is needed.

Lemma 9.1. *([24]) Let the undirected graph $\mathcal{G}$ is connected and $\mathcal{B} \neq 0$, then the disagreement vector $\boldsymbol{\delta}$ satisfies*

$$||\boldsymbol{\delta}(t)|| \leq \frac{||\boldsymbol{\xi}_1||}{\underline{\sigma}(\mathcal{L}+\mathcal{B})}, \tag{9.5}$$

where $\underline{\sigma}(\mathcal{L}+\mathcal{B})$ is the minimum singular value of the matrix $\mathcal{L}+\mathcal{B}$.

Due to Lemma 9.1, $\mathcal{H} = \mathcal{L}+\mathcal{B}$ is positive definite when the graph $\mathcal{G}$ is connected and $B \neq 0$. It is straightforward to see Eq. (9.5) from Eq. (9.4).

Remark 9.3. Lemma 9.1 implies that $\boldsymbol{\delta}$ is bounded as long as $\boldsymbol{\xi}_1$ is bounded. Moreover, $\boldsymbol{\xi}_1 \to \mathbf{0}$ implies $\boldsymbol{\delta} \to \mathbf{0}$. In other words, the synchronization of multi-agent outputs to the desired trajectory is achieved.

As $\xi_{i,1}$ reflects the distributed information of agent i through the network, the distributed control u_i will be essentially designed based upon $\xi_{i,1}$. Therefore, the distributed controller is proposed as follows:

Controller.

$$\begin{aligned}
\xi_{i,1} &= \sum_{j\in\mathcal{N}_i} a_{ij}(y_i - y_j) + b_i(y_i - y_d) & x_{i,2}^* &= -a_1\xi_{i,1} \\
\xi_{i,2} &= x_{i,2} - x_{i,2}^* & x_{i,3}^* &= -a_2\xi_{i,2} \\
\xi_{i,3} &= x_{i,3} - x_{i,3}^* & x_{i,4}^* &= -a_3\xi_{i,3} \\
&\vdots & &\vdots \\
\xi_{i,n} &= x_{i,n} - x_{i,n}^* & x_{i,n+1}^* &= -a_n\xi_{i,n} \\
u_i &= x_{i,n+1}^{*p} = -a_n^p\xi_{i,n}^p,\ \forall i = 1,\cdots,N
\end{aligned}, \tag{9.6}$$

where $a_1, \cdots, a_n$ are positive control gains.

With the above knowledge, we are ready to present the main result of this chapter as follows.

Theorem 9.1. *Given a network of N agents with dynamics Eq. (9.2) and a desired trajectory $y_d(t)$ under Assumptions 9.1–9.3, there is a distributed controller 1 in Eq. (9.6) such that the disagreement vector $\boldsymbol{\delta}(t)$ is UUB if the undirected graph $\mathcal{G}$ is connected and there exists at least one agent connected to the reference.*

Proof. **Step 1**: Define a Lyapunov function component V_1 as

$$V_1(\boldsymbol{\xi}_1) = \frac{1}{2}\boldsymbol{\xi}_1^T(\mathcal{L}+\mathcal{B})^{-1}\boldsymbol{\xi}_1 = \frac{1}{2}\sum_{i=1}^{N}\sum_{j=1}^{N} a_{ij}(x_{i,1} - x_{j,1})^2 + \frac{1}{2}\sum_{i=1}^{N} b_i(x_{i,1} - y_d)^2, \tag{9.7}$$

where $\boldsymbol{\xi}_1$ is defined in Eq. (9.4). It can be seen from Lemma 2.5 that V_1 is positive definite when the graph is connected and $\mathcal{B} \neq 0$. Then, the derivative of V_1 with respect to time t along Eq. (9.2) is

$$\dot{V}_1(\boldsymbol{\xi}_1) = \boldsymbol{\xi}_1^T\mathcal{H}^{-1}\dot{\boldsymbol{\xi}}_1 = \boldsymbol{\xi}_1^T\dot{\boldsymbol{\delta}} = \sum_{i=1}^{N}\xi_{i,1}(x_{i,2}^p + \varphi_{i,1}(x_{i,1}, x_{i,2}, t)) - \sum_{i=1}^{N}\xi_{i,1}\dot{y}_d(t) \tag{9.8}$$

due to Eq. (9.4). Define $x_{i,2}^* = -a_1\xi_{i,1}$ with $a_1 > 0$ and then Eq. (9.8) turns out to be

$$\begin{aligned}
\dot{V}_1(\boldsymbol{\xi}_1) &= \sum_{i=1}^{N}\xi_{i,1}x_{i,2}^{*p} + \sum_{i=1}^{N}\xi_{i,1}(x_{i,2}^p - x_{i,2}^{*p}) + \sum_{i=1}^{N}\xi_{i,1}(\varphi_{i,1}(x_{i,1}, x_{i,2}, t) - \dot{y}_d(t)) \\
&\leq -a_1^p\sum_{i=1}^{N}\xi_{i,1}^{p+1} + \sum_{i=1}^{N}\xi_{i,1}(x_{i,2}^p - x_{i,2}^{*p}) + \sum_{i=1}^{N}|\xi_{i,1}|(|\phi_1| + |v_d|).
\end{aligned} \tag{9.9}$$

Define $\xi_{i,2} = x_{i,2} - x_{i,2}^* = x_{i,2} + a_1\xi_{i,1}$ and notice that $x_{i,2}^p - {x_{i,2}^*}^p = (x_{i,2} - x_{i,2}^*)g(x_{i,2}, x_{i,2}^*)$ where $g(x_{i,2}, x_{i,2}^*)$ is a polynomial of $x_{i,2}$ and $x_{i,2}^*$ with the order of each term being $p-1$, it can be obtained from Lemmas 2.11 and 2.13 that

$$|g(x_{i,2}, x_{i,2}^*)| \leq \beta_1'(\xi_{i,2}^{p-1} + \xi_{i,1}^{p-1}), \tag{9.10}$$

where β_1' is a positive constant. Then, Eq. (9.9) can be rewritten as

$$\begin{aligned}
\dot{V}_1(\boldsymbol{\xi}_1) &\leq -a_1^p \sum_{i=1}^{N} \xi_{i,1}^{p+1} + \sum_{i=1}^{N} |\xi_{i,1}||\xi_{i,2}||g(x_{i,2}, x_{i,2}^*)| + \sum_{i=1}^{N} |\xi_{i,1}|(|\phi_1| + |v_d|) \\
&\leq -a_1^p \sum_{i=1}^{N} \xi_{i,1}^{p+1} + \beta'_1 \sum_{i=1}^{N} |\xi_{i,1}||\xi_{i,2}|(|\xi_{i,1}|^{p-1} + |\xi_{i,2}|^{p-1}) \\
&\quad + \sum_{i=1}^{N} |\xi_{i,1}|(|\phi_1^{\frac{1}{p}}|^p + |v_d^{\frac{1}{p}}|^p) \\
&\leq -a_1^p \sum_{i=1}^{N} \xi_{i,1}^{p+1} + \beta_{11} \sum_{i=1}^{N} \xi_{i,1}^{p+1} + \beta_{12} \sum_{i=1}^{N} \xi_{i,2}^{p+1} + \kappa_{1d}|v_d|^{\frac{p+1}{p}} + \kappa_{11}|\phi_1|^{\frac{p+1}{p}} \\
&= -k_{11} \sum_{i=1}^{N} \xi_{i,1}^{p+1} + k_{12} \sum_{i=1}^{N} \xi_{i,2}^{p+1} + \kappa_{1d}|v_d|^{\frac{p+1}{p}} + \kappa_{11}|\phi_1|^{\frac{p+1}{p}},
\end{aligned} \tag{9.11}$$

where a_1 is the positive control gain to be designed, $\beta_1', \beta_{11}, \beta_{12}, k_{11}, k_{12}, \kappa_{1d}, \kappa_{11}$ are positive constants. β_{11} can be set small by choosing β_{12}, κ_{1d} and κ_{11} to be large accordingly such that k_{11} and k_{12} can be positive.

Step 2: Choose the Lyapunov function component V_2 as

$$V_2(\boldsymbol{\xi}_1, \boldsymbol{\xi}_2) = V_1(\boldsymbol{\xi}_1) + \frac{1}{2} \sum_{i=1}^{N} \xi_{i,2}^2, \tag{9.12}$$

where $\boldsymbol{\xi}_2 = [\xi_{1,2}, \cdots, \xi_{N,2}]^T \in \mathbb{R}^N$ and note that $\xi_{i,2} = x_{i,2} - x_{i,2}^* = x_{i,2} + a_1\xi_{i,1}$. Then, the derivative of V_2 with respect to time t is

$$\begin{aligned}
\dot{V}_2(\boldsymbol{\xi}_1, \boldsymbol{\xi}_2) &= \dot{V}_1 + \sum_{i=1}^{N} \xi_{i,2}\dot{\xi}_{i,2} = \dot{V}_1 + \sum_{i=1}^{N} \xi_{i,2}(\dot{x}_{i,2} + a_1\dot{\xi}_{i,1}) \\
&= \dot{V}_1 + \sum_{i=1}^{N} \xi_{i,2}[{x_{i,3}^*}^p + (x_{i,3}^p - {x_{i,3}^*}^p) + \varphi_{i,2}(x_{i,1}, x_{i,2}, x_{i,3}, t)] + a_1 \sum_{i=1}^{N} \xi_{i,2}\dot{\xi}_{i,1}.
\end{aligned} \tag{9.13}$$

Define $\xi_{i,3} = x_{i,3} - x_{i,3}^*$ where $x_{i,3}^* = -a_2\xi_{i,2}$ and $a_2 > 0$, substituting Eq. (9.11) into Eq. (9.13) yields

$$\begin{aligned}
&\dot{V}_2(\boldsymbol{\xi}_1, \boldsymbol{\xi}_2) \\
&\leq -k_{11} \sum_{i=1}^{N} \xi_{i,1}^{p+1} + k_{12} \sum_{i=1}^{N} \xi_{i,2}^{p+1} + \kappa_{1d}|v_d|^{\frac{p+1}{p}} + \kappa_{11}|\phi_1|^{\frac{p+1}{p}} - a_2^p \sum_{i=1}^{N} \xi_{i,2}^{p+1} \\
&+ \sum_{i=1}^{N} |\xi_{i,2}||\phi_2^{\frac{1}{p}}|^p + \sum_{i=1}^{N} \xi_{i,2}\xi_{i,3}g(x_{i,3}, x_{i,3}^*)
\end{aligned}$$

$$
\begin{aligned}
&+a_1\sum_{i=1}^{N}\xi_{i,2}[\sum_{j=1}^{N}a_{ij}(x_{i,2}^p+\varphi_{i,1}-x_{j,2}^p-\varphi_{j,1})+b_i(x_{i,2}^p+\varphi_{i,1}-\dot{y}_d)]\\
&\le -k_{11}\sum_{i=1}^{N}\xi_{i,1}^{p+1}+k_{12}\sum_{i=1}^{N}\xi_{i,2}^{p+1}+\kappa_{1d}|v_d|^{\frac{p+1}{p}}+\kappa_{11}|\phi_1|^{\frac{p+1}{p}} \qquad (9.14)\\
&-a_2^p\sum_{i=1}^{N}\xi_{i,2}^{p+1}+\sum_{i=1}^{N}|\xi_{i,2}||\phi_2^{\frac{1}{p}}|^p+\sum_{i=1}^{N}\xi_{i,2}\xi_{i,3}g(x_{i,3},x_{i,3}^*)\\
&+a_1\sum_{i=1}^{N}\xi_{i,2}(\sum_{j=1}^{N}a_{ij}+b_i-\sum_{k=1}^{N}a_{ki})\varphi_{i,1}-a_1\sum_{i=1}^{N}b_i\xi_{i,2}\dot{y}_d\\
&+a_1\sum_{i=1}^{N}[(\sum_{j=1}^{N}a_{ij}+b_i)\xi_{i,2}-\sum_{k=1}^{N}a_{ki}\xi_{k,2}]x_{i,2}^p.
\end{aligned}
$$

Noting that $|x_{i,2}|^p=|\xi_{i,2}-a_1\xi_{i,1}|^p\le 2^{p-1}(|\xi_{i,2}|^p+a_1^p|\xi_{i,1}|^p)$ by Lemma 2.12 and $|g(x_{i,3},x_{i,3}^*)|\le\beta_2'(\xi_{i,3}^{p-1}+\xi_{i,2}^{p-1})$ similar to Eq. (9.10), we can obtain

$$
\begin{aligned}
\dot{V}_2(\boldsymbol{\xi}_1,\boldsymbol{\xi}_2)\le&-k_{11}\sum_{i=1}^{N}\xi_{i,1}^{p+1}+k_{12}\sum_{i=1}^{N}\xi_{i,2}^{p+1}-a_2^p\sum_{i=1}^{N}\xi_{i,2}^{p+1}\\
&+\beta'_2\sum_{i=1}^{N}|\xi_{i,2}||\xi_{i,3}|(\xi_{i,2}^{p-1}+\xi_{i,3}^{p-1})\\
&+2^{p-1}a_1(N+1)\sum_{i=1}^{N}(\sum_{j=1}^{N}|\xi_{j,2}|)(|a_1\xi_{i,1}|^p+|\xi_{i,2}|^p)+\kappa_{1d}|v_d|^{\frac{p+1}{p}}\\
&+\kappa_{11}|\phi_1|^{\frac{p+1}{p}}+\sum_{i=1}^{N}|\xi_{i,2}|(|\phi_2^{\frac{1}{p}}|^p+a_1|v_d^{\frac{1}{p}}|^p+a_1(N+1)|\phi_1^{\frac{1}{p}}|^p)\\
\le&-k_{11}\sum_{i=1}^{N}\xi_{i,1}^{p+1}+k_{12}\sum_{i=1}^{N}\xi_{i,2}^{p+1}+\beta_{21}\sum_{i=1}^{N}\xi_{i,1}^{p+1}+\beta_{22}\sum_{i=1}^{N}\xi_{i,2}^{p+1}\\
&+\beta_{23}\sum_{i=1}^{N}\xi_{i,3}^{p+1}-a_2^p\sum_{i=1}^{N}\xi_{i,2}^{p+1}+\kappa_{2d}|v_d|^{\frac{p+1}{p}}+\kappa_{21}|\phi_1|^{\frac{p+1}{p}}+\kappa_{22}|\phi_2|^{\frac{p+1}{p}}\\
\le&-k_{21}\sum_{i=1}^{N}\xi_{i,1}^{p+1}-k_{22}\sum_{i=1}^{N}\xi_{i,2}^{p+1}+k_{23}\sum_{i=1}^{N}\xi_{i,3}^{p+1}+\kappa_{2d}|v_d|^{\frac{p+1}{p}}\\
&+\kappa_{21}|\phi_1|^{\frac{p+1}{p}}+\kappa_{22}|\phi_2|^{\frac{p+1}{p}}
\end{aligned}
\tag{9.15}
$$

where a_2 is the positive control gain to be designed, $\beta_2{}'$, β_{21}, β_{22}, β_{23}, k_{21}, k_{22}, k_{23}, κ_{2d}, κ_{21}, κ_{22} are all positive constants. β_{21} can be set small while the corresponding β_{22} is large such that k_{21} is positive. Then, a_2 can be chosen large to dominate positive β_{22}, a_1 related to k_{12} such that k_{22} is positive.

Step m ($3\le m\le n-1$): Similarly, we define

$$
V_m(\boldsymbol{\xi}_1,\cdots,\boldsymbol{\xi}_m)=V_{m-1}(\boldsymbol{\xi}_1,\cdots,\boldsymbol{\xi}_{m-1})+\frac{1}{2}\sum_{i=1}^{N}\xi_{i,m}^2 \tag{9.16}
$$

with $\boldsymbol{\xi}_m = [\xi_{1,m}, \cdots, \xi_{N,m}]^T \in \mathbb{R}^N$. Define $\xi_{i,m} = x_{i,m} - x^*_{i,m}$ and $x^*_{i,m} = -a_{m-1}\xi_{i,m-1}$. It can be obtained that the derivative of Eq. (9.16) with respect to time t along Eq. (9.6) is

$$\begin{aligned}\dot{V}_m &= \dot{V}_{m-1} + \sum_{i=1}^{N} \xi_{i,m}\dot{\xi}_{i,m} \\ &= \dot{V}_{m-1} + \sum_{i=1}^{N} \xi_{i,m}[x^{*}_{i,m+1}{}^{p} + (x^{p}_{i,m+1} - x^{*}_{i,m+1}{}^{p}) + \varphi_{i,m}(x_{i,1}, \cdots, x_{i,m+1}, t)] \\ &\quad + a_{m-1}\sum_{i=1}^{N} \xi_{i,m}\dot{\xi}_{i,m-1},\end{aligned} \tag{9.17}$$

where $x^*_{i,m+1} = -a_m\xi_{i,m}$. Note that $a_{m-1}\sum_{i=1}^{N} |\xi_{i,m}||\dot{\xi}_{i,m-1}| \leq \beta_{m1}\sum_{i=1}^{N}\xi_{i,1}^{p+1} + \cdots + \beta_{mm}\sum_{i=1}^{N}\xi_{i,m}^{p+1} + \beta_{m,m+1}\sum_{i=1}^{N}\xi_{i,m+1}^{p+1} + \beta'_m(|\phi_1^{\frac{1}{p}}|^{p+1} + \cdots + |\phi_{m-1}^{\frac{1}{p}}|^{p+1}) + \beta''_m|v_d^{\frac{1}{p}}|^{p+1}$ with $\beta_{m1}\cdots\beta_{m,m+1}, \beta'_m, \beta''_m$ being positive constants. In a similar manner, the following inequality can be derived

$$\begin{aligned}\dot{V}_m(\boldsymbol{\xi}_1, \cdots, \boldsymbol{\xi}_m) \leq& -k_{m1}\sum_{i=1}^{N}\xi_{i,1}^{p+1} \cdots - k_{mm}\sum_{i=1}^{N}\xi_{i,m}^{p+1} + k_{m,m+1}\sum_{i=1}^{N}\xi_{i,m+1}^{p+1} \\ &+ \kappa_{md}|v_d|^{\frac{p+1}{p}} + \kappa_{m1}|\phi_1|^{\frac{p+1}{p}} + \cdots + \kappa_{mm}|\phi_m|^{\frac{p+1}{p}}\end{aligned} \tag{9.18}$$

with $k_{m1}, \cdots, k_{m,m+1}$ and $\kappa_{md}, \kappa_{m1}\cdots\kappa_{mm}$ being positive constants, and properly designed parameters $a_1, \cdots, a_m$. After the proof, we will explicitly explain how to design the control parameters.

Final step: Define

$$V(\boldsymbol{\xi}_1, \cdots, \boldsymbol{\xi}_n) = V_n = V_{n-1}(\boldsymbol{\xi}_1, \cdots, \boldsymbol{\xi}_{n-1}) + \frac{1}{2}\sum_{i=1}^{N}\xi_{i,n}^2 \tag{9.19}$$

with $\boldsymbol{\xi}_n = [\xi_{1,n}, \cdots, \xi_{N,n}]^T \in \mathbb{R}^N$. The derivative of Eq. (9.19) with respect to time t is

$$\begin{aligned}\dot{V}(\boldsymbol{\xi}_1, \cdots, \boldsymbol{\xi}_n) &= \dot{V}_{n-1} + \sum_{i=1}^{N}\xi_{i,n}\dot{\xi}_{i,n} \\ &= \dot{V}_{n-1} + \sum_{i=1}^{N}\xi_{i,n}(u_i + \varphi_{i,n} + \Delta_i + a_{n-1}\dot{\xi}_{i,n-1}) \\ &\leq -k_{n1}\sum_{i=1}^{N}\xi_{i,1}^{p+1} - k_{n2}\sum_{i=1}^{N}\xi_{i,2}^{p+1} \cdots - k_{nn}\sum_{i=1}^{N}\xi_{i,n}^{p+1} + \sum_{i=1}^{N}\xi_{i,n}(u_i - x^{*}_{i,n+1}{}^{p})\end{aligned}$$

$$+\kappa_{nd}|v_d|^{\frac{p+1}{p}}+\kappa'_{nd}|\bar{\Delta}|^{\frac{p+1}{p}}+\kappa_{n1}|\phi_1|^{\frac{p+1}{p}}+\cdots+\kappa_{nn}|\phi_n|^{\frac{p+1}{p}} \qquad (9.20)$$
$$\le -c'(\sum_{i=1}^{N}\xi_{i,1}^{p+1}+\sum_{i=1}^{N}\xi_{i,2}^{p+1}\cdots+\sum_{i=1}^{N}\xi_{i,n}^{p+1})+\varepsilon$$

when choosing the control as $u_i = {x^*_{i,n+1}}^p$, where $x^*_{i,n+1} = -a_n\xi_{i,n}$. The coefficients $k_{n1}\cdots, k_{nn}, \kappa_{n1},\cdots,\kappa_{nn},\kappa_{nd},\kappa'_{nd}$ are positive constants, $c' = \min\{k_{n1},\cdots,k_{nn}\}$ and $\varepsilon = \kappa_{nd}|v_d|^{\frac{p+1}{p}}+\kappa'_{nd}|\bar{\Delta}|^{\frac{p+1}{p}}+\kappa_{n1}|\phi_1|^{\frac{p+1}{p}}+\cdots+\kappa_{nn}|\phi_n|^{\frac{p+1}{p}}$ are both positive.

From the Lyapunov function (9.19) and the inequalities

$$\underline{\lambda}(\mathcal{H}^{-1})\sum_{i=1}^{N}\xi_{i,1}^2 \le \boldsymbol{\xi}_1^T\mathcal{H}^{-1}\boldsymbol{\xi}_1 \le \bar{\lambda}(\mathcal{H}^{-1})\sum_{i=1}^{N}\xi_{i,1}^2$$

we have

$$\frac{1}{2}[\underline{\lambda}(\mathcal{H}^{-1})\sum_{i=1}^{N}\xi_{i,1}^2+\cdots+\sum_{i=1}^{N}\xi_{i,n}^2] \le V(\boldsymbol{\xi}_1,\cdots,\boldsymbol{\xi}_n)$$
$$\le \frac{1}{2}[\bar{\lambda}(\mathcal{H}^{-1})\sum_{i=1}^{N}\xi_{i,1}^2+\cdots+\sum_{i=1}^{N}\xi_{i,n}^2]. \qquad (9.21)$$

Also from Lemma 2.11, when $p \ge 1$ we have

$$-\xi_{i,j}^{p+1} \le -\xi_{i,j}^2+(\frac{p-1}{p+1})(\frac{p+1}{2})^{-\frac{2}{p-1}}, \qquad (9.22)$$

where $a=1, b=1, x=\xi_{i,j}, y=1, m=2$ and $n=p-1$. Furthermore, Eqs. (9.20), (9.21) and (9.22) imply that

$$\dot{V}(\boldsymbol{\xi}_1,\cdots,\boldsymbol{\xi}_n) \le -\hat{c}V(\boldsymbol{\xi}_1,\cdots,\boldsymbol{\xi}_n)+\varepsilon', \qquad (9.23)$$

where $\hat{c} = 2c'/\max\{\bar{\lambda}(\mathcal{H}^{-1}),1\}$ and $\varepsilon' = \varepsilon + c'nN(\frac{p-1}{p+1})(\frac{p+1}{2})^{-\frac{2}{p-1}}$ are two positive constants.

Now define the set $\Omega_{\boldsymbol{\xi}} = \{\boldsymbol{\xi}_1,\cdots,\boldsymbol{\xi}_n | V(\boldsymbol{\xi}_1,\cdots,\boldsymbol{\xi}_n) \le \frac{\varepsilon'}{\hat{c}}\}$. Eq. (9.23) shows that $\dot{V}(\boldsymbol{\xi}_1,\cdots,\boldsymbol{\xi}_n) < 0$ once the errors are outside the compact set $\Omega_{\boldsymbol{\xi}}$. According to the standard Lyapunov theorem, we conclude that $\boldsymbol{\xi}_1\cdots\boldsymbol{\xi}_n$ are bounded and thus $\forall i$, $y_i - y_d$ is bounded under Eq. (9.6). To further explore the explicit bound of tracking error, the following inequality can be derived from Eq. (9.23)

$$V(\boldsymbol{\xi}_1(t),\cdots,\boldsymbol{\xi}_n(t)) \le V(\boldsymbol{\xi}_1(0),\cdots,\boldsymbol{\xi}_n(0))e^{-\hat{c}t}+\frac{\varepsilon'}{\hat{c}}(1-e^{-\hat{c}t}). \qquad (9.24)$$

By Lemma 9.1, the norm of disagreement vector $\boldsymbol{\delta}(t)$ is

$$||\boldsymbol{\delta}(t)|| \le \frac{||\boldsymbol{\xi}_1(t)||}{\underline{\sigma}(\mathcal{H})} \le \frac{1}{\underline{\sigma}(\mathcal{H})}\sqrt{\frac{2V(0)e^{-\hat{c}t}+2\frac{\varepsilon'}{\hat{c}}(1-e^{-\hat{c}t})}{\underline{\lambda}(\mathcal{H}^{-1})}}. \qquad (9.25)$$

As $t \to +\infty$,

$$||\delta(+\infty)|| \leq \frac{1}{\underline{\sigma}(\mathcal{H})}\sqrt{\frac{2\varepsilon'}{\hat{c}\underline{\lambda}(\mathcal{H}^{-1})}}, \tag{9.26}$$

which shows the tracking error is UUB and thus completes the proof. □

Theorem 9.1 reveals the existence of a cooperative tracking controller for power integrator systems and provides the explicit bound of tracking error. The distributed controller 1 in Eq. (9.6) is recursively designed and entirely based upon the state of agent i and the output of its neighbors, which only requires local information. Thus, it is in a distributed manner.

Remark 9.4. The control gains $a_1, \cdots, a_n$ actually provide the stability margin of the overall system regarding external disturbances and model uncertainties. The metric of choosing control gains $a_1, \cdots a_n$ can be guided as follows due to the recursive inherence:

- a_m $(1 \leq m \leq n-1)$ are chosen as positive constants that can be large enough to dominate $\beta_{1m}, \cdots, \beta_{mm}$ produced by applying Lemmas 2.2–2.4. Note that the coefficient b in Eq. (2.20) on the right-hand side can be set freely, so after a_m are set we choose relative small $\beta_{m+1,m}, \cdots, \beta_{nm}$ from step $m+1$ to step n.
- a_n is chosen such that $-a_n^p$ can cancel $\beta_{n-1,n}$ already produced in the previous step.

Note that β_{ij} are derived from the coefficients of the right-hand side in Lemma 2.11, such as b and $\frac{n}{m+n}(\frac{m+n}{m})^{-\frac{m}{n}} a^{\frac{m+n}{n}} b^{-\frac{m}{n}}$. Also, we know b can be set freely and thus by using the trade-off between them (i.e. when one is small and the other is large), the metric above is realizable. In other words, if every time we apply Lemma 2.11 to $\dot{V}_m$ with the same b, $a_1, \cdots, a_n$ can be written in a certain form. For the sake of simplicity and reality, we can select a_1 arbitrarily and a_2 much larger than a_1, and $a_3, \cdots, a_n$ in a similar manner.

Remark 9.5. It can be seen from Eq. (9.26) that the final tracking error can be minimized as small as possible by increasing $\hat{c}$ in the denominator. Noticing the relationship of $\hat{c}$ and control gains, it can be achieved by choosing large control gains $a_1, \cdots, a_n$ in the controller design. The transient performance Eq. (9.25) can also be improved by the trajectory initialization method in [4].

On the other hand, when the graph $\mathcal{G}$ is a balanced digraph and there is at least one link between followers and leader, the matrix $\mathcal{H} = \mathcal{L} + \mathcal{B}$ has the same positive definite property as in Lemma 2.5. For this reason, the controller in Theorem 9.1 can also be applied to the balanced digraph.

In what follows, as a special case of the proposed robust output tracking problem, the output consensus of networked power integrators systems to a constant reference is considered in order to further show the effectiveness of the proposed framework.

First, without consideration of external disturbances and model uncertainties, we have

$$\begin{cases} \dot{x}_{i,1} = x_{i,2}^p \\ \dot{x}_{i,2} = x_{i,3}^p \\ \vdots \\ \dot{x}_{i,n} = u_i \end{cases} . \tag{9.27}$$
$$y_i = x_{i,1}$$

If the constant reference is further assumed, we have the following corollary.

Corollary 9.1. *Given a network of N agents in Eq. (9.27) and a constant trajectory $y_d(t) = r$, the distributed controller 1 in Eq. (9.6) can synchronize each agent's output to the reference r, i.e., $\lim_{t\to\infty} x_{i,1}(t) = \lim_{t\to\infty} x_{j,1}(t) = r, \forall i, j \in \mathcal{V}$, if the undirected graph $\mathcal{G}$ is connected and there exists at least one agent connected to the reference.*

Proof. The proof is the same as the proof of Theorem 9.1 by removing the disturbance terms, and thus omitted. □

9.3 Numerical examples

This section will provide several numerical examples to illustrate the effectiveness of the proposed controller 1.

9.3.1 Perturbed case

Consider a network of 3 agents with the communication graph shown in Fig. 9.1. The weights of the edges are all set to 1 and thus

$$\mathcal{L} = \begin{bmatrix} 2 & -1 & -1 \\ -1 & 1 & 0 \\ -1 & 0 & 1 \end{bmatrix}, \mathcal{B} = \begin{bmatrix} 0 & 0 & 0 \\ 0 & 1 & 0 \\ 0 & 0 & 0 \end{bmatrix}$$

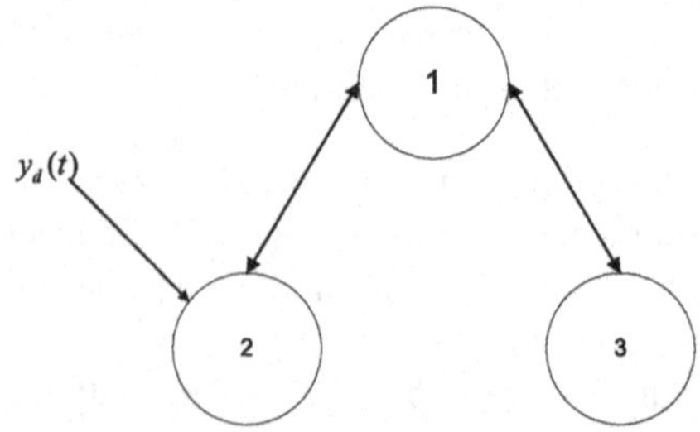

Figure 9.1 Communication graph.

The dynamics of agents are given as follows

$$\begin{cases} \dot{x}_{1,1} = x_{1,2}^3 - 0.08\frac{\sin(x_{1,2}^2)}{e^{-10t}+1} \\ \dot{x}_{1,2} = x_{1,3}^3 + \cos(t)\sin(x_{1,2}^2) \\ \dot{x}_{1,3} = u_1 + sat(randn(*), 0.02) \end{cases}$$
$$y_1 = x_{1,1},$$

$$\begin{cases} \dot{x}_{2,1} = x_{2,2}^3 - 0.1\frac{\sin(x_{2,2}^2)}{e^{-10t}+1} \\ \dot{x}_{2,2} = x_{2,3}^3 + \cos(t)\sin(x_{2,2}^2) \\ \dot{x}_{2,3} = u_2 + sat(randn(*), 0.02) \end{cases}$$
$$y_2 = x_{2,1},$$

$$\begin{cases} \dot{x}_{3,1} = x_{3,2}^3 - 0.12\frac{\sin(x_{3,2}^2)}{e^{-10t}+1} \\ \dot{x}_{3,2} = x_{3,3}^3 + \cos(t)\sin(x_{3,2}^2) \\ \dot{x}_{3,3} = u_3 + sat(randn(*), 0.02) \end{cases}$$
$$y_3 = x_{3,1},$$

where $sat(v, u)$ is the saturation function and $randn(*)$ is the standard Gaussian distributed random variable generated by Matlab®. Moreover, the desired trajectory is $y_d(t) = 1.32 + 2\sin(0.5t)$. Set the initial value for each agent to be

$$\begin{bmatrix} x_{1,1}(0) \\ x_{1,2}(0) \\ x_{1,3}(0) \end{bmatrix} = \begin{bmatrix} 2 \\ -0.1 \\ -0.05 \end{bmatrix}, \begin{bmatrix} x_{2,1}(0) \\ x_{2,2}(0) \\ x_{2,3}(0) \end{bmatrix} = \begin{bmatrix} 1.2 \\ -0.02 \\ 0.01 \end{bmatrix}, \begin{bmatrix} x_{3,1}(0) \\ x_{3,2}(0) \\ x_{3,3}(0) \end{bmatrix} = \begin{bmatrix} 1.6 \\ 0.05 \\ 0 \end{bmatrix}$$

and choose control gains as $a_1 = 5$, $a_2 = 30$, $a_3 = 600$ according to the selection rule in Remark 9.4. Then the simulation result can be seen in Fig. 9.2 by applying the distributed controller 1 in Eq. (9.6). It can be seen that all agents' output are synchronized to the desired time-varying trajectory with a bounded error. With the same initial value, increase the control gain as $a_1 = 10$, $a_2 = 50$, $a_3 = 850$. Fig. 9.3 shows that all agents are synchronized to the desired trajectory with a bounded error as well. Comparing Fig. 9.2 and Fig. 9.3, it can be seen that the tracking error in Fig. 9.3 is smaller than that of Fig. 9.2. The reason for this is that a larger control gain will yield a larger $\hat{c}$ such that the bound error will be smaller.

9.3.2 Perturbation-free case

In this subsection, we still consider three agents as depicted in Fig. 9.1. The weights of the edges are all set to 1. However, the dynamics of agents are given as

$$\begin{cases} \dot{x}_{i,1} = x_{i,2}^3 \\ \dot{x}_{i,2} = x_{i,3}^3 \\ \dot{x}_{i,3} = u_i \end{cases}, \forall i = 1, 2, 3$$
$$y_i = x_{i,1}$$

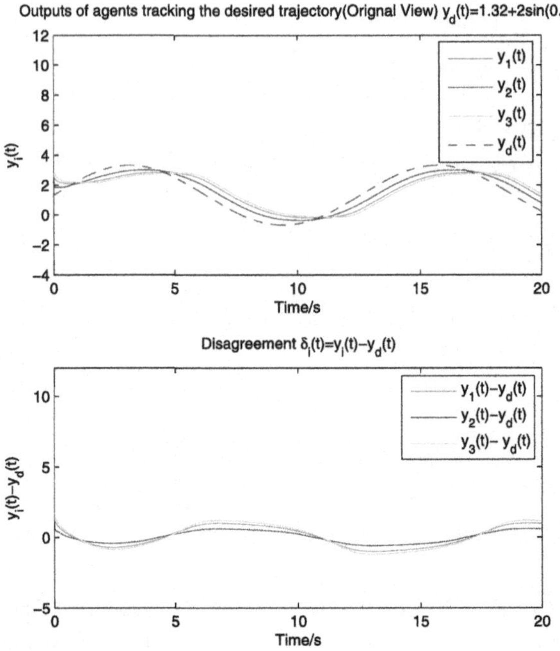

Figure 9.2 Small control gain case.

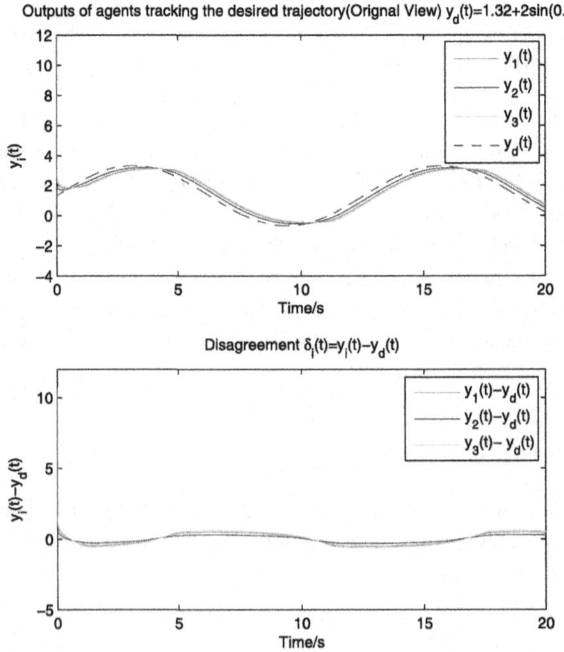

Figure 9.3 Large control gain case.

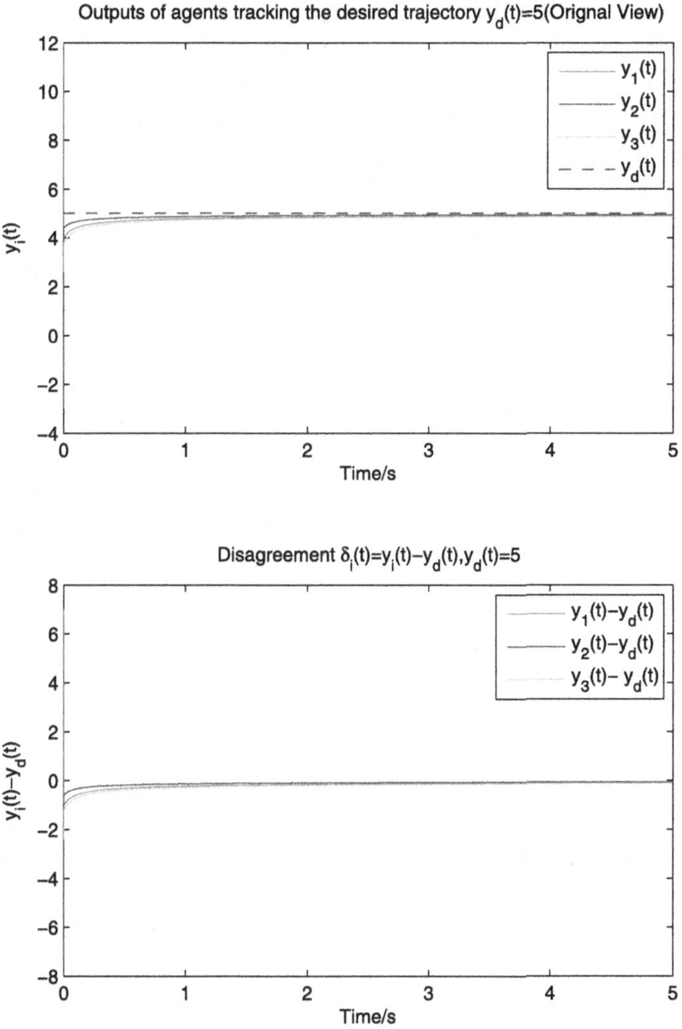

Figure 9.4 Perturbation-free case with controller in Corollary 2 when $p = 3$.

without consideration of disturbances and uncertainties. With the same initial conditions as those in the previous case and control gains being $a_1 = 10$, $a_2 = 50$, $a_3 = 850$, the simulation result of tracking a constant reference $y_d(t) = 5$ is given in Fig. 9.4 by applying the distributed controller 1 in Eq. (9.6). Fig. 9.4 shows that the proposed controller is able to drive multiple agents' outputs to track a constant desired value. It can be seen that without perturbation, all agents' outputs are synchronized to the desired value precisely. In other words, the distributed controller 1 in Eq. (9.6) can drive agents' outputs to any fixed value exactly. Moreover, due to $p \geq 1$, the derivatives varied tremendously, thus all agents converge to the reference rapidly.

9.4 Conclusion remarks

In this chapter, a distributed controller is designed recursively via backstepping technique for a network of nonlinear high-order power integrators to achieve the robust cooperative output tracking to an either time-varying or time-invariant trajectory. It is proven that the tracking error is ultimately uniformly bounded under some graph structural properties, i.e., the communication graph is undirected and connected and there is at least one follower that can receive the leader's information even under external disturbances and model uncertainties. Our future work will focus on relaxing the restriction on identical odd integers p and on the case with switching topologies.

Robust cooperative control of networked negative-imaginary systems

10

10.1 Problem formulation

10.1.1 NI system definition

We consider the first order transfer function $P_1(s) = \frac{b}{\tau s+1}$ where $b > 0$, $\tau > 0$, and the frequency characteristics of the transfer function can be described by the Bode diagram or the Nyquist diagram, as shown in Fig. 10.1. Similarly, for the second order transfer function $P_2(s) = \frac{\psi_i^2}{s^2+2\zeta_i\omega_i s+\omega_i^2}$ where $\zeta_i \geq 0$, $\omega_i > 0$, and the Bode and Nyquist diagrams are illustrated in Fig. 10.2.

It can be observed from Fig. 10.1 and Fig. 10.2 that the phase angle and imaginary part of the transfer functions satisfy $\angle P_i(j\omega) \in (0, -\pi)$, $\omega \in (0, \infty)$ and $\mathrm{Im}\left[P_i(j\omega)\right] < 0$, $\omega \in (0, \infty)$, $i = 1, 2$, respectively, which define the negative-imaginary (NI) property for a linear SISO system. In addition, a general definition for a MIMO NI system is given as follows.

Definition 10.1 ([109]). A square, real, rational, proper transfer function $P(s)$ is NI if the following conditions are satisfied:

1. $P(s)$ has no pole in $\mathrm{Re}[s] > 0$;
2. $\forall \omega > 0$ such that $j\omega$ is not a pole of $P(s)$, $j(P(j\omega) - P(j\omega)^*) \geq 0$;

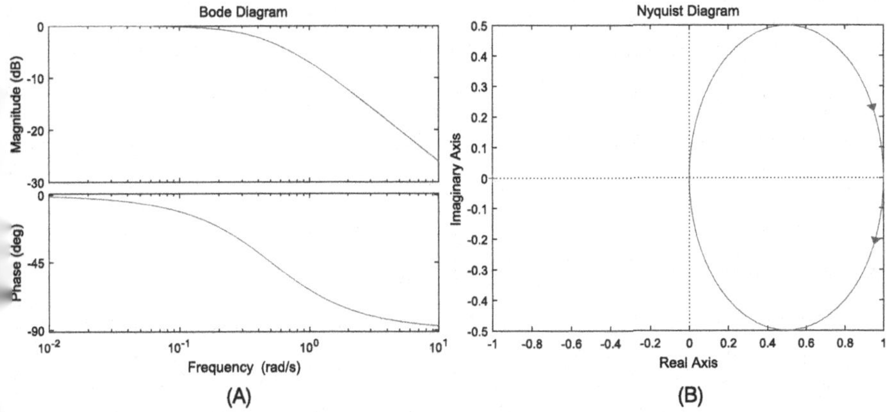

Figure 10.1 The frequency characteristics of $G_1(s) = \frac{1}{2s+1}$. (A) Bode diagram. (B) Nyquist diagram.

Cooperative Control of Multi-Agent Systems. https://doi.org/10.1016/B978-0-12-820118-3.00021-1

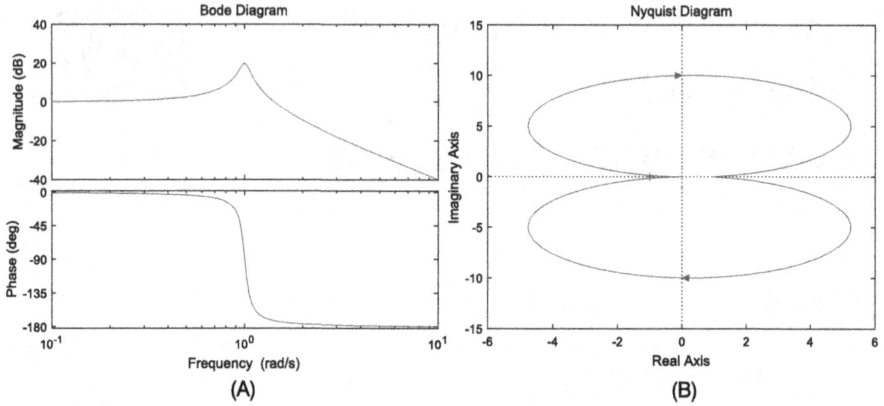

Figure 10.2 The frequency characteristics of $G_2(s) = \frac{1}{s^2+0.1s+1}$. (A) Bode diagram. (B) Nyquist diagram.

3. If $s = j\omega_0$ with $\omega_0 > 0$ is a pole of $P(s)$, then it is a simple pole and the residue matrix $K = \lim_{s \to j\omega_0} (s - j\omega_0) jP(s)$ is Hermitian and positive semi-definite;
4. If $s = 0$ is a pole of $P(s)$, then $\lim_{s\to 0} s^k P(s) = 0$, $\forall k \geq 3$ and $P_2 = \lim_{s\to 0} s^2 P(s)$ is Hermitian and positive semi-definite.

In Definition 10.1, condition 1 indicates that NI systems are stable in the view of the Lyapunov theory. Conditions 2 and 3 together define the phase frequency characteristics of the NI systems. For SISO systems, $j\left[G(j\omega) - G^*(j\omega)\right] = -2\,\mathrm{Im}[G(j\omega)] \geq 0$, i.e., $\mathrm{Im}\left[G_i(j\omega)\right] \leq 0$, which can be used to verify whether a transfer function is NI or not. In addition, the definition of a strictly negative imaginary (SNI) system is given as follows.

Definition 10.2 ([83]). A square, real, rational, proper transfer function $P_s(s)$ is SNI if the following conditions are satisfied:

1. $P_s(s)$ has no pole in $\mathrm{Re}[s] \geq 0$;
2. $\forall \omega > 0$, $j(P_s(j\omega) - P_s(j\omega)^*) > 0$.

According to the definition, transfer functions $G_1(s)$ and $G_2(s)$ considered at the beginning of this section are SNI. $G_3(s) = \frac{2s^2+s+1}{(s^2+2s+5)(s+1)(2s+1)}$ shown in Fig. 10.3 is NI, but not SNI.

10.1.2 Typical results for NI systems

The next question is how to judge whether a transfer function is NI? Currently, there are four ways: The first one is to use the definition directly. This method is effective for a simple SISO system, such as a first-order or second-order system. This method is not feasible for a MIMO system, a high-order system and a system whose transfer function has poles on the imaginary axis. The second one is to use the Bode diagram

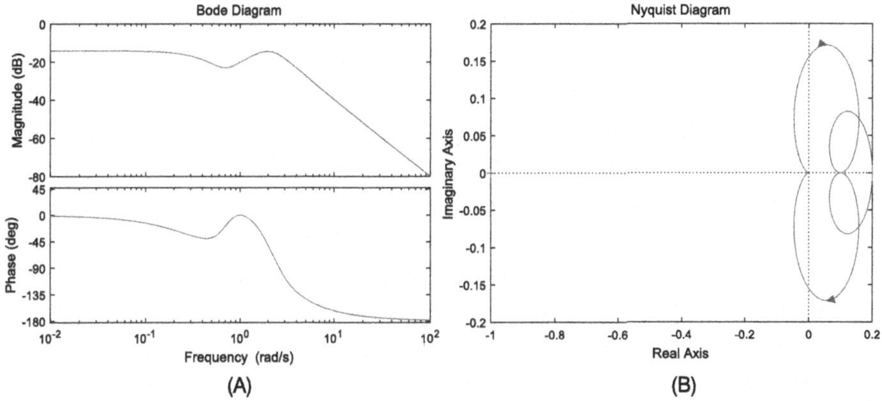

Figure 10.3 The frequency characteristics of $G_3(s) = \frac{2s^2+s+1}{(s^2+2s+5)(s+1)(2s+1)}$. (A) Bode diagram. (B) Nyquist diagram.

or the Nyquist diagram to judge. The method is very visual and intuitive, but only can be applied to a SISO system without a pole on the imaginary axis. The third one is based on the 'enumeration' numerical verification, which tries to enumerate all the $\omega \in (0, \infty)$ to examine whether the condition in the definition is satisfied. Since all of them cannot be enumerated and poles on the imaginary axis cannot be effectively processed, this method cannot guarantee the reliability theoretically. The fourth one is to judge the NI property of the transfer function according to its state space realization. Such results are easy verified numerically, and the reliability can be guaranteed, which is shown in the following lemma.

Lemma 10.1 (NI Lemma). *Let (A, B, C, D) be the minimum realization of the transfer function $P(s)$, where $A \in \mathbb{R}^{n\times n}$, $B \in \mathbb{R}^{n\times m}$, $C \in \mathbb{R}^{m\times n}$, $D \in \mathbb{R}^{m\times m}$, $m \leq n$. $P(s)$ is NI if and only if*

1. $\det(A) \neq 0$, $D = D^T$;
2. *There exists symmetric positive definite matrix $Y \in \mathbb{R}^{n\times n}$ satisfying $AY + YA^T \leq 0$, $B + AYC^T = 0$.*

The above lemma, also known as NI lemma, provides a method based on the state space realization to determine whether a dynamic system satisfies the NI property. The condition 1 of Lemma 10.1 is easy to verify, and the feasible solution set of condition 2 is a convex set. When the system order is not high, it can be solved directly using MATLAB®'s YALMIP toolbox. When the system order is high, it can be solved using a numerical iterative algorithm. For a nonminimal realization, the two conditions in Lemma 10.1 are sufficient conditions for determining a NI system.

The stability of a NI system connected by positive feedback is called NI systems theory. Considering the stability issues of the interconnect system as shown in Fig. 10.4, the following stability conditions can be established.

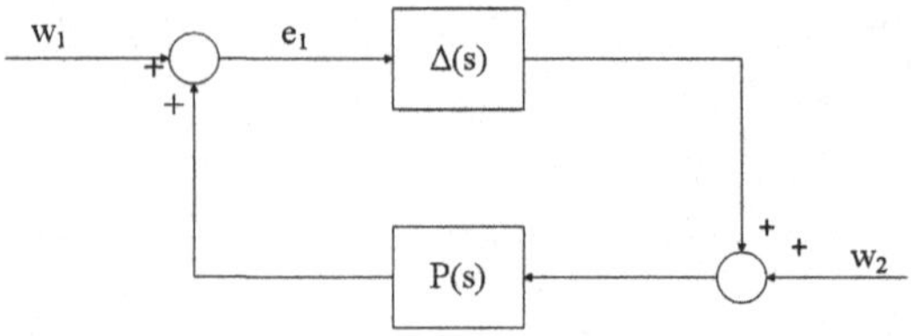

Figure 10.4 The stability of NI systems.

Lemma 10.2 (NI system theorem). *Let $\Delta(s)$ be a strictly NI system, $P(s)$ be a NI system, $\Delta(\infty)P(\infty)=0$, $\Delta(\infty)\geq 0$. Then the interconnect system shown in Fig. 10.4 is stable if and only if $\lambda_{\max}(\Delta(0)P(0))<1$.*

Lemma 10.2 shows that if a system has NI property, then the stability of interconnected system depends on the gains of the systems at frequency of zero and infinity. In particular, when the system is a strictly NI system, the stability of system depends only on the gain at zero frequency.

Examples of NI systems can be found in many literatures, including single integrator systems, double integrator systems, second order systems (such as those found in undamped and damped flexible structures or inertial systems), etc. The motion loop of aircraft system can be modeled as the double integrator system.

10.2 Robust consensus control for multi-NI systems

10.2.1 Homogeneous NI systems

In this section, a homogeneous network of NI systems and a fixed communication topology are assumed. The i-th NI system is described in the s-domain as

$$\boldsymbol{y}_i = P(s)\boldsymbol{u}_i,\ i=1,\cdots,n, \tag{10.1}$$

where $P(s)$ is the transfer function (generally MIMO). $\boldsymbol{y}_i\in\mathbb{R}^{m\times 1}$ and $\boldsymbol{u}_i\in\mathbb{R}^{m\times 1}$ are the output and input of the system with the dimension $m\geq 1$. $n>1$ is the number of agents. Since $P(s)$ is in general a MIMO plant, the Laplacian matrix describing the network interconnection is modified via the Kronecker product to $\mathcal{L}_n\otimes I_m$ and the total networked plant under consideration is depicted in Fig. 10.5 with

$$\tilde{\boldsymbol{y}} = \bar{P}(s)\boldsymbol{u} = (\mathcal{L}_n\otimes I_m)(I_n\otimes P(s))\boldsymbol{u} = (\mathcal{L}_n\otimes P(s))\boldsymbol{u}, \tag{10.2}$$

where $\bar{P}(s)$ is the augmented plant, $\boldsymbol{y}=[\boldsymbol{y}_1^T,\cdots,\boldsymbol{y}_n^T]^T\in\mathbb{R}^{nm\times 1}$ and $\boldsymbol{u}=[\boldsymbol{u}_1^T,\cdots,\boldsymbol{u}_n^T]^T\in\mathbb{R}^{nm\times 1}$.

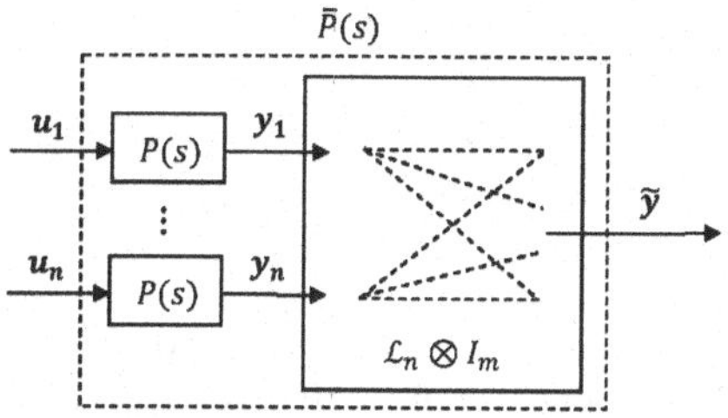

Figure 10.5 Networked NI system.

10.2.2 Robust output feedback consensus algorithm for homogeneous multi-agent systems

In general, the robust output feedback consensus is defined as follows.

Definition 10.3. A distributed output feedback control law achieves *robust output feedback consensus* for a network of systems when:

1. output consensus is achieved, i.e., $\boldsymbol{y}_i \to \boldsymbol{y}_{ss}$, $\forall i \in \{1, \cdots, n\}$ for a family of plant dynamics with no external disturbance, where $\boldsymbol{y}_{ss}$ is the final convergence trajectory;
2. $\boldsymbol{y}_{ss}$ is perturbed by additive $\mathfrak{L}_2[0, \infty]$ signals when $\mathfrak{L}_2[0, \infty]$ disturbances are present on both input and output.

It can be seen that the output $\boldsymbol{y}$ reaches consensus when $\tilde{\boldsymbol{y}} \to \boldsymbol{0}$ via the properties of the Laplacian matrix in Chapter 2. This formulation actually converts the output consensus problem to an internal stability problem which is easier to tackle and investigate the robustness property via standard control theoretic methods. We now impose the following standing assumption.

Assumption 10.1. $\mathcal{G}$ is undirected and connected.

The following preliminary lemmas are needed.

Lemma 10.3 ([84]). *Let λ_j and γ_k, $j = 1, \cdots, n$, $k = 1, \cdots, m$, be eigenvalues of matrices $\Lambda_{n\times n}$ and $\Gamma_{m\times m}$ respectively. Then the eigenvalues of $\Lambda \otimes \Gamma$ are $\lambda_j \gamma_k$.*

Note that Lemma 10.3 also applies to singular values [84].

Lemma 10.4. *Given $\Lambda \in \mathbb{R}^{n\times n}$ and $\Gamma \in \mathbb{R}^{m\times m}$, we have*

$$\mathbb{N}(\Lambda \otimes \Gamma) = \{\boldsymbol{a} \otimes \boldsymbol{b} : \boldsymbol{b} \in \mathbb{R}^{m\times 1}, \boldsymbol{a} \in \mathbb{N}(\Lambda)\} \cup \{\boldsymbol{c} \otimes \boldsymbol{d} : \boldsymbol{c} \in \mathbb{R}^{n\times 1}, \boldsymbol{d} \in \mathbb{N}(\Gamma)\},$$

where $\mathbb{N}(\Lambda \otimes \Gamma)$ denotes the null space of the matrix $\Lambda \otimes \Gamma$.

Proof. The proof simply follows from the definition of the null space and the properties of the Kronecker product. □

The following lemma states that the augmented networked plant $\bar{P}(s) = \mathcal{L}_n \otimes P(s)$ is NI if and only if every single system $P(s)$ is NI.

Lemma 10.5. *$\bar{P}(s)$ is NI if and only if $P(s)$ is NI.*

Proof. First note that $\mathcal{L}_n \geq 0$ due to Assumption 10.1 in Eq. (10.2). Then, the sufficiency and necessity are straightforward by applying Lemma 10.3 to Definition 10.1. □

Since Lemma 10.5 requires positive semi-definiteness of $\mathcal{L}_n$, this work cannot be applied to directed graphs. The output $\tilde{\boldsymbol{y}} \to \boldsymbol{0}$ if internal stability is achieved for $\bar{P}(s)$ with some controller. From [83], [195] and [109], the following internal stability results are summarized.

Lemma 10.6. *Given an NI transfer function $P(s)$ and an SNI transfer function $P_s(s)$ with $P_2 = \lim_{s\to 0} s^2 P(s)$, $P_1 = \lim_{s\to 0} s(P(s) - \frac{P_2}{s^2})$ and $P_0 = \lim_{s\to 0}(P(s) - \frac{P_2}{s^2} - \frac{P_1}{s})$, the positive feedback interconnection $[P(s), P_s(s)]$ is internally stable if and only if any of the following conditions is satisfied:*

1. *$\bar{\lambda}(P(0)P_s(0)) < 1$ when $P(s)$ has no pole(s) at the origin, $P(\infty)P_s(\infty) = 0$ and $P_s(\infty) \geq 0$;*
2. *$J^T P_s(0) J < 0$ when $P(s)$ has pole(s) at the origin and is strictly proper, $P_2 \neq 0$, $P_1 = 0$, $\mathbb{N}(P_2) \subseteq \mathbb{N}(P_0^T)$, where $P_2 = JJ^T$ with J having full column rank;*
3. *$F_1^T P_s(0) F_1 < 0$ when $P(s)$ has pole(s) at the origin and is strictly proper, $P_2 = 0$, $P_1 \neq 0$, $\mathbb{N}(P_1^T) \subseteq \mathbb{N}(P_0^T)$, where $P_1 = F_1 V_1^T$ with F_1 and V_1 having full column rank and $V_1^T V_1 = I$.*

Note that the above result is actually a robust stability result because an NI plant $P(s)$ can be perturbed by any unmodeled dynamics $\Delta(s)$ such that the perturbed plant $P_\Delta(s)$ which then replaces the nominal plant $P(s)$ in Lemma 10.6 retains the NI system property and still fulfills any one of the conditions in Lemma 10.6. Similarly, $P_s(s)$ can be perturbed to any SNI controller subject to 1), 2), 3). Henceforth, we do not distinguish between $P(s)$ and $P_\Delta(s)$ for simplicity of notation, though it is stressed that $P(s)$ could be the resulting perturbed dynamics of some simpler nominal plant. There is clearly a huge class of permissible dynamic perturbations to the nominal dynamics as conditions 1), 2) and 3) impose a restriction on $P(s)$ only at the frequency $\omega = 0$ or on the associated residues of $P(s)$ at $\omega = 0$ and the NI class has no gain or order restriction [83]. A few examples of permissible perturbations are additive perturbations where the uncertainty is also NI [83], feedback perturbations where both systems in the feedback interconnection are NI [137] and more general perturbations based on Redheffer star products and linear fractional transformations [137]. For example, $\frac{1}{s+5}$ and $\frac{(2s^2+s+1)}{(s^2+2s+5)(s+1)(2s+1)}$ are both NI with the same D.C. gain.

Now, we are ready to state the first main result of this chapter.

Theorem 10.1. *Given a graph $\mathcal{G}$ which satisfies Assumption 10.1 and models the communication links for networked homogeneous NI systems. Given any SNI control law $P_s(s)$, robust output feedback consensus is achieved via the protocol*

$$\boldsymbol{U}_{cs} = \bar{P}_s(s)\tilde{\boldsymbol{y}} = C_{cs}(s)\boldsymbol{y} = (\mathcal{L}_n \otimes P_s(s))\boldsymbol{y} \tag{10.3}$$

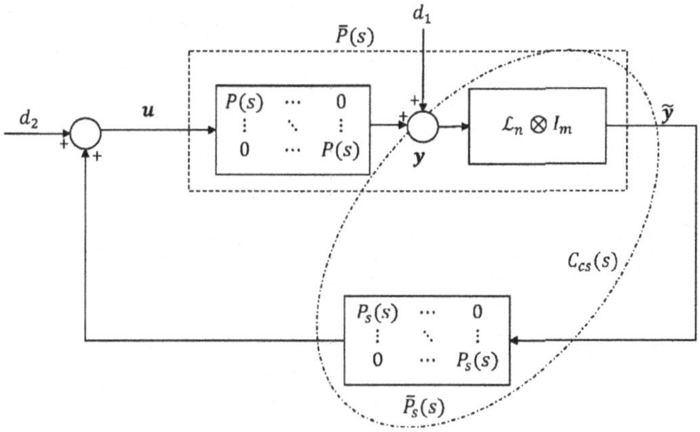

Figure 10.6 Closed-loop system with SNI controllers.

shown in Fig. 10.6, or in a distributed manner for each agent i,

$$\boldsymbol{u}_i = P_s(s)\sum_{j=1}^{n} a_{ij}(\boldsymbol{y}_i - \boldsymbol{y}_j), \tag{10.4}$$

under any external disturbances $\boldsymbol{d}_1, \boldsymbol{d}_2 \in \mathfrak{L}_2[0,\infty)$ *and any model uncertainty which retains the NI system property of the perturbed plant* $P(s)$ *if and only if* $P(s)$ *and* $P_s(s)$ *satisfy 1), 2) or 3) in Lemma 10.6 except that*

$$\bar{\lambda}(P(0)P_s(0)) < \frac{1}{\bar{\lambda}(\mathcal{L}_n)} \tag{10.5}$$

replaces $\bar{\lambda}(P(0)P_s(0)) < 1$ *in case 1).*

Proof. Before presenting the consensus result, let us first prove the internal stability of $[\bar{P}(s), \bar{P}_s(s)]$ using Lemma 10.6. From Fig. 10.6, we have $\bar{P}(s) = \mathcal{L}_n \otimes P(s)$ which has been shown to be NI in Lemma 10.5 and it is straightforward to see that $\bar{P}_s(s) = I_n \otimes P_s(s)$ is SNI since $P_s(s)$ is SNI.

($\Leftarrow$) Sufficiency: From Lemma 10.6, we can conclude that $[\bar{P}(s), \bar{P}_s(s)]$ is internally stable since:

1. When $P(s)$ has no pole(s) at the origin, $\bar{P}(s)$ has no pole(s) at the origin as well. Also, $\bar{P}(\infty)\bar{P}_s(\infty) = (\mathcal{L}_n \otimes P(\infty))(I_n \otimes P_s(\infty)) = \mathcal{L}_n \otimes (P(\infty)P_s(\infty)) = 0$ and $\bar{P}_s(\infty) = \mathcal{L}_n \otimes P_s(\infty) \geq 0$ due to Lemma 10.3 as well as the pair of $[P(s), P_s(s)]$ satisfies condition 1) of Lemma 10.6. Finally, $\bar{\lambda}(\bar{P}(0)\bar{P}_s(0)) = \bar{\lambda}(\mathcal{L}_n \otimes (P(0)P_s(0))) < 1$ since $\bar{\lambda}(P(0)P_s(0)) < \frac{1}{\bar{\lambda}(\mathcal{L}_n)}$ due to Lemma 10.3.
2. When $P(s)$ has pole(s) at the origin, $\bar{P}(s)$ has pole(s) at the origin as well. In the case of $P_2 \neq 0$, $P_1 = 0$, it is straightforward to see that $\mathbb{N}(\mathcal{L}_n \otimes P_2) = \mathbb{N}(\bar{P}_2) \subseteq \mathbb{N}(\bar{P}_0^T) = \mathbb{N}(\mathcal{L}_n \otimes P_0^T)$ due to Lemma 10.4 and $\mathbb{N}(P_2) \subseteq \mathbb{N}(P_0^T)$. Furthermore, $\bar{P}_2 = \mathcal{L}_n \otimes P_2 = (J_{\mathcal{L}}J_{\mathcal{L}}^T) \otimes (JJ^T) = (J_{\mathcal{L}} \otimes J)(J_{\mathcal{L}} \otimes J)^T = \bar{J}\bar{J}^T$

since $\mathcal{L}_n$ and P_2 are both Hermitian and positive semi-definite, where $J_{\mathcal{L}}$ has full column rank being $n-1$. With the definition of $\bar{J} = J_{\mathcal{L}} \otimes J$, we have $\bar{J}^T \bar{P}_s(0)\bar{J} = (J_{\mathcal{L}} \otimes J)^T(I_n \otimes P_s(0))(J_{\mathcal{L}} \otimes J) = (J_{\mathcal{L}}^T I_n J_{\mathcal{L}}) \otimes (J^T P_s(0)J) = (J_{\mathcal{L}}^T J_{\mathcal{L}}) \otimes (J^T P_s(0)J) < 0$ since $J_{\mathcal{L}}^T J_{\mathcal{L}} > 0$ (with full rank of $n-1$) as well as Lemma 10.3 and condition 2) of Lemma 10.6.

3. The case of $P_2 = 0$, $P_1 \neq 0$ follows in the similar manner as case 2) by noting that $\bar{F}_1 = J_{\mathcal{L}} \otimes F_1$.

($\Rightarrow$) Necessity is trivial by reversing the above arguments.

The internal stability of $[\bar{P}(s), \bar{P}_s(s)]$ implies output consensus when $\boldsymbol{d}_1 = \boldsymbol{d}_2 = \boldsymbol{0}$, by noting that $\tilde{\boldsymbol{y}} \to \boldsymbol{0} \Longleftrightarrow \boldsymbol{y} \to \boldsymbol{1}_n \otimes \boldsymbol{y}_{ss}$, i.e., $\boldsymbol{y}_i \to \boldsymbol{y}_{ss} \in \mathbb{R}^{m\times 1}$, which is the null space of $\mathcal{L}_n \otimes I_m$ when $\mathcal{G}$ is undirected and connected.

Robustness to model uncertainty which retains the NI property of $P(s)$ is assured as the result is applicable to any NI plant $P(s)$. Furthermore, the external disturbances $\boldsymbol{d}_2, \boldsymbol{d}_1$ in Fig. 10.6 on input $\boldsymbol{u}$ and output $\boldsymbol{y}$ are equivalent to $\boldsymbol{d}_2$, $(\mathcal{L}_n \otimes I_m)\boldsymbol{d}_1$ on input $\boldsymbol{u}$ and output $\tilde{\boldsymbol{y}}$, which is a subset of $\mathfrak{L}_2$ disturbances. Hence, the control protocol (10.3) or (10.4) will achieve a perturbed $\mathfrak{L}_2$ consensus signal on output $\boldsymbol{y}$ (due to superposition principle of linear systems) for all $\mathfrak{L}_2$ disturbances $\boldsymbol{d}_1, \boldsymbol{d}_2$. □

Remark 10.1. It can be seen that the condition in inequality (10.5) is stricter than that in the inequality of case 1) of Lemma 10.6 due to the network interconnection. If originally $P_s(0)$ was such that $0 < \bar{\lambda}(P(0)P_s(0)) < 1$, the controller $P_s(0)$ needs to be tuned for smaller eigenvalues in order to satisfy inequality (10.5). On the other hand, if $\bar{\lambda}(P(0)P_s(0)) < 0$, there is no need to tune further.

From Fig. 10.6 and [216], it is convenient to define the input loop transfer matrix, $L_i = -(I_n \otimes P_s(s))(\mathcal{L}_n \otimes P(s)) = -\mathcal{L}_n \otimes (P_s(s)P(s))$, and output loop transfer matrix, $L_o = -(\mathcal{L}_n \otimes P(s))(I_n \otimes P_s(s)) = -\mathcal{L}_n \otimes (P(s)P_s(s))$, respectively. The input and output sensitivity matrices are defined as $S_i = (I + L_i)^{-1}$ and $S_o = (I + L_o)^{-1}$. If the closed-loop system is internally stable, the following equations hold:

$$\begin{aligned} \tilde{\boldsymbol{y}} &= S_o(\mathcal{L}_n \otimes I_m)\boldsymbol{d}_1 + S_o(\mathcal{L}_n \otimes P(s))\boldsymbol{d}_2 \\ \boldsymbol{u} &= S_i(\mathcal{L}_n \otimes P_s(s))\boldsymbol{d}_1 + S_i\boldsymbol{d}_2. \end{aligned} \tag{10.6}$$

Good robustness to high frequency unmodeled dynamics is given by the condition in [216]:

$$\begin{aligned} &\overline{\sigma}(-\mathcal{L}_n \otimes P(j\omega)P_s(j\omega)) \ll 1,\ \overline{\sigma}(-\mathcal{L}_n \otimes P_s(j\omega)P(j\omega)) \ll 1 \text{ and} \\ &\overline{\sigma}(-I_m \otimes P_s(j\omega)) \ll M \\ \Longleftrightarrow &\overline{\sigma}(P(j\omega)P_s(j\omega)) \ll \frac{1}{\overline{\sigma}(\mathcal{L}_n)},\ \overline{\sigma}(P_s(j\omega)P(j\omega)) \ll \frac{1}{\overline{\sigma}(\mathcal{L}_n)} \text{ and} \\ &\overline{\sigma}(P_s(j\omega)) \ll \frac{M}{\overline{\sigma}(\mathcal{L}_n)} \end{aligned} \tag{10.7}$$

where M is sufficiently small and $\overline{\sigma}(\mathcal{L}_n) = \bar{\lambda}(\mathcal{L}_n)$ for the undirected and connected graph.

Remark 10.2. Inequality (10.7) implies that the robust condition for networked systems is always more stringent than that for a single system by noting that $\overline{\sigma}(\mathcal{L}_n) = \overline{\lambda}(\mathcal{L}_n) > 1$ ([57]), which reaffirms the result of [89].

10.2.3 Convergence set study

Section 10.2.2 provides a class of general robust output feedback consensus protocols that guarantee the convergence of the NI systems' outputs $\boldsymbol{y}_i$ under external disturbances as well as the NI model uncertainty. This section mainly aims at investigating the steady state nominal values of $\boldsymbol{y}_{ss}$ under the proposed output feedback consensus protocol. In order to specify the exact convergence set, the external disturbances and model uncertainty will not be considered in this section.

Given a minimal realization of the i-th NI plant $P(s)$,

$$\begin{cases} \dot{\boldsymbol{x}}_i^{p\times 1} = A^{p\times p}\boldsymbol{x}_i^{p\times 1} + B^{p\times m}\boldsymbol{u}_i^{m\times 1} \\ \boldsymbol{y}_i^{m\times 1} = C^{m\times p}\boldsymbol{x}_i^{p\times 1} + D^{m\times m}\boldsymbol{u}_i^{m\times 1} \end{cases}, \quad i = 1, \cdots, n, \tag{10.8}$$

and a minimal realization of the i-th SNI controller $P_s(s)$,

$$\begin{cases} \dot{\bar{\boldsymbol{x}}}_i^{q\times 1} = \bar{A}^{q\times q}\bar{\boldsymbol{x}}_i^{q\times 1} + \bar{B}^{q\times m}\bar{\boldsymbol{u}}_i^{m\times 1} \\ \bar{\boldsymbol{y}}_i^{m\times 1} = \bar{C}^{m\times q}\bar{\boldsymbol{x}}_i^{q\times 1} + \bar{D}^{m\times m}\bar{\boldsymbol{u}}_i^{m\times 1} \end{cases}, \quad i = 1, \cdots, n, \tag{10.9}$$

where p and q are the dimensions of the states of the NI plant and the SNI controller, respectively. The closed-loop system of Fig. 10.6 is given as

$$\begin{bmatrix} \dot{\bar{\boldsymbol{x}}} \\ \dot{\boldsymbol{x}} \end{bmatrix} = \begin{bmatrix} I_n \otimes \bar{A} + \mathcal{L}_n \otimes \bar{B}D\bar{C} & \mathcal{L}_n \otimes \bar{B}C \\ I_n \otimes B\bar{C} & I_n \otimes A + \mathcal{L}_n \otimes B\bar{D}C \end{bmatrix} \begin{bmatrix} \bar{\boldsymbol{x}} \\ \boldsymbol{x} \end{bmatrix} \triangleq \Psi \begin{bmatrix} \bar{\boldsymbol{x}} \\ \boldsymbol{x} \end{bmatrix}. \tag{10.10}$$

The spectrum of Ψ is of importance since it will determine the equilibria. In particular, in this work, the eigenvalues of Ψ on the imaginary axis will determine the steady-state behavior. To this end, the following lemma is given to characterize the spectrum of Ψ.

Lemma 10.7. *Let $\lambda_{\mathcal{L}}^i$ be the i-th eigenvalue of $\mathcal{L}_n$ associated with eigenvector $\boldsymbol{v}_{\mathcal{L}}^i$. The spectrum of Ψ is given by the union of spectra of the following matrices:*

$$\psi_i = \begin{bmatrix} \bar{A} + \lambda_{\mathcal{L}}^i \bar{B}D\bar{C} & \lambda_{\mathcal{L}}^i \bar{B}C \\ B\bar{C} & A + \lambda_{\mathcal{L}}^i B\bar{D}C \end{bmatrix}, \quad i = 1, \cdots, n.$$

Furthermore, let $[\boldsymbol{v}_1^{i\,T} \;\; \boldsymbol{v}_2^{i\,T}]^T$ be an eigenvector of ψ_i. Then the corresponding eigenvector of Ψ is $\begin{bmatrix} \boldsymbol{v}_{\mathcal{L}}^i \otimes \boldsymbol{v}_1^i \\ \boldsymbol{v}_{\mathcal{L}}^i \otimes \boldsymbol{v}_2^i \end{bmatrix}$.

Proof. Let λ_{ψ_i} be the eigenvalue of ψ_i and

$$\Psi \begin{bmatrix} v^i_{\mathcal{L}} \otimes v^i_1 \\ v^i_{\mathcal{L}} \otimes v^i_2 \end{bmatrix} = \begin{bmatrix} v^i_{\mathcal{L}} \otimes (\bar{A}v^i_1 + \lambda^i_{\mathcal{L}}\bar{B}D\bar{C}v^i_1 + \lambda^i_{\mathcal{L}}\bar{B}Cv^i_2) \\ v^i_{\mathcal{L}} \otimes (B\bar{C}v^i_1 + Av^i_2 + \lambda^i_{\mathcal{L}}B\bar{D}Cv^i_2) \end{bmatrix}$$
$$= \begin{bmatrix} v^i_{\mathcal{L}} \otimes \lambda_{\psi_i} v^i_1 \\ v^i_{\mathcal{L}} \otimes \lambda_{\psi_i} v^i_2 \end{bmatrix} = \lambda_{\psi_i} \begin{bmatrix} v^i_{\mathcal{L}} \otimes v^i_1 \\ v^i_{\mathcal{L}} \otimes v^i_2 \end{bmatrix}$$

which shows that λ_{ψ_i} is also an eigenvalue of Ψ with the associated eigenvector being $\begin{bmatrix} v^i_{\mathcal{L}} \otimes v^i_1 \\ v^i_{\mathcal{L}} \otimes v^i_2 \end{bmatrix}$. □

It is well known in [148] that there is only one zero eigenvalue in $\mathcal{L}_n$, $\lambda^i_{\mathcal{L}} = 0$, when the graph $\mathcal{G}$ satisfies Assumption 10.1. In this case, ψ_i has eigenvalues λ_A and $\lambda_{\bar{A}}$ associated with eigenvectors $\begin{bmatrix} \mathbf{0} \\ v_A \end{bmatrix}$ and $\begin{bmatrix} v_{\bar{A}} \\ (\lambda_{\bar{A}}I_n - A)^{-1}B\bar{C}v_{\bar{A}} \end{bmatrix}$ respectively since $\psi_i = \begin{bmatrix} \bar{A} & 0 \\ B\bar{C} & A \end{bmatrix}$, where λ_A and $\lambda_{\bar{A}}$ are the eigenvalues of A and $\bar{A}$, v_A and $v_{\bar{A}}$ are the corresponding eigenvectors of A and $\bar{A}$, respectively. This also shows that eigenvalues of Ψ include λ_A and $\lambda_{\bar{A}}$ with the associated eigenvectors being $\begin{bmatrix} \mathbf{0} \\ \mathbf{1} \otimes v_A \end{bmatrix}$ and $\begin{bmatrix} \mathbf{1} \otimes v_{\bar{A}} \\ \mathbf{1} \otimes (\lambda_{\bar{A}}I_n - A)^{-1}B\bar{C}v_{\bar{A}} \end{bmatrix}$. It is worth noting that the invertibility of $A - \lambda_{\bar{A}}I_n$ follows since an SNI controller can always be chosen such that $\lambda_{\bar{A}} \neq \lambda_A$.

In the case of $\lambda^i_{\mathcal{L}} > 0$ and $\det(A) \neq 0$, it can be shown in a similar manner as in Theorem 5 of [83] that

$$\psi_i = \begin{bmatrix} \bar{A} + \lambda^i_{\mathcal{L}}\bar{B}D\bar{C} & \lambda^i_{\mathcal{L}}\bar{B}C \\ B\bar{C} & A + \lambda^i_{\mathcal{L}}B\bar{D}C \end{bmatrix} = \begin{bmatrix} \bar{A} & 0 \\ B\bar{C} & A \end{bmatrix} + \lambda^i_{\mathcal{L}} \begin{bmatrix} \bar{B} \\ B\bar{D} \end{bmatrix} \begin{bmatrix} D\bar{C} & C \end{bmatrix} = \Phi T \tag{10.11}$$

where $T = \begin{bmatrix} \bar{Y}^{-1} - \lambda^i_{\mathcal{L}}\bar{C}^* D\bar{C} & -\lambda^i_{\mathcal{L}}\bar{C}^* C \\ -C^*\bar{C} & Y^{-1} - \lambda^i_{\mathcal{L}}C^*\bar{D}C \end{bmatrix}$ and $\Phi = \begin{bmatrix} \bar{A}\bar{Y} & 0 \\ 0 & AY \end{bmatrix}$. ψ_i is Hurwitz if and only if $\bar{\lambda}(P(0)P_s(0)) < \frac{1}{\lambda^i_{\mathcal{L}}}$, which coincides with the condition in Theorem 10.1 when $\lambda^i_{\mathcal{L}} = \bar{\lambda}(\mathcal{L}_n)$.

In the case of $\lambda^i_{\mathcal{L}} > 0$ and $\det(A) = 0$, it can be verified in a similar manner as [109] that

$$\psi_i = \begin{bmatrix} \bar{A} & \lambda^i_{\mathcal{L}}\bar{B}C \\ B\bar{C} & A + \lambda^i_{\mathcal{L}}B\bar{D}C \end{bmatrix} \tag{10.12}$$

due to $D = 0$. ψ_i is also Hurwitz when the conditions 2) and 3) in Lemma 10.6 hold.

One direct observation from the above analysis: the number of eigenvalues of Ψ on the imaginary axis is equal to the number of eigenvalues of A on the imaginary axis and all of the other eigenvalues lie in the open left half plane (OLHP) since $\bar{A}$ is Hurwitz [195]. Thus, the steady state of the closed-loop system (10.10) in general depends only on the eigenvalues of A on the imaginary axis as shown in the following theorem.

Theorem 10.2. *Given the closed-loop system in Eq. (10.10), the steady state can be expressed in the general form*

$$\begin{bmatrix} \bar{\boldsymbol{x}}(t) \\ \boldsymbol{x}(t) \end{bmatrix} \xrightarrow{t \to \infty} \begin{bmatrix} \boldsymbol{w}_j, & \cdots, & \boldsymbol{w}_k^g \end{bmatrix} e^{J't} \begin{bmatrix} \boldsymbol{v}_j^T \\ \vdots \\ \boldsymbol{v}_k^{gT} \end{bmatrix} \begin{bmatrix} \bar{\boldsymbol{x}}(0) \\ \boldsymbol{x}(0) \end{bmatrix}, \tag{10.13}$$

where J' is the Jordan block associated with n_0 eigenvalues of Ψ on the imaginary axis denoted by λ_A, $\boldsymbol{w}_j$ and $\boldsymbol{v}_j$ are the right and left eigenvectors of Ψ associated with λ_A given by

$$\boldsymbol{w}_j = \begin{bmatrix} \mathbf{0} \\ \mathbf{1} \otimes \boldsymbol{v}_A^r \end{bmatrix}, \boldsymbol{v}_j = \begin{bmatrix} \mathbf{1} \otimes (\frac{1}{n}(\lambda_A I_q - \bar{A})^{-1} \bar{C}^T B^T \boldsymbol{v}_A^l) \\ \mathbf{1} \otimes \frac{1}{n} \boldsymbol{v}_A^l \end{bmatrix}, \tag{10.14}$$

$\forall j = 1, \cdots, n_0 - (n_a - n_g)$, where n_a and n_g denote the algebraic and geometric multiplicity of λ_A respectively. $\boldsymbol{v}_A^r$, $\boldsymbol{v}_A^l$ are the right and left eigenvectors of A associated with λ_A. Moreover, in the case when $n_a > n_g$, $\boldsymbol{w}_k^g$ and $\boldsymbol{v}_k^g$ are the generalized right and left eigenvectors given by

$$\boldsymbol{w}_k^g = \begin{bmatrix} \mathbf{0} \\ \mathbf{1} \otimes \boldsymbol{v}_A^{r_g} \end{bmatrix}, \boldsymbol{v}_k^g = \begin{bmatrix} \mathbf{1} \otimes (\frac{1}{n}(\lambda_A I_q - \bar{A})^{-1} \bar{C}^T B^T \boldsymbol{v}_A^{l_g}) \\ \mathbf{1} \otimes \frac{1}{n} \boldsymbol{v}_A^{l_g} \end{bmatrix}, \tag{10.15}$$

where $k = 1, \cdots, n_a - n_g$, $\boldsymbol{v}_A^{r_g}$ and $\boldsymbol{v}_A^{l_g}$ are the generalized right and left eigenvectors of A associated with λ_A.

Proof. It is straightforward that

$$\begin{bmatrix} \bar{\boldsymbol{x}}(t) \\ \boldsymbol{x}(t) \end{bmatrix} = e^{\Psi t} \begin{bmatrix} \bar{\boldsymbol{x}}(0) \\ \boldsymbol{x}(0) \end{bmatrix} = P e^{Jt} P^{-1} \begin{bmatrix} \bar{\boldsymbol{x}}(0) \\ \boldsymbol{x}(0) \end{bmatrix} \xrightarrow{t \to \infty} P \begin{bmatrix} e^{J't} & 0 \\ 0 & 0 \end{bmatrix} P^{-1} \begin{bmatrix} \bar{\boldsymbol{x}}(0) \\ \boldsymbol{x}(0) \end{bmatrix},$$

where $J'_{r \times r}$ is the Jordan block associated with n_0 eigenvalues on the imaginary axis. Also, $P = [\boldsymbol{w}_1, \cdots, \boldsymbol{w}_{n_0}, \cdots, \boldsymbol{w}_{(p+q)n}]$, where $\boldsymbol{w}_i$ is the right eigenvector of Ψ and $P^{-1} = [\boldsymbol{v}_1, \cdots, \boldsymbol{v}_{n_0}, \cdots, \boldsymbol{v}_{(p+q)n}]^T$, where $\boldsymbol{v}_i$ is the left eigenvector of Ψ.

It can be found, without loss of generality, that the right and left eigenvectors of Ψ associated with the eigenvalues on imaginary axis are given in Eq. (10.14).

Thereby, the steady state generally converges to

$$\begin{bmatrix}\bar{\boldsymbol{x}}(t)\\ \boldsymbol{x}(t)\end{bmatrix}\xrightarrow{t\to\infty}\begin{bmatrix}\boldsymbol{w}_1, & \cdots, & \boldsymbol{w}_{n_0}\end{bmatrix}e^{J't}\begin{bmatrix}\boldsymbol{v}_1^T\\ \vdots\\ \boldsymbol{v}_{n_0}^T\end{bmatrix}\begin{bmatrix}\bar{\boldsymbol{x}}(0)\\ \boldsymbol{x}(0)\end{bmatrix}. \tag{10.16}$$

However, in the case when $n_a > n_g$, the generalized right and left eigenvectors are given in Eq. (10.15). Thus, the steady state converges to Eq. (10.13) instead of Eq. (10.16). □

Next, convergence sets of several special cases of NI systems are given in detail.

Corollary 10.1. *In the case when the NI plant is a single-integrator, i.e.,* $\dot{x}_i = u_i$, $y_i = x_i$, *the convergence set of Eq. (10.10) is* $y_{ss} = -\bar{C}\bar{A}^{-T}\cdot \text{ave}(\bar{\boldsymbol{x}}(0)) + \text{ave}(\boldsymbol{x}(0))$.

Proof. The convergence set can be obtained by noting the eigenvectors $\boldsymbol{w}_j = \begin{bmatrix}\boldsymbol{0}^T & \boldsymbol{1}_n^T\end{bmatrix}^T$, $\boldsymbol{v}_j = \begin{bmatrix}-\frac{1}{n}\bar{C}\bar{A}^{-T}\boldsymbol{1}_n^T & \frac{1}{n}\boldsymbol{1}_n^T\end{bmatrix}^T$ and applying Eq. (10.16) in Theorem 10.2. □

Corollary 10.2. *In the case when the NI plant is a double-integrator, i.e.,* $\dot{\xi}_i = \zeta_i$, $\dot{\zeta}_i = u_i$, $y_i = \xi_i$, *the convergence set of Eq. (10.10) is* $y_{ss} = -\bar{C}\bar{A}^{-T}\cdot \text{ave}(\bar{\boldsymbol{x}}(0))t + \text{ave}(\boldsymbol{\xi}(0)) + \text{ave}(\boldsymbol{\zeta}(0))t$.

Proof. For double-integrator plants, $n_a = 2 > 1 = n_g$ for $\lambda(A) = 0$. The convergence set is straightforward by noting that $\boldsymbol{w}_j = \begin{bmatrix}\boldsymbol{0}^T & \boldsymbol{1}_n^T & \boldsymbol{0}^T\end{bmatrix}^T$, $\boldsymbol{v}_j = \begin{bmatrix}\boldsymbol{0}^T & \frac{1}{n}\boldsymbol{1}_n^T & \boldsymbol{0}^T\end{bmatrix}^T$, $\boldsymbol{w}_k^g = \begin{bmatrix}\boldsymbol{0}^T & \boldsymbol{0}^T & \boldsymbol{1}_n^T\end{bmatrix}^T$, $\boldsymbol{v}_k^g = \begin{bmatrix}-\frac{1}{n}\bar{C}\bar{A}^{-T}\boldsymbol{1}_n^T & \boldsymbol{0}^T & \frac{1}{n}\boldsymbol{1}_n^T\end{bmatrix}^T$ after rearranging $\boldsymbol{x} = [\boldsymbol{\xi}^T\ \boldsymbol{\zeta}^T]^T$ and applying Eq. (10.13). □

Corollary 10.3. *In the case when the NI plant is a damped flexible structure, the convergence set of Eq. (10.10) is* $y_{ss} = 0$.

Proof. This is straightforward and thus omitted. □

10.2.4 Numerical examples

In this section, numerical examples of typical NI systems are given to illustrate the main results of this technical note. A scenario of 3 NI systems is considered and the communication graph $\mathcal{G}$ is given as in Fig. 10.7. Therefore, the Laplacian matrix of $\mathcal{G}$ can be derived according to the definition:

$$\mathcal{L}_3 = \begin{bmatrix}1 & -1 & 0\\ -1 & 2 & -1\\ 0 & -1 & 1\end{bmatrix}$$

10.2.4.1 Multiple single-integrator systems

Suppose that the NI systems have identical single-integrator dynamics as shown in Corollary 10.1 with the initial condition being $\boldsymbol{x}(0) = [1\ 2\ 3]^T$. The SNI controller is designed as indicated in Theorem 10.1 to be $\bar{A} = -2$, $\bar{B} = 1$, $\bar{C} = 1$, $\bar{D} = -1$, with the initial condition being $\bar{\boldsymbol{x}}(0) = [0.1\ 0.2\ 0.3]^T$. Without considering disturbances firstly, it can be verified as Corollary 10.1 that $y_{ss} = -\bar{C}\bar{A}^{-T} \cdot \text{ave}(\bar{\boldsymbol{x}}(0)) + \text{ave}(\boldsymbol{x}(0)) = \frac{1}{2} * 0.2 + 2 = 2.1$, which is shown at the top left of Fig. 10.8. If external disturbances are inserted, robust output feedback consensus is also achieved with the steady state consensus value perturbed by filtered disturbances as shown at the top right of Fig. 10.8. The robust performance of the control law can be improved by tuning the SNI controller to, for example, $\bar{D} = -5$, what is shown in the bottom left and right of Fig. 10.8, respectively.

One may notice that when the initial condition of the controller $\bar{\boldsymbol{x}}(0)$ is set to $\mathbf{0}$ (a reasonable choice as the controller is determined by the designer), the convergence set naturally becomes the centroid of the initial pattern, i.e., $y_{ss} = \text{ave}(\boldsymbol{x}(0))$, which in turn implies that the result for the average consensus protocol in [148] is a special case of the proposed result. Alternatively, the desired convergence point can be chosen by properly initializing the SNI controller, which can be seen as a more general result.

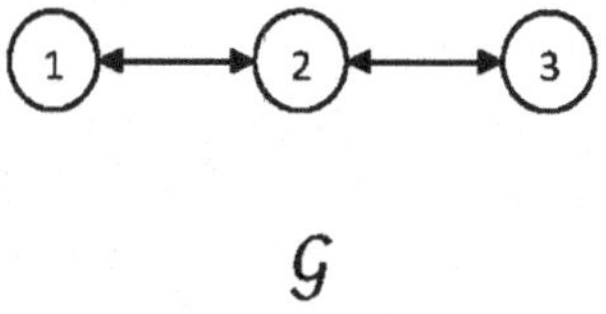

Figure 10.7 Communication topology $\mathcal{G}$ and the associated Laplacian matrix.

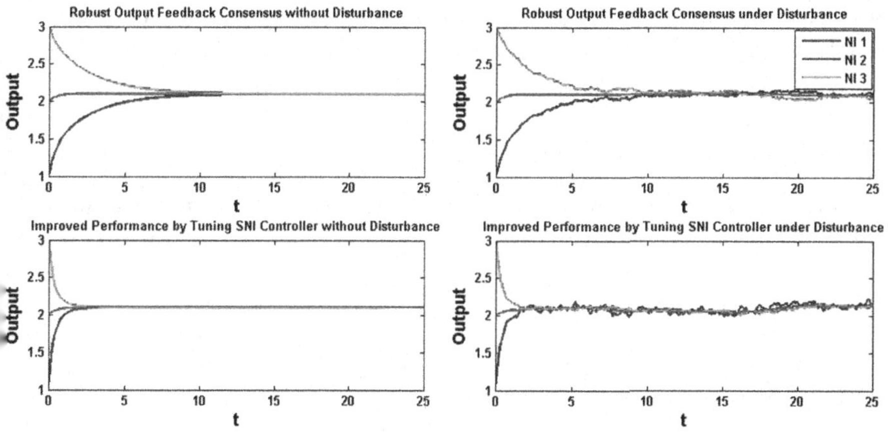

Figure 10.8 Robust output consensus for networked single-integrator systems.

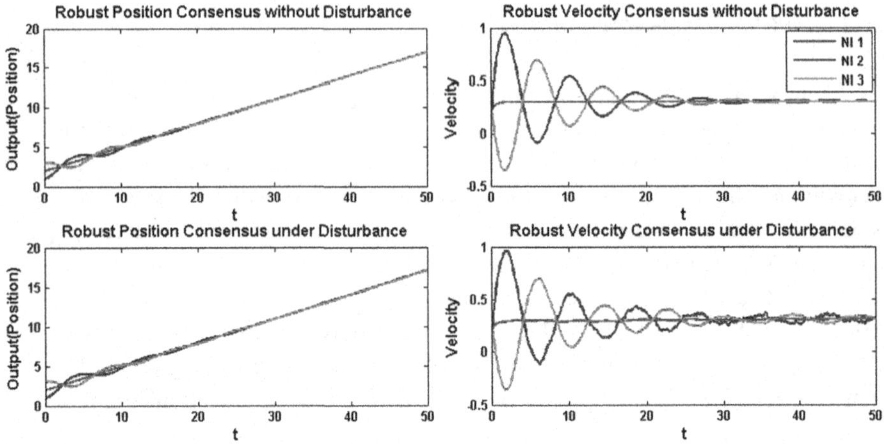

Figure 10.9 Robust output consensus for networked double-integrator systems.

10.2.4.2 *Multiple double-integrator systems*

Suppose that the NI systems have identical double-integrator dynamics as shown in Corollary 10.2 with the initial conditions being $\boldsymbol{\xi}(0) = [1\ 2\ 3]^T$, $\boldsymbol{\zeta}(0) = [0.1\ 0.2\ 0.3]^T$. The same SNI controller can be adopted as in Subsection 10.2.4.1. Without considering disturbances firstly, it can be verified using Corollary 10.2 that $y_{ss} = \xi_i(\infty) = -\bar{C}\bar{A}^{-T} \cdot \text{ave}(\bar{\boldsymbol{x}}(0)) + \text{ave}(\boldsymbol{\xi}(0)) + \text{ave}(\boldsymbol{\zeta}(0))t = \frac{1}{2} * 0.2 + 2 + 0.2t = 2.1 + 0.2t$ and $\zeta_i(\infty) = -\bar{C}\bar{A}^{-T} \cdot \text{ave}(\bar{\boldsymbol{x}}(0)) + \text{ave}(\boldsymbol{\zeta}(0)) = \frac{1}{2} * 0.2 + 0.2 = 0.3$, which is exactly as shown at the top of Fig. 10.9. If the same disturbances as in Subsection 10.2.4.1 are inserted, output consensus is also achieved with the steady state values perturbed by filtered disturbances as shown at the bottom of Fig. 10.9. Again, appropriate choices of the SNI controller can be made to minimize the effects of external disturbances, which is omitted here due to the page limitations.

One can also choose the initial condition of the controller to be $\bar{\boldsymbol{x}}(0) = \mathbf{0}$ to obtain the natural convergence set as $y_{ss} = \xi_{ss} = \text{ave}(\boldsymbol{\xi}(0)) + \text{ave}(\boldsymbol{\zeta}(0))t$ and $\zeta_{ss} = \text{ave}(\boldsymbol{\zeta}(0))$. The same conclusion can hence be drawn as in Subsection 10.2.4.1.

10.2.4.3 *Multiple flexible structures systems*

Suppose that the NI systems are damped flexible structures as shown, for example, in Fig. 2 of [83]: $M\ddot{\boldsymbol{x}}_i + C\dot{\boldsymbol{x}}_i + K\boldsymbol{x}_i = \boldsymbol{u}_i$, $\boldsymbol{y}_i = \boldsymbol{x}_i$, $i = 1, \cdots, 3$ where $\boldsymbol{x}_i = [x_i^{1^T}\ x_i^{2^T}]^T$, $\boldsymbol{u}_i = [u_i^{1^T}\ u_i^{2^T}]^T$, $M = \text{diag}(m_1, m_2)$, $C = \begin{bmatrix} c_1 + c & -c \\ -c & c_2 + c \end{bmatrix}$, $K = \begin{bmatrix} k_1 + k & -k \\ -k & k_2 + k \end{bmatrix}$ with $m_1 = 1$, $m_2 = 0.5$, $k_1 = k_2 = k = 1$ and $c_1 = c_2 = c = 0.1$. The initial conditions are given as

$$\boldsymbol{x}(0) = [1\ 2\ 3\ 4\ 5\ 6]^T \quad \text{and} \quad \dot{\boldsymbol{x}}(0) = [0.1\ 0.2\ 0.3\ 0.4\ 0.5\ 0.6]^T.$$

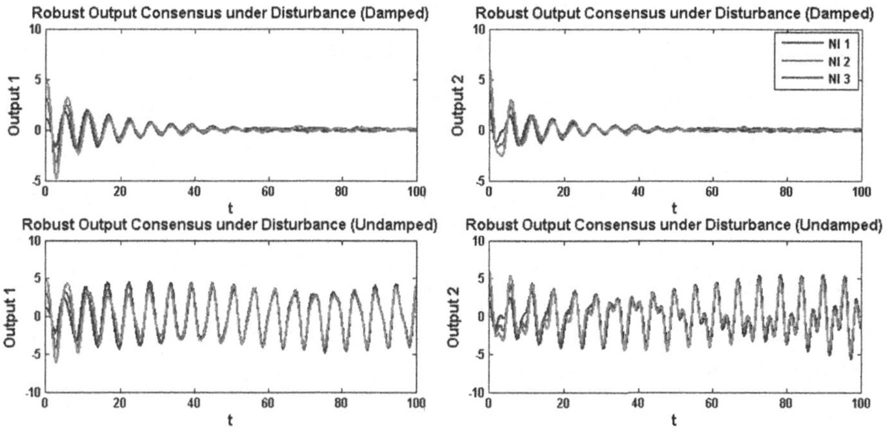

Figure 10.10 Robust output consensus for networked flexible structures.

The SNI controller can be designed as indicated in Theorem 10.1 to be $\bar{A} = -4I_2$, $\bar{B} = I_2$, $\bar{C} = I_2$, $\bar{D} = 0_2$ since $\bar{\lambda}(P(0)) = 1$ and thus $\bar{\lambda}(P(0)P_s(0)) = \frac{1}{4} < \frac{1}{\bar{\lambda}(\mathcal{L}_n)} = \frac{1}{3}$ with the initial condition being $[0\,0\,0\,0\,0\,0]^T$. Robust output feedback consensus can be achieved as shown at the top of Fig. 10.10 under external disturbances, which also validates Corollary 10.3.

If the NI systems are considered as undamped flexible structures as shown in Fig. 2 of [195], which correspond to the above damped flexible structure dynamics without the damping term C, robust output feedback consensus can be achieved as shown at the bottom of Fig. 10.10 under external disturbances.

10.3 Robust consensus control of heterogeneous multi-NI systems

10.3.1 Heterogeneous NI systems

For multiple heterogeneous NI systems (in general MIMO) with $n > 1$ agents, the transfer function of agent $i \in \{1, \cdots, n\}$ is given as

$$\hat{\boldsymbol{y}}_i = \hat{P}_i(s)\hat{\boldsymbol{u}}_i, \tag{10.17}$$

where $\hat{\boldsymbol{y}}_i \in \mathbb{R}^{m_i \times 1}$ and $\hat{\boldsymbol{u}}_i \in \mathbb{R}^{m_i \times 1}$ are the output and input of agent i respectively. In order to deal with the consensus of different dimensional inputs/outputs, $\hat{P}_i(s)$ can be padded with zeros up to $m = \max_{i=1}^{n}\{m_i\}$ and the locations of padding zeros depend on which output needs to be coordinated, for instance, $P_i(s) = \begin{bmatrix} \hat{P}_i(s) & 0 \\ 0 & 0 \end{bmatrix}$ has dimension of m such that the first m_i outputs are to be coordinated, or

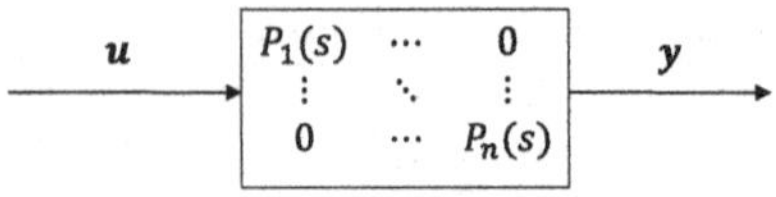

Figure 10.11 Multiple Heterogeneous NI Plants.

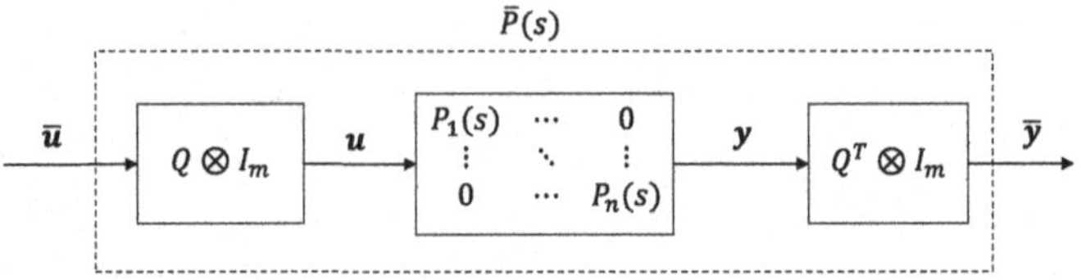

Figure 10.12 Overall network plant.

$P_i(s) = \begin{bmatrix} 0 & 0 \\ 0 & \hat{P}_i(s) \end{bmatrix}$ also has dimension of m, but now the last m_i outputs are to be coordinated instead. Accordingly, the input $\hat{\boldsymbol{u}}_i$ and output $\hat{\boldsymbol{y}}_i$ are extended to be $\boldsymbol{u}_i = \begin{bmatrix} \hat{\boldsymbol{u}}_i^T & 0 \end{bmatrix}^T$ or $\begin{bmatrix} 0 & \hat{\boldsymbol{u}}_i^T \end{bmatrix}^T \in \mathbb{R}^{m \times 1}$, and $\boldsymbol{y}_i = \begin{bmatrix} \hat{\boldsymbol{y}}_i^T & 0 \end{bmatrix}^T$ or $\begin{bmatrix} 0 & \hat{\boldsymbol{y}}_i^T \end{bmatrix}^T \in \mathbb{R}^{m \times 1}$, respectively. Note that interleaving zero rows and corresponding columns within $\hat{P}_i(s)$ is also permissible. It can be easily seen that the above manipulation would preserve the NI property by checking the definition. Therefore, without loss of generality, the overall plant can be described as in Fig. 10.11 where $\boldsymbol{y} = \begin{bmatrix} \boldsymbol{y}_1^T & , \cdots, & \boldsymbol{y}_n^T \end{bmatrix}^T \in \mathbb{R}^{nm \times 1}$ and $\boldsymbol{u} = \begin{bmatrix} \boldsymbol{u}_1^T & , \cdots, & \boldsymbol{u}_n^T \end{bmatrix}^T \in \mathbb{R}^{nm \times 1}$.

We now define robust output feedback consensus as follows.

Definition 10.4. A distributed output feedback control law achieves robust output feedback consensus for a network of systems if for a family of plant dynamics and for all $\mathfrak{L}_2[0, \infty)$ disturbances on the plant input and/or plant output, $\boldsymbol{y}_i - \boldsymbol{y}_{ss} \in \mathfrak{L}_2[0, \infty)$ $\forall i \in \{1, \cdots, n\}$. Here $\boldsymbol{y}_{ss}$ is the final convergence trajectory, which can be a function of time depending on the plant and controller dynamics.

Remark 10.3. Note that since transfer functions in $\mathcal{RH}_\infty$ map $\mathfrak{L}_2[0, \infty)$ to $\mathfrak{L}_2[0, \infty)$ and due to the superposition theorem of linear systems [216], if there are no disturbances, then $\boldsymbol{y}_i - \boldsymbol{y}_{ss} \to \boldsymbol{0}$ $\forall i \in \{1, \cdots, n\}$ in Definition 10.1 retrieving the typical consensus meaning in the literature.

Observe that if one was to construct the overall networked plant dynamics involving the heterogeneous multiple agents $P_i(s)$ and the communications graph was represented by the Laplacian matrix $\mathcal{L}_n$ as $(\mathcal{L}_n \otimes I_m) \cdot \operatorname{diag}_{i=1}^{n}\{P_i(s)\}$, then the overall networked plant would be not NI any more due to asymmetry despite each heterogeneous agent was individually NI. This would then make NI systems theory inapplicable. Instead, we can utilize the incidence matrix $\mathcal{Q}$ instead of $\mathcal{L}_n$ to reformulate the overall networked plant as shown in Fig. 10.12. The reason for adopting the incidence matrix $\mathcal{Q}$ before and after the plant dynamics instead of the Laplacian matrix $\mathcal{L}_n = \mathcal{Q}\mathcal{Q}^T$ totally before or after the plant dynamics is to guarantee that the resultant controller is distributed and only uses local information. This will be explained in more detail later in this section. The augmented system can be derived as

$$\bar{\mathbf{y}} = \bar{P}(s)\bar{\mathbf{u}} = (\mathcal{Q}^T \otimes I_m)\operatorname*{diag}_{i=1}^{n}\{P_i(s)\}(\mathcal{Q} \otimes I_m)\bar{\mathbf{u}} \tag{10.18}$$

where $\bar{\mathbf{y}} = \begin{bmatrix}\bar{\mathbf{y}}_1^T & ,\cdots, & \bar{\mathbf{y}}_l^T\end{bmatrix}^T \in \mathbb{R}^{lm\times 1}$ and $\bar{\mathbf{u}} = \begin{bmatrix}\bar{\mathbf{u}}_1^T & ,\cdots, & \bar{\mathbf{u}}_l^T\end{bmatrix}^T \in \mathbb{R}^{lm\times 1}$ are the output and input vectors for the overall system. It can be concluded that the overall system $\bar{P}(s)$ is still NI due to the following lemmas.

Lemma 10.8. $\operatorname*{diag}_{i=1}^{n}\{P_i(s)\}$ *is NI if and only if* $P_i(s)$ *are all NI* $\forall i \in \{1, \cdots, n\}$.

The proof of Lemma 10.8 is straightforward from the definition of NI systems. The same argument also applies for SNI functions. The following lemma which is also straightforward from the definition of NI systems allows further manipulation.

Lemma 10.9. *Given any NI MIMO* $P(s)$, *we have that* $\bar{P}(s) = FP(s)F^*$ *is NI for any constant matrix* F.

The output $\mathbf{y} \in \mathbb{R}^{nm\times 1}$ reaches consensus (i.e. $\mathbf{y}_i = \mathbf{y}_j\ \forall i, j$) when $\bar{\mathbf{y}} \to \mathbf{0} \in \mathbb{R}^{lm\times 1}$ by noticing the properties of the incidence matrix $\mathcal{Q}$ given in Eq. (10.4). This formulation converts the output consensus problem to an equivalent internal stability problem. Then robustness properties can be studied via standard control theoretic methods to yield robust consensus results. We now impose additional assumption throughout the rest of this chapter.

Assumption 10.2. Let $\Delta_i(s)\ \forall i \in \{1, \cdots, n\}$ be arbitrary SNI systems satisfying $\bar{\lambda}(\Delta_i(0)) < \mu$, $\Delta_i(\infty) = 0\ \forall i \in \{1, \cdots, n\}$, and $0 < \mu \in \mathbb{R}$.

10.3.2 Robust output feedback consensus algorithm for heterogeneous multi-agent systems

In this section, the robust output feedback consensus will be discussed in two directions: NI plants without or with free body dynamics to cover all the heterogeneous types of NI systems.

10.3.2.1 NI plants without free body dynamics

In this subsection, NI plants without free body dynamics will be firstly considered, which also means that $\hat{P}_i(s)$ has no poles at the origin. The following lemmas are needed.

Lemma 10.10 ([56]). *For* $M \in \mathbb{R}^{n\times m}$, *we have* $\bar{\lambda}(MM^T) = \bar{\lambda}(M^T M)$.

Lemma 10.11. *Assume* M *is Hermitian with* $\bar{\lambda}(M) \geq 0$ *and* $N \geq 0$, *we have* $\bar{\lambda}(MN) \leq \bar{\lambda}(M)\bar{\lambda}(N)$.

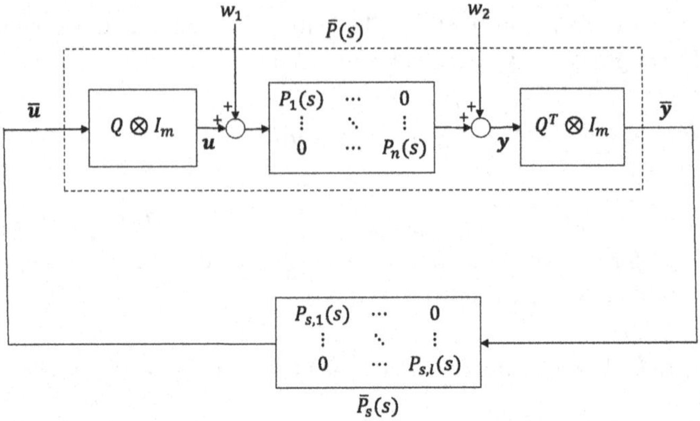

Figure 10.13 Positive feedback interconnection with SNI compensators through the network topology.

Proof. Since $M \leq \bar{\lambda}(M)I$, we obtain $N^{\frac{1}{2}}MN^{\frac{1}{2}} \leq \bar{\lambda}(M)N$. With the condition of $\bar{\lambda}(M) \geq 0$, $N^{\frac{1}{2}}MN^{\frac{1}{2}} \leq \bar{\lambda}(M)N \leq \bar{\lambda}(M)\bar{\lambda}(N)I$. Thus, $\bar{\lambda}(MN) = \bar{\lambda}(N^{\frac{1}{2}}MN^{\frac{1}{2}}) \leq \bar{\lambda}(M)\bar{\lambda}(N)$. □

Lemma 10.12 ([83,195]). *Given an NI transfer function $P(s)$ and an SNI function $P_s(s)$ with $P(s)$ having no pole(s) at the origin, $P(\infty)P_s(\infty) = 0$ and $P_s(\infty) \geq 0$. $[P(s), P_s(s)]$ is internally stable if and only if $\bar{\lambda}(P(0)P_s(0)) < 1$.*

Next we present the following result with the definition of $\bar{P}_s(s) = \operatorname{diag}_{j=1}^{l}\{P_{s,j}(s)\}$ where $P_{s,j}(s)$ are arbitrary SNI compensators.

Theorem 10.3. *Given a graph $\mathcal{G}$ with the incidence matrix $\mathcal{Q}$, satisfying Assumption 10.1 and modeling the communication links among multiple NI agents $\hat{P}_i(s)$ with no pole(s) at the origin which are appropriately padded with rows and columns of zeros to give $P_i(s)$ in Fig. 10.13. The robust output feedback consensus is achieved via the output feedback control law*

$$\boldsymbol{u} = (\mathcal{Q} \otimes I_m)\bar{P}_s(s)(\mathcal{Q}^T \otimes I_m)\boldsymbol{y} \tag{10.19}$$

(or in a distributed manner for agent i via

$$\boldsymbol{u}_i = \sum_{k=1}^{n} a_{ik} P_{s,j}(s)(\boldsymbol{y}_i - \boldsymbol{y}_k), \tag{10.20}$$

where a_{ik} are the elements of the adjacency matrix[1] *and j is the edge connecting vertex i to vertex k) under any external disturbances $w_1 \in \operatorname{Im}_{\mathfrak{L}_2}(\mathcal{Q} \otimes I_m)$ and $w_2 \in \mathfrak{L}_2$*

[1] See [148] for definition.

if $\exists i \in \{1, \cdots, n\} : \bar{\lambda}(P_i(0)) \geq 0$ *and* $\forall i \in \{1, \cdots, n\},\ j \in \{1, \cdots, l\}$ *all the following conditions hold:*

$$\bar{\lambda}(P_i(0))\bar{\lambda}(P_{s,j}(0)) < \frac{1}{\bar{\lambda}(\mathcal{L}_n)}, \tag{10.21}$$

$P_i(\infty)P_{s,j}(\infty) = 0$ *(where* i *is the vertex of edge* j*) and* $P_{s,j}(\infty) \geq 0$*. The output feedback consensus control law (10.19) will be robust to all model uncertainty* $\Delta_i(s), i \in \{1, \cdots, n\}$ *satisfying Assumption 10.2 if the D.C. gain of the SNI compensator* $\bar{P}_s(s)$ *is tuned more stringently such that* $\forall i \in \{1, \cdots, n\},\ j \in \{1, \cdots, l\}$

$$\bar{\lambda}(P_i(0)) + \mu < \frac{1}{\bar{\lambda}(\mathcal{L}_n)\bar{\lambda}(P_{s,j}(0))}. \tag{10.22}$$

Proof. From Fig. 10.13, Lemmas 10.8 and 10.9, it can be seen that $\bar{P}(s)$ is NI without pole(s) at the origin and $\bar{P}_s(s)$ is SNI. Applying Lemma 10.11, we obtain

$$\begin{aligned}
&\bar{\lambda}(\bar{P}(0)\bar{P}_s(0)) \\
=&\bar{\lambda}((\mathcal{Q}^T \otimes I_m)\operatorname{diag}_{i=1}^{n}\{P_i(0)\}(\mathcal{Q} \otimes I_m)\operatorname{diag}_{j=1}^{l}\{P_{s,j}(0)\}) \\
\leq&\bar{\lambda}((\mathcal{Q}^T \otimes I_m)\operatorname{diag}_{i=1}^{n}\{P_i(0)\}(\mathcal{Q} \otimes I_m)) \max_{j=1}^{l}\{\bar{\lambda}(P_{s,j}(0))\} \\
\leq&\max_{i=1}^{n}\{\bar{\lambda}(P_i(0))\}\bar{\lambda}(\mathcal{Q}^T\mathcal{Q}) \max_{j=1}^{l}\{\bar{\lambda}(P_{s,j}(0))\} \quad (\text{since } \exists i : \bar{\lambda}(P_i(0)) \geq 0) \\
=&\max_{i=1}^{n}\{\bar{\lambda}(P_i(0))\} \max_{j=1}^{l}\{\bar{\lambda}(P_{s,j}(0))\}\bar{\lambda}(\mathcal{L}_n) \qquad (\text{by Lemma 10.10})
\end{aligned}$$

since $\bar{\lambda}(\bar{P}(0)) \geq 0$ (because $\exists i : \bar{\lambda}(P_i(0)) \geq 0$) and $P_{s,j}(0) > P_{s,j}(\infty) \geq 0\ \forall j \in \{1, \cdots, l\}$ (due to Lemma 2 in [83] with the assumption of $P_{s,j}(\infty) \geq 0$). Thus, since $\exists i \in \{1, \cdots, n\} : \bar{\lambda}(P_i(0)) \geq 0$ and $\forall i = 1, \cdots, n$ and $j = 1, \cdots, l$, all of the following hold: $\bar{\lambda}(P_i(0))\bar{\lambda}(P_{s,j}(0)) < \frac{1}{\bar{\lambda}(\mathcal{L}_n)}$, $P_i(\infty)P_{s,j}(\infty) = 0$ (where i is the vertex of edge j) and $P_{s,j}(\infty) \geq 0$, $[\bar{P}(s), \bar{P}_s(s)]$ is internally stable via the NI systems theory in Lemma 10.12. This then implies nominal output consensus when the disturbances $\boldsymbol{w}_1$ and $\boldsymbol{w}_2$ are set to zero by noting that $\bar{\boldsymbol{y}} \to \boldsymbol{0} \Leftrightarrow \boldsymbol{y} - \boldsymbol{1}_n \otimes \boldsymbol{y}_{ss} \to \boldsymbol{0}$, i.e., $\boldsymbol{y}_i - \boldsymbol{y}_{ss} \to \boldsymbol{0}$ since the graph $\mathcal{G}$ is undirected and connected.

In addition, the internal stability of $[\bar{P}(s), \bar{P}_s(s)]$ and the superposition principle of linear systems ([216]) guarantee that $\boldsymbol{y}_i \to \boldsymbol{y}_{ss} + \delta$ with $\delta \in \mathfrak{L}_2$ for all $\mathfrak{L}_2$ exogenous signal injections perturbing signals $\bar{\boldsymbol{u}}$ and $\bar{\boldsymbol{y}}$, which in turn means that any $w_1 \in \operatorname{Im}_{\mathfrak{L}_2}(\mathcal{Q} \otimes I_m)$ and any $w_2 \in \mathfrak{L}_2$ can be injected in Fig. 10.13. Hence, the control protocol (10.19) will achieve a perturbed $\mathfrak{L}_2$ consensus signal on output $\boldsymbol{y}$ (due to the superposition principle of linear systems) for all disturbances $w_1 \in \operatorname{Im}_{\mathfrak{L}_2}(\mathcal{Q} \otimes I_m)$ and $w_2 \in \mathfrak{L}_2$.

Additive model uncertainties $\Delta_i(s)\ \forall i \in \{1, \cdots, n\}$ satisfying Assumption 10.2 can be dealt with as in [163], which is shown in Fig. 10.14.

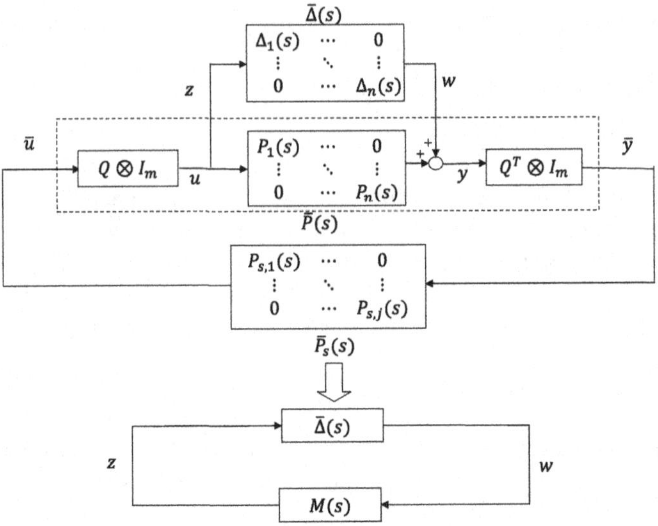

Figure 10.14 Robustness to model uncertainty via the NI system theory.

Fig. 10.14 (top) can be manipulated to Fig. 10.14 (bottom) with $M(s) = (\mathcal{Q} \otimes I_m)\bar{P}_s(s)(I - \bar{P}(s)\bar{P}_s(s))^{-1}(\mathcal{Q}^T \otimes I_m)$. The internal stability already yields $M(s) \in \mathcal{RH}_\infty$ and $M(s)$ is NI via Theorem 6 in [50] in general, or via Theorem 6 in [137] when $P_i(s)$ have no poles on the imaginary axis. This NI system $M(s)$ is connected with $\bar{\Delta}(s)$ which fulfills Assumption 10.2. Now

$$
\begin{aligned}
&\bar{\lambda}(\Delta(0)M(0)) \\
\leq &\bar{\lambda}(\Delta(0))\bar{\lambda}[(\mathcal{Q} \otimes I_m)\bar{P}_s(0)(I - \bar{P}(0)\bar{P}_s(0))^{-1}(\mathcal{Q}^T \otimes I_m)] \\
\leq &\mu\bar{\lambda}(\mathcal{L}_n)\bar{\lambda}[\bar{P}_s(0)(I - \bar{P}(0)\bar{P}_s(0))^{-1}] \\
\leq &\frac{\mu\bar{\lambda}(\mathcal{L}_n)\bar{\lambda}(\bar{P}_s(0))}{1 - \bar{\lambda}(\bar{P}(0)\bar{P}_s(0))} \\
\leq &\frac{\mu\bar{\lambda}(\mathcal{L}_n)\max_{j=1}^{l}\{\bar{\lambda}(P_{s,j}(0))\}}{1 - \max_{i=1}^{n}\{\bar{\lambda}(P_i(0))\}\bar{\lambda}(\mathcal{L}_n)\max_{j=1}^{l}\{\bar{\lambda}(P_{s,j}(0))\}}
\end{aligned}
$$

It is then clear that inequality (10.22) guarantees $\bar{\lambda}(\Delta(0)M(0)) < 1$ which in turn implies robust stability for all uncertainties that satisfy Assumption 10.2. □

Remark 10.4. Inequality (10.21) only provides a sufficient condition due to the heterogeneousness of systems. It also implies $\bar{\lambda}(P_i(0)P_{s,j}(0)) \leq \bar{\lambda}(P_i(0))\bar{\lambda}(P_{s,j}(0)) < \frac{1}{\bar{\lambda}(\mathcal{L}_n)}$ due to Lemma 10.11, which gives a more stringent condition than that of internal stability of $[P_i(s), P_{s,j}(s)]$, i.e., $\bar{\lambda}(P_i(0)P_{s,j}(0)) < 1$ (on noting that $\bar{\lambda}(\mathcal{L}_n) > 1$

due to [57]). This coincides with the engineering intuition since the stability condition for networked systems is always more stringent than that of single agent system ([89]).

Remark 10.5. Since we assume that $\exists i : \bar{\lambda}(P_i(0)) \geq 0$ and since values of i such that $\bar{\lambda}(P_i(0)) \leq 0$ automatically fulfill inequality (10.21), only values of i such that $\bar{\lambda}(P_i(0)) > 0$ need to be checked. For values of i such that $\bar{\lambda}(P_i(0)) > 0$, the D.C. gain of the SNI controllers always need to be tuned for small eigenvalues in order to satisfy inequality (10.21). The SNI control synthesis for robust performance is beyond the scope of this chapter. Interested readers are referred to [164,163].

Remark 10.6. There is clearly a huge class of permissible dynamic perturbations to the nominal dynamics as Assumption 10.2 only imposes a restriction on $\Delta_i(s)$ at the frequency $\omega = 0$ and $\omega = \infty$ and the SNI class has no gain (as long as it is finite gain) or order restriction [83]. The result in Theorem 10.3 is for additive perturbations, but similar analysis can be performed for other types of perturbations that preserve the NI class. A few examples of permissible perturbations that preserve the NI class include additive perturbations where uncertainty is also NI [83], feedback perturbations where both systems in the feedback interconnection are NI [137] and more general perturbations based on Redheffer star products and linear fractional transformations [50]. For example, $\frac{1}{s+5}$ and $\frac{(2s^2+s+1)}{(s^2+2s+5)(s+1)(2s+1)}$ are SNI with the same D.C. gain.

10.3.2.2 NI plants with free body dynamics

In this subsection, we will consider more general NI plants by including free body dynamics (i.e. poles at the origin) under the assumption of strict properness, i.e. $P_i(\infty) = 0$. Hence, this subsection covers the cases where the NI plant has poles at the origin. The NI class restricts the number of such poles at the origin to be at most 2. The following residue matrices carry information about the properties of the free body motion for the NI system $\boldsymbol{y} = P(s)\boldsymbol{u}$ where $P(s) \in \mathbb{R}^{m\times m}$ ([109]): $P_2 = \lim_{s\to 0} s^2 P(s)$, $P_1 = \lim_{s\to 0} s(P(s) - \frac{P_2}{s^2})$, $P_0 = \lim_{s\to 0}(P(s) - \frac{P_2}{s^2} - \frac{P_1}{s})$. It can be observed that $P_1 = 0$, $P_2 = 0$ means there is no free body dynamics, $P_1 \neq 0$, $P_2 = 0$ means there is free body dynamics with 1 pole at the origin, $P_2 \neq 0$ means there is free body dynamics with 2 poles at the origin. Then, we can define the Hankel matrix Γ as $\Gamma = \begin{bmatrix} P_1 & P_2 \\ P_2 & 0 \end{bmatrix} \in \mathbb{R}^{2m\times 2m}$, where $P_1, P_2 \in \mathbb{R}^{m\times m}$. In the subsection, we assume $\Gamma \neq 0$ since either $P_1 \neq 0$ or $P_2 \neq 0$. Then Γ can be decomposed by Singular Value Decomposition (SVD) as

$$\Gamma = \begin{bmatrix} U_1 & U_2 \end{bmatrix} \begin{bmatrix} S & 0 \\ 0 & 0 \end{bmatrix} \begin{bmatrix} V_1^T \\ V_2^T \end{bmatrix} = U_1 S V_1^T = H V_1^T = \begin{bmatrix} H_1 \\ H_2 \end{bmatrix} V_1^T,$$

where the diagonal matrix $S > 0$, the matrices $\begin{bmatrix} U_1 & U_2 \end{bmatrix}$ and $\begin{bmatrix} V_1 & V_2 \end{bmatrix}$ are orthogonal, $H = U_1 S \in \mathbb{R}^{2m\times \tilde{n}}$, $H_1, H_2 \in \mathbb{R}^{m\times \tilde{n}}$ and the matrices H and V_1 have orthogonal columns. Then, the matrix $H_1^T H_2$ can be further decomposed by SVD as $H_1^T H_2 = \hat{U}\hat{S}\hat{V}^T = \hat{U} \begin{bmatrix} S_1 & 0 \\ 0 & 0 \end{bmatrix} \begin{bmatrix} \hat{V}_1^T \\ \hat{V}_2^T \end{bmatrix}$, where $\hat{U}, \hat{V} = \begin{bmatrix} \hat{V}_1 & \hat{V}_2 \end{bmatrix}^T \in \mathbb{R}^{\tilde{n}\times \tilde{n}}$ are orthogonal matrices, $\hat{V}_2 \in \mathbb{R}^{\tilde{n}\times\tilde{n}}$ and the diagonal matrix $S_1 > 0$. Also, define $F = H_1\hat{V}_2 \in \mathbb{R}^{m\times\tilde{n}}$ and $N_f = P_s(0) - P_s(0)F(F^T P_s(0)F)^{-1}F^T P_s(0)$, where $P_s(s)$ is an SNI controller. When $P_2 \neq 0$, define $N_2 = P_s(0) - P_s(0)J(J^T P_s(0)J)^{-1}J^T P_s(0)$, where J is a full column rank matrix satisfying $JJ^T = P_2$. When $P_2 = 0$ and $P_1 \neq 0$, P_1 can be decomposed by SVD as $P_1 = \begin{bmatrix}\tilde{U}_1 & \tilde{U}_2\end{bmatrix}\begin{bmatrix}S_2 & 0\\ 0 & 0\end{bmatrix}\begin{bmatrix}V_1^T\\ V_2^T\end{bmatrix} = F_1 V_1^T$, where $\begin{bmatrix}\tilde{U}_1 & \tilde{U}_2\end{bmatrix}$ and $\begin{bmatrix}V_1 & V_2\end{bmatrix}$ are orthogonal matrices, the diagonal matrix $S_2 > 0$ and F_1 and V_1 have orthogonal columns. Then, define $N_1 = P_s(0) - P_s(0)F_1(F_1^T P_s(0)F_1)^{-1}F_1^T P_s(0)$. Next, the internal stability of $[P(s), P_s(s)]$ with free body dynamics can be summarized in the following lemma.

Lemma 10.13 ([109]). *Let $P(s)$ be a strictly proper NI plant and $P_s(s)$ be an SNI controller.*

1. *Suppose $P_2 \neq 0$, N_f is sign definite and $F^T P_s(0)F$ is nonsingular. Then, $[P(s), P_s(s)]$ is internally stable if and only if $F^T P_s(0)F < 0$ and either*

$$I - N_f^{\frac{1}{2}} P_0 N_f^{\frac{1}{2}} - N_f^{\frac{1}{2}} P_1 J(J^T J)^{-2} J^T P_1^T N_f^{\frac{1}{2}} > 0 \tag{10.23}$$

when $N_f \geq 0$ or

$$\det(I + \tilde{N}_f P_0 \tilde{N}_f + \tilde{N}_f P_1 J(J^T J)^{-2} J^T P_1^T \tilde{N}_f) \neq 0 \tag{10.24}$$

when $N_f \leq 0$ where $\tilde{N}_f = (-N_f)^{\frac{1}{2}}$.
If furthermore $P_1 = 0$, N_2 is sign definite and $J^T P_s(0)J$ is nonsingular, the necessary and sufficient conditions for the internal stability of $[P(s), P_s(s)]$ reduce to $J^T P_s(0)J < 0$ and either $I - N_2^{\frac{1}{2}} P_0 N_2^{\frac{1}{2}} > 0$ when $N_2 \geq 0$ or $\det(I + \tilde{N}_2 P_0 \tilde{N}_2) \neq 0$ when $N_2 \leq 0$ where $\tilde{N}_2 = (-N_2)^{\frac{1}{2}}$.
If additionally $\mathrm{Ker}(P_2) \subseteq \mathrm{Ker}(P_0^T)$, the necessary and sufficient condition for the internal stability of $[P(s), P_s(s)]$ reduces to $J^T P_s(0)J < 0$. When $P_2 > 0$, the necessary and sufficient condition for the internal stability of $[P(s), P_s(s)]$ reduces to $P_s(0) < 0$.

2. *Suppose $P_2 = 0$, $P_1 \neq 0$, N_1 is sign definite and $F_1^T P_s(0)F_1$ is nonsingular. Then $[P(s), P_s(s)]$ is internally stable if and only if $F_1^T P_s(0)F_1 < 0$ and either $I - N_1^{\frac{1}{2}} P_0 N_1^{\frac{1}{2}} > 0$ when $N_1 \geq 0$ or $\det(I + \tilde{N}_1 P_0 \tilde{N}_1) \neq 0$ when $N_1 \leq 0$ where $\tilde{N}_1 = (-N_1)^{\frac{1}{2}}$.*
If furthermore $\mathrm{Ker}(P_1^T) \subseteq \mathrm{Ker}(P_0^T)$, the necessary and sufficient condition for the internal stability of $[P(s), P_s(s)]$ reduces to $F_1^T P_s(0)F_1 < 0$, when P_1 is invertible, the necessary and sufficient condition for the internal stability of $[P(s), P_s(s)]$ reduces to $P_s(0) < 0$.

Next we present the second main result of this chapter with the following notation: $\bar{P}_2 = \lim_{s\to 0} s^2 \bar{P}(s)$, $\bar{P}_1 = \lim_{s\to 0} s(\bar{P}(s) - \frac{\bar{P}_2}{s^2})$, and $\bar{P}_0 = \lim_{s\to 0}(\bar{P}(s) - \frac{\bar{P}_2}{s^2} - \frac{\bar{P}_1}{s})$.

Theorem 10.4. *Given a graph $\mathcal{G}$ with the incidence matrix $\mathcal{Q}$, satisfying Assumption 10.1 and modeling the communication links among multiple strictly proper NI agents $\hat{P}_i(s)$ (allowing possible poles at the origin) which are appropriately extended to $P_i(s)$ as in Fig. 10.13, robust output feedback consensus is achieved via the feedback control law in Eq. (10.19) or Eq. (10.20) under any external disturbances $w_1 \in \mathrm{Im}_{\mathfrak{L}_2}(\mathcal{Q} \otimes I_m)$ and $w_2 \in \mathfrak{L}_2$ as well as under any model uncertainty $\Delta_i(s), i \in \{1, \cdots, n\}$, satisfying Assumption 10.2 if and only if the necessary and sufficient conditions in Lemma 10.13 are satisfied for $[\bar{P}(s), \bar{P}_s(s)]$.*

Proof. Lemma 10.13 guarantees the internal stability of $[\bar{P}(s), \bar{P}_s(s)]$. The nominal output consensus is then achieved without considering the external disturbances w_1 and w_2 via the internal stability as discussed in the proof of Theorem 10.3. Then, similar analysis as in the proof of Theorem 10.3 guarantees robustness against both external disturbances as well as additive SNI model uncertainty. □

One could inquire whether the conditions in Lemma 10.13 simplify or not in some cases. The answer is positive as we present next.

Theorem 10.5. *Given a graph $\mathcal{G}$ with the incidence matrix $\mathcal{Q}$, satisfying Assumption 10.1 and modeling the communication links among n_2 strictly proper NI agents $\hat{P}_i(s)$ (allowing possible poles at the origin) with double poles at the origin (i.e. no single pole at the origin) and n_1 (at least 1) agents without free body dynamics (i.e. without poles at the origin) in Fig. 10.13, a necessary and sufficient condition for the robust output feedback consensus via the feedback control law in Eq. (10.19) or Eq. (10.20) under external disturbances $w_1 \in \mathrm{Im}_{\mathfrak{L}_2}(\mathcal{Q} \otimes I_m)$ and $w_2 \in \mathfrak{L}_2$ and under any model uncertainty satisfying Assumption 10.2 is*

$$J_{n_2}^T \mathcal{L}_{e,11} J_{n_2} < 0,$$

where $J_{n_2} \triangleq \operatorname{diag}_{i=1}^{n_2}\{J_{2,i}\}$ with $J_{2,i}$ being full column rank matrices satisfying $J_{2,i} J_{2,i}^T = \lim_{s\to 0} s^2 P_i(s) \neq 0$ for n_2 agents and $\mathcal{L}_{e,11} \in \mathbb{R}^{n_2 m \times n_2 m}$ is a part of the weighted Laplacian matrix constructed as follows:

$$\mathcal{L}_e = \begin{bmatrix} \mathcal{L}_{e,11} & \mathcal{L}_{e,12} \\ \mathcal{L}_{e,12}^T & \mathcal{L}_{e,22} \end{bmatrix} = \chi(\mathcal{Q} \otimes I_m) \operatorname{diag}_{j=1}^{l}\{P_{s,j}(0)\}(\mathcal{Q}^T \otimes I_m)\chi^T$$

where χ is a permutation matrix such that

$$\chi \lim_{s\to 0} s^2 \operatorname{diag}_{i=1}^{n}\{P_i(s)\}\chi^T = \begin{bmatrix} \operatorname{diag}_{i=1}^{n_2}\{P_{2,i}\} & 0 \\ 0 & 0_{mn_1 \times mn_1} \end{bmatrix}$$

and $P_{s,j}(s)\ \forall j \in \{1, \cdots, l\}$ are SNI compensators.

Proof. It can be seen that this case corresponds to $\bar{P}_2 \neq 0$, $\bar{P}_1 = 0$ and $\mathrm{Ker}(\bar{P}_2) \subseteq \mathrm{Ker}(\bar{P}_0^T)$ in Theorem 10.4. The necessary and sufficient condition in this case is

$\bar{J}^T \bar{P}_s(0)\bar{J} < 0$ due to Lemma 10.13, where $\bar{P}_2 = \bar{J}\bar{J}^T$ with $\bar{J}$ being full column rank. Since

$$\begin{aligned}
\bar{P}_2 &= \lim_{s\to 0} s^2 \bar{P}(s) \\
&= \lim_{s\to 0} (\mathcal{Q}^T \otimes I_m) s^2 \operatorname*{diag}_{i=1}^{n}\{P_i(s)\}(\mathcal{Q}\otimes I_m) \\
&= (\mathcal{Q}^T \otimes I_m)\chi^T \begin{bmatrix} \operatorname*{diag}_{i=1}^{n_2}\{P_{2,i}\} & 0 \\ 0 & 0_{mn_1\times mn_1}\end{bmatrix} \chi(\mathcal{Q}\otimes I_m) \\
&= (\mathcal{Q}^T \otimes I_m)\chi^T \underbrace{\begin{bmatrix} \operatorname*{diag}_{i=1}^{n_2}\{J_{2,i}\} \\ 0 \end{bmatrix}}_{J_P} \underbrace{\begin{bmatrix} \operatorname*{diag}_{i=1}^{n_2}\{J_{2,i}^T\} & 0 \end{bmatrix}}_{J_P^T} \chi(\mathcal{Q}\otimes I_m) \\
&= \bar{J}\bar{J}^T,
\end{aligned}$$

where χ is a permutation matrix which also has a representation of $\chi = (\Upsilon \otimes I_m)$ with Υ also being a permutation matrix, $P_{2,i} = \lim_{s\to 0} s^2 P_i(s) \neq 0$ for the n_2 agents $P_s(s)$ that have a double pole at the origin and $J_{2,i} J_{2,i}^T = P_{2,i}$ where $J_{2,i}$ is a full column rank matrix. It can be seen that J_P has at least m zero rows due to the existence of at least one $m \times m$ agent without free body dynamics. Thus

$$\begin{aligned}
\bar{J} &= (\mathcal{Q}^T \otimes I_m)\chi^T J_P = (\mathcal{Q}^T \Upsilon^T \otimes I_m) J_P \\
&= (\begin{bmatrix} \hat{\mathcal{Q}}_1^T & \hat{\mathcal{Q}}_2^T \end{bmatrix} \otimes I_m) J_P = (\hat{\mathcal{Q}}_1^T \otimes I_m) \operatorname*{diag}_{i=1}^{n_2}\{J_{2,i}\},
\end{aligned}$$

where $\hat{\mathcal{Q}}_1$, $\hat{\mathcal{Q}}_2$ are row submatrices of $\Upsilon\mathcal{Q} = \begin{bmatrix} \hat{\mathcal{Q}}_1 \\ \hat{\mathcal{Q}}_2 \end{bmatrix}$. According to Assumption 10.1, the rank property of $\mathcal{Q}$ and the invertibility of permutation matrix Υ, the removal of $\hat{\mathcal{Q}}_2 \otimes I_m$ yields a full column rank of $\hat{\mathcal{Q}}_1^T$ and thus $\operatorname{rank}(\bar{J}) = \operatorname{rank}(\operatorname*{diag}_{i=1}^{n_2}\{J_{2,i}\})$, which also implies that $\bar{J}$ is full column rank due to the full column rank matrix $J_{2,i}$. Then, with the notation of $J_{n_2} \triangleq \operatorname*{diag}_{i=1}^{n_2}\{J_{2,i}\}$,

$$\begin{aligned}
& \bar{J}^T \bar{P}_s(0)\bar{J} < 0 \\
\Leftrightarrow & \begin{bmatrix} J_{n_2}^T & 0 \end{bmatrix} \chi(\mathcal{Q}\otimes I_m) \operatorname*{diag}_{j=1}^{l}\{P_{s,j}(0)\}(\mathcal{Q}^T \otimes I_m)\chi^T \begin{bmatrix} J_{n_2} \\ 0 \end{bmatrix} < 0 \\
\Leftrightarrow & \begin{bmatrix} J_{n_2}^T & 0 \end{bmatrix} \mathcal{L}_e \begin{bmatrix} J_{n_2} \\ 0 \end{bmatrix} = \begin{bmatrix} J_{n_2}^T & 0 \end{bmatrix} \begin{bmatrix} \mathcal{L}_{e,11} & \mathcal{L}_{e,12} \\ \mathcal{L}_{e,12}^T & \mathcal{L}_{e,22} \end{bmatrix} \begin{bmatrix} J_{n_2} \\ 0 \end{bmatrix} < 0 \\
\Leftrightarrow & J_{n_2}^T \mathcal{L}_{e,11} J_{n_2} < 0
\end{aligned}$$

Similar steps as in the proof of Theorem 10.3 then shows that robustness against external disturbances as well as an additive SNI model uncertainty also holds here. □

The above theorem gives a necessary and sufficient condition for robust output feedback consensus directly on the graph information and on the D.C. gain of the SNI controllers. The edge weights (i.e. the D.C. gains of the SNI controllers that are connected with agents that have a double pole at the origin) are important in determining the sign definiteness of $J_{n2}^T \mathcal{L}_{e,11} J_{n2}$, in other words, the internal stability of the networked system. The D.C. gains of the remaining SNI controllers are irrelevant and can be freely chosen as long as they are nonsingular.

Remark 10.7. When the SNI controllers are homogeneous, the consensus law (10.19) simplifies to $\boldsymbol{u} = (\mathcal{Q} \otimes I_m)(I_n \otimes P_s(s))(\mathcal{Q}^T \otimes I_m)\boldsymbol{y} = \mathcal{L}_n \otimes P_s(s)\boldsymbol{y}$, or in a distributed manner, $\boldsymbol{u}_i = P_s(s)\sum_{k=1}^n a_{ik}(\boldsymbol{y}_i - \boldsymbol{y}_k)$. It can be seen that this captures the main result of [183] in the homogeneous plant case but also generalizes the results to the heterogeneous plant case. In the case of heterogeneous SNI controllers, the controller is given by $\boldsymbol{u} = (\mathcal{Q} \otimes I_m)\bar{P}_s(s)(\mathcal{Q}^T \otimes I_m)\boldsymbol{y} = \bar{\mathcal{L}}_e(s)\boldsymbol{y}$, which can be interpreted as a weighted graph $\mathcal{G}$ with the edges weighted by the controller transfer functions $P_{s,j}(s)$, $j = 1, \cdots, l$, or in a distributed manner: $\boldsymbol{u}_i = \sum_{k=1}^n a_{ik} P_{s,j}(s)(\boldsymbol{y}_i - \boldsymbol{y}_k)$, where j is the edge connecting vertices i and k. The above facts give a nice intuitive interpretation and explain why we adopt the incidence matrix for the distributed property rather than the Laplacian matrix as indicated earlier.

10.3.3 Numerical examples

Two cases are given to illustrate the main results of this chapter, Theorems 10.3 and 10.4 respectively. The first case considers multiple NI systems without poles at the origin but allowing the plants to be biproper, while the second case is to show the more general case by including NI systems with poles at the origin but under the requirement of strictly proper plants.

10.3.3.1 Two lightly damped and one undamped flexible structures

Fig. 2 in [83] depicts a flexible structure that can also be studied in this chapter. The dynamics can be expressed as $M_i\ddot{\boldsymbol{x}}_i + C_i\dot{\boldsymbol{x}}_i + K_i\boldsymbol{x}_i = \boldsymbol{u}_i$, $\boldsymbol{y}_i = \boldsymbol{x}_i$, $i \in \{1, \cdots, 3\}$, where $\boldsymbol{x}_i = \begin{bmatrix} x_{i,1} \\ x_{i,2} \end{bmatrix}$, $\boldsymbol{u}_i = \begin{bmatrix} u_{i,1} \\ u_{i,2} \end{bmatrix}$, $M_i = \begin{bmatrix} m_{i,1} & 0 \\ 0 & m_{i,2} \end{bmatrix}$, $C_i = \begin{bmatrix} c_{i,1} + c_i & -c_i \\ -c_i & c_{i,2} + c_i \end{bmatrix}$, $K_i = \begin{bmatrix} k_{i,1} + k_i & -k_i \\ -k_i & k_{i,2} + k_i \end{bmatrix}$. The undamped flexible structure is given by letting the damped term $C_i = 0$. The parameters are given as follows.

System 1: $k_1 = k_{1,1} = k_{1,2} = 0.5$, $c_1 = c_{1,1} = c_{1,2} = 0.2$, $m_{1,1} = m_{1,2} = 1$ with the initial condition of $\begin{bmatrix} 0.5 & 0.1 & 1 & 0.2 \end{bmatrix}^T$

System 2: $k_2 = k_{2,1} = k_{2,2} = 1$, $c_2 = c_{2,1} = c_{2,2} = 0.1$, $m_{2,1} = 1$, $m_{2,2} = 0.5$ with the initial condition of $\begin{bmatrix} 1 & 0.1 & 1.5 & 0.2 \end{bmatrix}^T$.

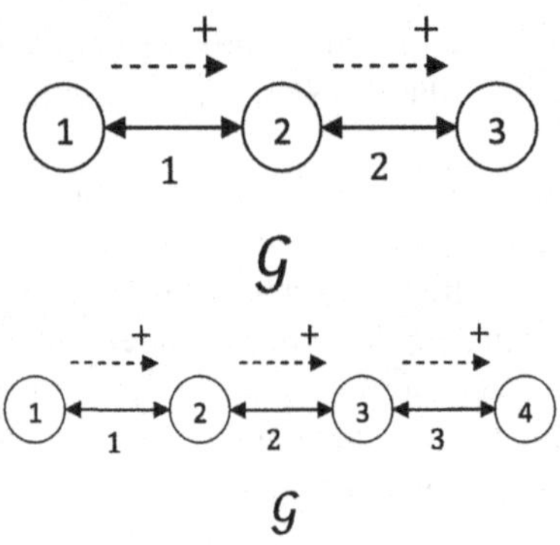

Figure 10.15 Graph for 3 and 4 NI systems.

System 3: $k_3 = k_{3,1} = k_{3,2} = 1$, $c_3 = c_{3,1} = c_{3,2} = 0$, $m_{3,1} = 1$, $m_{3,2} = 0.5$ with the initial condition of $\begin{bmatrix}1.5 & 0.1 & 2 & 0.2\end{bmatrix}^T$.

The communication topology is given in Fig. 10.15 and thus $\mathcal{Q} = \begin{bmatrix}1 & 0\\ -1 & 1\\ 0 & -1\end{bmatrix}$ and $\mathcal{L}_3 = \begin{bmatrix}1 & -1 & 0\\ -1 & 2 & -1\\ 0 & -1 & 1\end{bmatrix}$. It can easily be seen that $\max_{i=1}^{3} \bar{\lambda}(P_i(0)) = 2 > 0$. Both the SNI controllers are chosen as $\frac{1}{s+8}$ with an initial condition of -1 such that $\bar{\lambda}(P_i(0))\bar{\lambda}(P_{s,j}(0)) = 2 * \frac{1}{8} = \frac{1}{4} < \frac{1}{3} = \frac{1}{\bar{\lambda}(\mathcal{L}_n)}$, $\forall i \in \{1, \cdots, n\}$, $j \in \{1, \cdots, l\}$. In addition, $\forall i \in \{1, \cdots, n\}$, $j \in \{1, \cdots, l\}$, $P_i(\infty)P_{s,j}(\infty) = 0$, $P_{s,j}(\infty) = 0$, which all satisfy the assumptions of Theorem 10.3. First, without disturbances and the model uncertainty, the nominal output feedback consensus is achieved via the control law in Eq. (10.19) or Eq. (10.20) as shown in the top two figures of Fig. 10.16. Then introducing an additive SNI model uncertainty, for example given by $\frac{1}{s+4}$ as well as $\mathfrak{L}_2$ external disturbances, the robust output feedback consensus is also achieved as shown in the bottom two figures of Fig. 10.16.

10.3.3.2 One single integrator, one double integrator, one undamped and one lightly damped flexible structure

In order to illustrate Theorem 10.4, systems with free body dynamics will be included in this example. Therefore, a complicated case containing 1 single integrator, 1 double integrator, 1 MIMO undamped and 1 MIMO lightly damped flexible structure is considered in this example. For consistency of dimension, the single integrator and the double integrator are extended as follows: diag$\{\frac{1}{s}, 0\}$ and diag$\{0, \frac{1}{s^2}\}$, which also means that the output of the single integrator will be coordinated with first outputs of both the undamped and the lightly damped flexible structures, while the output of the double integrator will be coordinated with second outputs of both the undamped and the lightly damped flexible structures. The parameters of all NI systems are as follows.

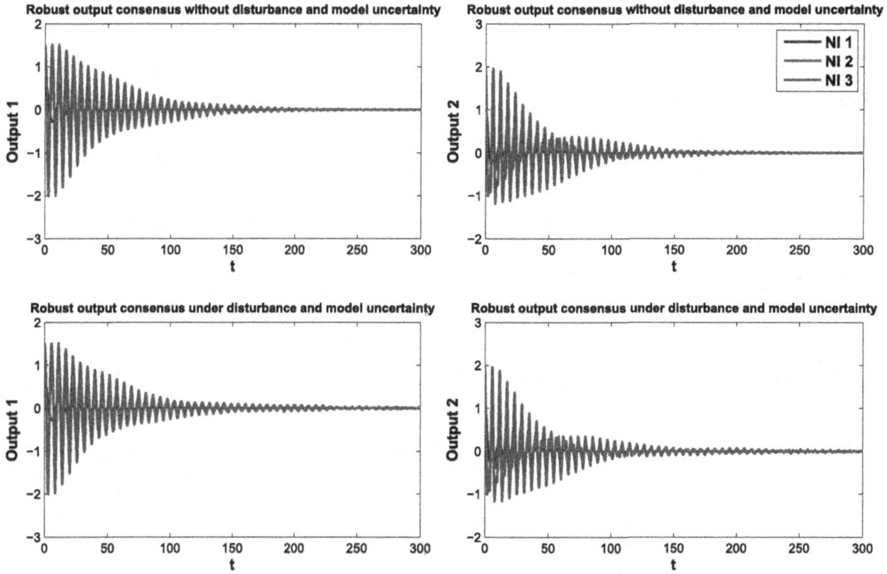

Figure 10.16 Robust output consensus of heterogeneous NI systems.

System 1: $\frac{1}{s^2}$ with the initial condition of $\begin{bmatrix}1 & 0.1\end{bmatrix}^T$.
System 2: $\frac{1}{s}$ with the initial condition of 2.
System 3: $k_3 = k_{3,1} = k_{3,2} = 1$, $c_3 = c_{3,1} = c_{3,2} = 0$, $m_{3,1} = 1$, $m_{3,2} = 0.5$ with the initial condition of $\begin{bmatrix}3 & 0.1 & 3 & 0.2\end{bmatrix}^T$.
System 4: $k_4 = k_{4,1} = k_{4,2} = 1$, $c_4 = c_{4,1} = c_{4,2} = 0.1$, $m_{4,1} = 1$, $m_{4,2} = 0.5$ with the initial condition of $\begin{bmatrix}4 & 0.1 & 4 & 0.2\end{bmatrix}^T$.

The communication topology is given in Fig. 10.15 and thus $\mathcal{Q} = \begin{bmatrix} 1 & 0 & 0 \\ -1 & 1 & 0 \\ 0 & -1 & 1 \\ 0 & 0 & -1 \end{bmatrix}$ and $\mathcal{L}_4 = \begin{bmatrix} 1 & -1 & 0 & 0 \\ -1 & 2 & -1 & 0 \\ 0 & -1 & 2 & 1 \\ 0 & 0 & -1 & 1 \end{bmatrix}$. All three SNI controllers are chosen as $-\frac{s+1}{s+2}$ with an initial condition of 0.1. Through the calculation process discussed earlier in Subsection 10.3.2.2, the inequality condition (10.24) can be verified as $\det(I + \tilde{N}_f \bar{P}_0 \tilde{N}_f + \tilde{N}_f \bar{P}_1 J (J^T J)^{-2} J^T \bar{P}_1^T \tilde{N}_f) = 3.7813 \neq 0$, which indicates the internal stability of $[\bar{P}(s), \bar{P}_s(s)]$. Firstly, without disturbances and the model uncertainty, the nominal output feedback consensus is achieved via the output feedback control law (10.19) or (10.20) as shown in the top two figures of Fig. 10.17. If the same external disturbances and model uncertainty as in Subsection 10.3.3.1 are inserted, robust output feedback consensus is also achieved via output feedback control law Eq. (10.19) or Eq. (10.20)

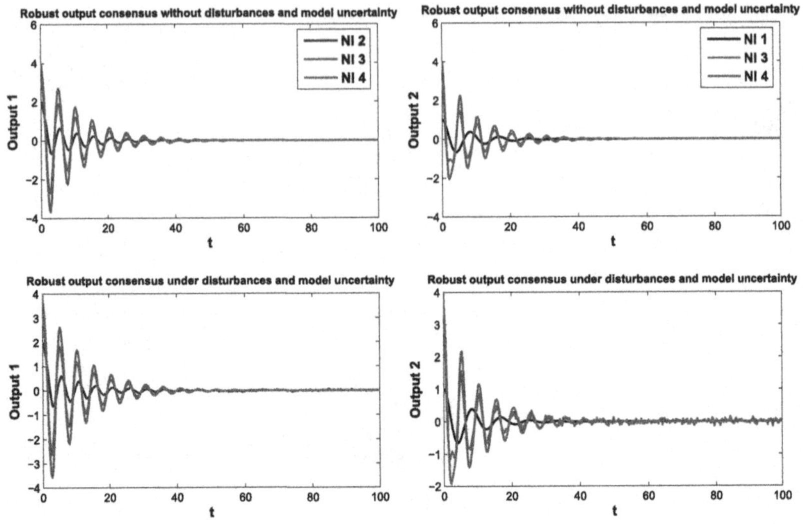

Figure 10.17 Robust output consensus of heterogeneous NI systems.

as shown in the bottom two figures of Fig. 10.17. The left two figures of Fig. 10.17 indicate that the output of the single integrator (System 2) is coordinated with the first output of the undamped flexible structure (System 3) and the first output of the lightly damped flexible structure (System 4) even under external disturbances and the model uncertainty. Similarly, the right two figures of Fig. 10.17 indicate that the output of the double integrator (System 1) is coordinated with the second output of the undamped flexible structure (System 3) and the second output of the lightly damped flexible structure (System 4) even under external disturbances and the model uncertainty.

10.4 Extension to robust cooperative control

In this section, we exploit the proposed consensus results for cooperative tracking to obtain a robust cooperative control framework for multiple NI systems. An famous rendezvous problem is presented to show the effectiveness of the proposed framework. Other cooperative problems can be obtained analogously by adapting the proposed consensus algorithm. The objective of cooperative tracking is to achieve the convergence of all agents' outputs to a predefined constant reference while the objective of rendezvous is its direct application for all the agents to converge to a predefined point. To solve this problem, let us first define a matrix $\mathcal{B}$ to express the connections between agents following reference [122]: $\mathcal{B} = \operatorname{diag}\{b_i\}_{i=1}^{n}$, where $b_i = 1$ if agent i is connected with a reference, otherwise $b_i = 0$. From $\mathcal{L}_n = \mathcal{Q}\mathcal{Q}^T$, we can similarly decompose $\mathcal{B}$ as $\mathcal{Q}_b\mathcal{Q}_b^T$ where $\mathcal{Q}_b$ is a full column rank matrix. It can be seen that $\mathcal{Q}_b$ is defined analogously as $\mathcal{Q}$ by $\mathcal{Q}_b = \left[\beta_1 \quad , \cdots , \beta_{l_b}\right]$, where l_b is the number of agents connected to the reference and β_{j_b} $\forall j_b \in \{1, \cdots, l_b\}$ is a vector in $\mathbb{R}^n$ with the i-th element being a 1 if agent i is connected to the reference. Then, the incidence matrix $\mathcal{Q}$ can be augmented with $\mathcal{Q}_b$ to give $\begin{bmatrix}\mathcal{Q} & \mathcal{Q}_b\end{bmatrix}$ which also shows the additional links from the reference to agents. The augmented matrix $\begin{bmatrix}\mathcal{Q} & \mathcal{Q}_b\end{bmatrix}$ still guarantees that $(\begin{bmatrix}\mathcal{Q} & \mathcal{Q}_b\end{bmatrix}^T \otimes I_m)\bar{P}(s)(\begin{bmatrix}\mathcal{Q} & \mathcal{Q}_b\end{bmatrix} \otimes I_m)$ is NI in Fig. 10.12 and thus the NI system theory can be applied to the robust output feedback consensus problem as shown in the previous section. The main result of this section is given next.

Theorem 10.6. *Given a graph $\mathcal{G}$ with the incidence matrix $\mathcal{Q}$, satisfying Assumption 10.1 and modeling the communication links among multiple NI agents $\hat{P}_i(s)$ which are appropriately extended to $P_i(s)$ as in Fig. 10.13, the robust cooperative output tracking of a constant reference $\boldsymbol{r}$ is achieved via the output feedback control law*

$$\boldsymbol{u}_{ct} = (\begin{bmatrix}\mathcal{Q} & \mathcal{Q}_b\end{bmatrix} \otimes I_m)\bar{P}_s(s)(\begin{bmatrix}\mathcal{Q} & \mathcal{Q}_b\end{bmatrix}^T \otimes I_m)(\boldsymbol{y} - \mathbf{1}_n \otimes \boldsymbol{r}), \tag{10.25}$$

or in a distributed manner $\forall i \in \{1, \cdots, n\}$ by

$$\boldsymbol{u}_i^{ct} = \sum_{k=1}^{n} a_{ik} P_{s,j}(s)(\boldsymbol{y}_i - \boldsymbol{y}_k) + b_i P_{s,j_b}(s)(\boldsymbol{y}_i - \boldsymbol{r}) \tag{10.26}$$

where $j \in \{1, \cdots, l\}$ is the number of the edge connecting agents i and k, and $j_b \in \{1, \cdots, l_b\}$ is the number of the link connecting reference $\boldsymbol{r}$ to agent i, under any external disturbances $w_1 \in \mathrm{Im}_{\mathfrak{L}_2}(\mathcal{Q} \otimes I_m)$ and $w_2 \in \mathfrak{L}_2$ as well as under any model uncertainty satisfying Assumption 10.2 if and only if the relevant conditions in Theorem 10.3 or Lemma 10.13 are satisfied for $[\bar{P}(s), \bar{P}_s(s)]$.

Proof. Given a constant reference $\boldsymbol{r}$, the matrix $\begin{bmatrix}\mathcal{Q} & \mathcal{Q}_b\end{bmatrix}^T$ has the property that

$$\bar{\boldsymbol{y}} = (\begin{bmatrix}\mathcal{Q}^T \\ \mathcal{Q}_b^T\end{bmatrix} \otimes I_m)(\boldsymbol{y} - \mathbf{1}_n \otimes \boldsymbol{r}) = \begin{bmatrix}(\mathcal{Q}^T \otimes I_m)\boldsymbol{y} \\ (\mathcal{Q}_b^T \otimes I_m)(\boldsymbol{y} - \mathbf{1}_n \otimes \boldsymbol{r})\end{bmatrix}$$

since $(\mathcal{Q}^T \otimes I_m)(\mathbf{1}_n \otimes \boldsymbol{r}) = \mathbf{0}$ due to the null space property of the incidence matrix in Section 1. Therefore, the internal stability guaranteed by Theorem 10.3 or Lemma 10.13 leads to $\bar{\boldsymbol{y}} \to \mathbf{0}$, which is equivalent to $\boldsymbol{y}_i \to \boldsymbol{y}_k$ $\forall i \neq k \in \{1, \cdots, n\}$ due to $(\mathcal{Q}^T \otimes I_m)\boldsymbol{y} \to \mathbf{0}$ and $\boldsymbol{y}_{j_b} \to \boldsymbol{r}$ $\forall j_b \in \{1, \cdots, l_b\}$ due to $(\mathcal{Q}_b^T \otimes I_m)(\boldsymbol{y} - \mathbf{1}_n \otimes \boldsymbol{r}) \to \mathbf{0}$. This then implies that the robust cooperative output tracking is achieved, i.e., $\boldsymbol{y}_i = \boldsymbol{y}_k = \boldsymbol{r}$ $\forall i \in \{1, \cdots, n\}$ and $k \in \{1, \cdots, n\}/\{i\}$ via arguments like those in the proof of Theorem 10.3. □

As a direct application of the cooperative tracking, the robust rendezvous is shown in a 3D scenario. We use plants composed of $\frac{1}{s^2}I_3$, $\frac{1}{s^2}I_3$, $\frac{1}{s}I_3$ with the initial conditions in 3 axis being $\begin{bmatrix}1 & 0.1 & 1 & 1 & 1 & -0.1\end{bmatrix}^T$, $\begin{bmatrix}2 & 0.2 & 2 & -1 & 2 & -0.2\end{bmatrix}^T$ and $\begin{bmatrix}0.3 & 0.5 & 0.8\end{bmatrix}^T$ respectively. The predefined rendezvous point is $\begin{bmatrix}1 & -0.5 & 0\end{bmatrix}^T$. The graph is exactly the same as in Fig. 10.15 and the reference only sends

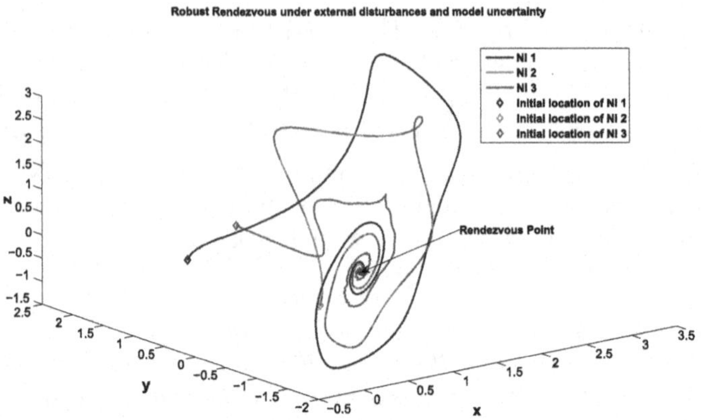

Figure 10.18 3D robust rendezvous of heterogeneous NI systems.

information to system 3, which gives $\mathcal{B} = \text{diag}\{0, 0, 1\}$ and $\mathcal{Q}_b = \begin{bmatrix} 0 & 0 & 1 \end{bmatrix}^T$. Comparing with the cases in Theorem 10.4, it can be seen that this is the case when $P_1 \neq 0$ and $P_2 \neq 0$. Therefore, Lemma 10.13 is used and it is easy to see that $N_f \leq 0$ and $\det(I + \tilde{N}_f P_0 \tilde{N}_f + \tilde{N}_f P_1 J (J^T J)^{-2} J^T P_1^T \tilde{N}_f) = 1 \neq 0$. The robust rendezvous is then guaranteed via the cooperative tracking controller (10.25) or (10.26) in Theorem 10.6. It can be seen from Fig. 10.18 that the cooperative tracking control law (10.25) or (10.26) is able to drive the systems from the initial positions marked as diamonds to the final rendezvous point even under the same external disturbances and model uncertainty as in Subsection 10.3.3.1.

10.5 Conclusion remarks

The robust cooperative control for multi-NI systems is proposed via the NI systems theory. The robust output feedback consensus against external disturbances and the NI model uncertainty is studied first and then the results are exploited for a cooperative tracking to derive a cooperative control framework. The key contributions of this chapter can be summarized as: (1) the cooperative control which is robust to exogenous disturbances and the SNI model uncertainty for a general heterogeneous network of MIMO NI systems under any undirected and connected graph; (2) only exploiting the output feedback information in contrast to a full state information commonly used in the literature; (3) providing a whole class of cooperative control laws, i.e. SNI controllers, that can be tuned for performance, and characterizing conditions that can be easily checked for the robust output feedback consensus; (4) showing how consensus and cooperative control problems can exploit powerful internal stability and robust stability results available in the literature.

Bibliography

[1] A. Abdessameud, A. Tayebi, Formation control of VTOL unmanned aerial vehicles with communication delays, Automatica 47 (11) (2011) 2383–2394.

[2] H. Ando, Y. Oasa, I. Suzuki, M. Yamashita, Distributed memoryless point convergence algorithm for mobile robots with limited visibility, IEEE Transactions on Robotics and Automation 15 (5) (1999) 818–828.

[3] Z. Artstein, Linear systems with delayed controls: a reduction, IEEE Transactions on Automatic Control 27 (4) (1982) 869–879.

[4] K.J. Aström, G.C. Goodwin, P.R. Kumar, Adaptive Control, Filtering, and Signal Processing, Springer-Verlag, 1995.

[5] J. Back, J. Kim, A disturbance observer based practical coordinated tracking controller for uncertain heterogeneous multi-agent systems, International Journal of Robust and Nonlinear Control 25 (14) (2015) 2254–2278.

[6] A. Bahr, J.J. Leonard, M.F. Fallon, Cooperative localization for autonomous underwater vehicles, The International Journal of Robotics Research 28 (6) (2009) 714–728.

[7] T. Balch, How multirobot systems research will accelerate our understanding of social animal behavior, in: Proceedings of the IEEE, vol. 94, IEEE, 2006, pp. 1445–1463.

[8] T. Balch, R.C. Arkin, Behavior-based formation control for multirobot teams, IEEE Transactions on Robotics and Automation 14 (6) (1998) 926–939.

[9] H.T. Banks, M.Q. Jacobs, M.R. Latina, The synthesis of optimal controls for linear, time-optimal problems with retarded controls, Journal of Optimization Theory and Applications 8 (5) (1971) 319–366.

[10] D. Bauso, L. Giarre, R. Pesenti, Mechanism design for optimal consensus problems, in: Proceedings of the 45th IEEE Conference on Decision and Control, IEEE, 2006, pp. 3381–3386.

[11] D.S. Bernstein, Nonquadratic cost and nonlinear feedback control, International Journal of Robust and Nonlinear Control 3 (3) (1993) 211–229.

[12] E. Camponogara, D. Jia, B.H. Krogh, S. Talukdar, Distributed model predictive control, IEEE Control Systems Magazine 22 (1) (2002) 44–52.

[13] W. Cao, J. Zhang, W. Ren, Leader–follower consensus of linear multi-agent systems with unknown external disturbances, Systems & Control Letters 82 (2015) 64–70.

[14] Y. Cao, W. Ren, Containment control with multiple stationary or dynamic leaders under a directed interaction graph, in: Proceedings of the 48th IEEE Conference on Decision and Control and the 28th Chinese Control Conference, 2009.

[15] Y. Cao, W. Ren, Optimal linear-consensus algorithms: an LQR perspective, IEEE Transactions on Systems, Man and Cybernetics. Part B. Cybernetics 40 (3) (2009) 819–830.

[16] Y. Cao, D. Stuart, W. Ren, Z. Meng, Distributed containment control for multiple autonomous vehicles with double-integrator dynamics: algorithms and experiments, IEEE Transactions on Control Systems Technology 19 (4) (2011) 929–938.

[17] Y. Cao, W. Ren, M. Egerstedt, Distributed containment control with multiple stationary or dynamic leaders in fixed and switching directed networks, Automatica 48 (8) (2012) 1586–1597.

[18] Y. Cao, W. Yu, W. Ren, G. Chen, An overview of recent progress in the study of distributed multi-agent coordination, IEEE Transactions on Industrial Informatics 9 (1) (2013) 427–438.
[19] F. Chen, W. Ren, Z. Lin, Multi-agent coordination with cohesion, dispersion, and containment control, in: Proceedings of the 2011 American Control Conference, 2010.
[20] M. Chen, W. Chen, Disturbance-observer-based robust control for time delay uncertain systems, International Journal of Control, Automation, and Systems 8 (2) (2010) 445–453.
[21] L. Consolini, F. Morbidi, D. Prattichizzo, Leader-follower formation control of nonholonomic mobile robots with input constraints, Automatica 44 (5) (2008) 1343–1349.
[22] P. Cortes, J. Rodriguez, C. Silva, A. Flores, Delay compensation in model predictive current control of a three-phase inverter, IEEE Transactions on Industrial Electronics 59 (2) (2012) 1323–1325.
[23] J. Cortés, S. Martinéz, F. Bullo, Robust rendezvous for mobile autonomous agents via proximity graphs in arbitrary dimensions, IEEE Transactions on Automatic Control 51 (8) (2006) 1289–1298.
[24] A. Das, F.L. Lewis, Distributed adaptive control for synchronization of unknown nonlinear networked systems, Automatica 46 (12) (2010) 2014–2021.
[25] M.E. Dehshalie, M.B. Menhaj, M. Karrari, Fault tolerant cooperative control for affine multi-agent systems: an optimal control approach, Journal of the Franklin Institute 356 (3) (2019) 1360–1378.
[26] J. Delvenne, R. Carli, S. Zampieri, Optimal strategies in the average consensus problem, in: 2007 46th IEEE Conference on Decision and Control, Dec 2007, pp. 2498–2503.
[27] J.C. Derenick, J.R. Spletzer, Convex optimization strategies for coordinating large-scale robot formations, IEEE Transactions on Robotics 23 (6) (2007) 1252–1259.
[28] J.P. Desai, J.P. Ostrowski, V. Kumar, Modeling and control of formations of nonholonomic mobile robots, IEEE Transactions on Robotics and Automation 17 (6) (2001) 905–908.
[29] D.V. Dimarogonas, M. Egerstedt, K.J. Kyriakopoulos, A leader-based containment control strategy for multiple unicycles, in: Proceedings of the 45th IEEE Conference on Decision and Control, 2006.
[30] D.V. Dimarogonas, M. Egerstedt, K.J. Kyriakopoulos, Further results on the stability of distance-based multi-robot formations, in: Proceedings of the 2009 American Control Conference, 2009.
[31] L. Ding, Q. Han, L.Y. Wang, E. Sindi, Distributed cooperative optimal control of DC microgrids with communication delays, IEEE Transactions on Industrial Informatics 14 (9) (2018) 3924–3935.
[32] Z. Ding, Consensus output regulation of a class of heterogeneous nonlinear systems, IEEE Transactions on Automatic Control 58 (10) (2013) 2648–2653.
[33] Z. Ding, Consensus control of a class of Lipschitz nonlinear systems, International Journal of Control 87 (11) (2014) 2372–2382.
[34] Z. Ding, Adaptive consensus output regulation of a class of nonlinear systems with unknown high-frequency gain, Automatica 51 (2015) 348–355.
[35] Z. Ding, Consensus disturbance rejection with disturbance observers, IEEE Transactions on Industrial Electronics 62 (9) (2015) 5829–5837.
[36] Z. Ding, Z. Lin, Truncated state prediction for control of Lipschitz nonlinear systems with input delay, in: Proceeding of IEEE 53rd Annual Conference on Decision and Control (CDC), IEEE, 2014, pp. 1966–1971.
[37] Zhengtao Ding, Nonlinear and Adaptive Control Systems, IET, London, 2013.

[38] W. Dong, J.A. Farrell, Cooperative control of multiple nonholonomic mobile agents, IEEE Transactions on Automatic Control 53 (6) (2008) 1434–1448.
[39] X. Dong, J. Xi, G. Lu, Formation control for high-order linear time-invariant multiagent systems with time delays, IEEE Transactions on Control of Network Systems 1 (3) (2014) 232–240.
[40] H. Du, S. Li, P. Shi, Robust consensus algorithm for second-order multi-agent systems with external disturbances, International Journal of Control 85 (12) (2012) 1913–1928.
[41] H.B. Duan, S.Q. Liu, Non-linear dual-mode receding horizon control for multiple unmanned air vehicles formation flight based on chaotic particle swarm optimisation, IET Control Theory & Applications 4 (11) (2010) 2565–2578.
[42] H.B. Duan, Q.N. Luo, Y.X. Yu, Trophallaxis network control approach to formation flight of multiple unmanned aerial vehicles, Science China. Technological Sciences 56 (5) (2013) 1066–1074.
[43] Z.S. Duan, G.R. Chen, L. Huang, Disconnected synchronized regions of complex dynamical networks, IEEE Transactions on Automatic Control 54 (4) (2009) 845–849.
[44] W.B. Dunbar, A distributed receding horizon control algorithm for dynamically coupled nonlinear systems, in: Proceedings of the 44th IEEE Conference on Decision and Control, IEEE, 2005, pp. 6673–6679.
[45] W.B. Dunbar, R.M. Murray, Model predictive control of coordinated multi-vehicle formations, in: Proceedings of the 41st IEEE Conference on Decision and Control, 2002, vol. 4, IEEE, 2002, pp. 4631–4636.
[46] W.B. Dunbar, R.M. Murray, Receding horizon control of multi-vehicle formations: a distributed implementation, in: 2004 43rd IEEE Conference on Decision and Control (CDC) (IEEE Cat. No. 04CH37601), vol. 2, IEEE, 2004, pp. 1995–2002.
[47] K. Engelborghs, M. Dambrine, D. Roose, Limitations of a class of stabilization methods for delay systems, IEEE Transactions on Automatic Control 46 (2) (2001) 336–339.
[48] J.A. Fax, R.M. Murray, Graph Laplacians and stabilization of vehicle formations, 2001, pp. 1–11.
[49] J.A. Fax, R.M. Murray, Information flow and cooperative control of vehicle formations, IEEE Transactions on Automatic Control 49 (9) (2004) 1465–1476.
[50] A. Ferrante, A. Lanzon, L. Ntogramatzidis, Foundations of not necessarily rational negative imaginary systems theory: relations between classes of negative imaginary and positive real systems, IEEE Transactions on Automatic Control 61 (10) (2016) 3052–3057, see also technical report at arXiv:1412.5709v1.
[51] G. Ferrari-Trecate, M. Egerstedt, A. Buffa, M. Ji, Laplacian Sheep: A Hybrid, Stop-Go Policy for Leader-Based Containment Control, Springer, 2006.
[52] G. Ferrari-Trecate, L. Galbusera, M.P.E. Marciandi, R. Scattolini, Model predictive control schemes for consensus in multi-agent systems with single-and double-integrator dynamics, IEEE Transactions on Automatic Control 54 (11) (2009) 2560–2572.
[53] D. Fox, J. Ko, K. Konolige, B. Limketkai, D. Schulz, B. Stewart, Distributed multirobot exploration and mapping, Proceedings of the IEEE 94 (7) (2006) 1325–1339.
[54] R. Ghrist, S.M. Lavalle, Nonpositive curvature and Pareto optimal coordination of robots, SIAM Journal on Control and Optimization 45 (5) (2006) 1697–1713.
[55] L. Giovanini, J. Balderud, R. Katebi, Autonomous and decentralized mission planning for clusters of UUVs, International Journal of Control 80 (7) (2007) 1169–1179.
[56] G.H. Golub, C.F. Van Loan, Matrix Computations, JHU Press, Baltimore, MD, USA, 2012.
[57] R. Grone, R. Merris, The Laplacian spectrum of a graph II, SIAM Journal on Discrete Mathematics 7 (2) (1994) 221–229.

[58] K. Gu, J. Chen, V.L. Kharitonov, Stability of Time-Delay Systems, Springer Science & Business Media, 2003.

[59] K. Gu, S.-I. Niculescu, Survey on recent results in the stability and control of time-delay systems, Journal of Dynamic Systems, Measurement, and Control 125 (2) (2003) 158–165.

[60] Y. Gu, B. Seanor, G. Campa, M.R. Napolitano, L. Rowe, S. Gururajan, S. Wan, Design and flight testing evaluation of formation control laws, IEEE Transactions on Control Systems Technology 14 (6) (Nov 2006) 1105–1112.

[61] D. Han, G. Chesi, Robust consensus for uncertain multi-agent systems with discrete-time dynamics, International Journal of Robust and Nonlinear Control 24 (13) (2014) 1858–1872.

[62] D. Han, G. Chesi, Y.S. Hung, Robust consensus for a class of uncertain multi-agent dynamical systems, IEEE Transactions on Industrial Informatics 9 (1) (Feb 2013) 306–312.

[63] Y. He, Q.-G. Wang, C. Lin, M. Wu, Delay-range-dependent stability for systems with time-varying delay, Automatica 43 (2) (2007) 371–376.

[64] Y. Hong, J. Hu, L. Gao, Tracking control for multi-agent consensus with an active leader and variable topology, Automatica 42 (7) (2006) 1177–1182.

[65] R.A. Horn, C.R. Johnson, Matrix Analysis, Cambridge University Press, 2012.

[66] H. Hu, Biologically inspired design of autonomous robotic fish at Essex, in: IEEE SMC UK-RI Chapter Conference, on Advances in Cybernetic Systems, 2006.

[67] J. Hu, M. Prandini, C. Tomlin, Conjugate points in formation constrained optimal multi-agent coordination: a case study, SIAM Journal on Control and Optimization 45 (6) (2007) 2119–2137.

[68] Y. Hu, J. Lam, J. Liang, Consensus of multi-agent systems with Luenberger observers, Journal of the Franklin Institute 350 (9) (2013) 2769–2790.

[69] Alberto Isidori, Nonlinear Control Systems, 3rd edition, Springer-Verlag, Berlin, 1995.

[70] A. Jadbabaie, J. Lin, A.S. Morse, Coordination of groups of mobile autonomous agents using nearest neighbor rules, IEEE Transactions on Automatic Control 48 (6) (2003) 988–1001.

[71] M. Ji, G. Ferrari-Trecate, M. Egerstedt, A. Buffa, Containment control in mobile networks, IEEE Transactions on Automatic Control 53 (8) (2008) 1972–1975.

[72] F. Jiang, L. Wang, G. Xie, Consensus of high-order dynamic multi-agent systems with switching topology and time-varying delays, Journal of Control Theory and Applications 8 (1) (2010) 52–60.

[73] B. Johansson, A. Speranzon, M. Johansson, K.H. Johansson, On decentralized negotiation of optimal consensus, Automatica 44 (4) (2008) 1175–1179.

[74] G.A. Kaminka, R. Schechter-Glick, V. Sadov, Using sensor morphology for multirobot formations, IEEE Transactions on Robotics 24 (2) (2008) 271–282.

[75] T. Keviczky, F. Borrelli, K. Fregene, D. Godbole, G.J. Balas, Decentralized receding horizon control and coordination of autonomous vehicle formations, IEEE Transactions on Control Systems Technology 16 (1) (2007) 19–33.

[76] H. Kim, H. Shim, J. Seo, Output consensus of heterogeneous uncertain linear multi-agent systems, IEEE Transactions on Automatic Control 56 (1) (2011) 200–206.

[77] Y. Kim, M. Mesbahi, On maximizing the second smallest eigenvalue of a state-dependent graph Laplacian, in: Proceedings of the 2005, American Control Conference, 2005, IEEE, 2005, pp. 99–103.

[78] L. Kocarev, P. Amato Chaos, Synchronization in power-law networks, IEEE Transactions on Automatic Control 15 (2) (2005) 1–7.

[79] M. Krstic, Delay Compensation for Nonlinear, Adaptive, and PDE Systems, Springer, 2009.

[80] Y. Kuwata, A. Richards, T. Schouwenaars, J.P. How, Distributed robust receding horizon control for multivehicle guidance, IEEE Transactions on Control Systems Technology 15 (4) (2007) 627–641.

[81] W. Kwon, A. Pearson, Feedback stabilization of linear systems with delayed control, IEEE Transactions on Automatic Control 25 (2) (1980) 266–269.

[82] G. Lafferriere, A. Williams, J. Caughman, J.J.P. Veerman, Decentralized control of vehicle formations, Systems & Control Letters 54 (9) (2005) 899–910.

[83] A. Lanzon, I.R. Petersen, Stability robustness of a feedback interconnection of systems with negative imaginary frequency response, IEEE Transactions on Automatic Control 53 (4) (2008) 1042–1046.

[84] A.J. Laub, Matrix Analysis for Scientists and Engineers, Society for Industrial and Applied Mathematics, Philadelphia, PA, 2005.

[85] J. Lavaei, A. Momeni, A.G. Aghdam, A model predictive decentralized control scheme with reduced communication requirement for spacecraft formation, IEEE Transactions on Control Systems Technology 16 (2) (2008) 268–278.

[86] S. Li, J. Yang, W. Chen, X. Chen, Disturbance Observer-Based Control: Methods and Applications, CRC Press, 2014.

[87] Z. Li, Z. Duan, Cooperative Control of Multi-Agent Systems: A Consensus Region Approach, CRC Press, 2014.

[88] Z. Li, Z. Duan, G. Chen, On H_∞ and H_2 performance regions of multi-agent systems, Automatica 47 (4) (2011) 797–803.

[89] Z. Li, Z. Duan, G. Chen, L. Huang, Consensus of multi-agent systems and synchronization of complex networks: a unified viewpoint, IEEE Transactions on Circuits and Systems I: Regular Papers 57 (1) (2010) 213–224.

[90] Z. Li, H. Ishiguro, Consensus of linear multi-agent systems based on full-order observer, Journal of the Franklin Institute 351 (2) (2014) 1151–1160.

[91] Z. Li, X. Liu, M. Fu, L. Xie, Global H_∞ consensus of multi-agent systems with Lipschitz non-linear dynamics, IET Control Theory & Applications 6 (13) (2012) 2041–2048.

[92] Z. Li, W. Ren, X. Liu, M. Fu, Consensus of multi-agent systems with general linear and Lipschitz nonlinear dynamics using distributed adaptive protocols, IEEE Transactions on Automatic Control 58 (7) (2013) 1786–1791.

[93] Z. Li, W. Ren, X. Liu, M. Fu, Distributed containment control for multi-agent systems with general linear dynamics in the presence of multiple leaders, International Journal of Robust and Nonlinear Control 23 (5) (2013) 534–547.

[94] J. Liang, Z. Wang, Y. Liu, X. Liu, State estimation for two-dimensional complex networks with randomly occurring nonlinearities and randomly varying sensor delays, International Journal of Robust and Nonlinear Control 24 (1) (2014) 18–38.

[95] K.Y. Liang, S. Van de Hoef, H. Terelius, V. Turri, B. Besselink, J. Mårtensson, K.H. Johansson, Networked control challenges in collaborative road freight transport, European Journal of Control (2016).

[96] F. Liao, Y. Lu, H. Liu, Cooperative optimal preview tracking control of continuous-time multi-agent systems, International Journal of Control 89 (10) (2016) 2019–2028.

[97] J. Lin, A.S. Morse, B.D.O. Anderson, The multi-agent rendezvous problem–Part 1: the synchronous case, SIAM Journal on Control and Optimization 46 (6) (2007) 2096–2119.

[98] J. Lin, A.S. Morse, B.D.O. Anderson, The multi-agent rendezvous problem–Part 2: the asynchronous case, SIAM Journal on Control and Optimization 46 (6) (2007) 2120–2147.

[99] P. Lin, Y. Jia, Robust H_∞ consensus analysis of a class of second-order multi-agent systems with uncertainty, IET Control Theory & Applications 4 (3) (2010) 487–498.

[100] P. Lin, Y. Jia, L. Li, Distributed robust H_∞ consensus control in directed networks of agents with time-delay, Systems & Control Letters 57 (8) (2008) 643–653.

[101] Z. Lin, Low Gain Feedback, Springer, London, 1999.

[102] Z. Lin, H. Fang, On asymptotic stabilizability of linear systems with delayed input, IEEE Transactions on Automatic Control 52 (6) (2007) 998–1013.

[103] Z. Lin, B. Francis, M. Maggiore, Necessary and sufficient graphical conditions for formation control of unicycles, IEEE Transactions on Automatic Control 50 (1) (2005) 121–127.

[104] Y. Liu, Y. Jia, Robust H_∞ consensus control of uncertain multi-agent systems with time delays, International Journal of Control, Automation, and Systems 9 (6) (2011) 1086–1094.

[105] Y. Liu, Y. Jia, H_∞ consensus control for multi-agent systems with linear coupling dynamics and communication delays, International Journal of Systems Science 43 (1) (2012) 50–62.

[106] J. Lu, J. Cao, D.W.C. Ho, Adaptive stabilization and synchronization for chaotic Lur'e systems with time-varying delay, IEEE Transactions on Circuits and Systems I: Regular Papers 55 (5) (2008) 1347–1356.

[107] W. Luand, T. Chen, New approach to synchronization analysis of linearly coupled ordinary differential systems, Physica D 213 (2) (2006) 214–230.

[108] D.J. Stilwell, M. Porfiri, E.M. Bollt, Synchronization in random weighted directed networks, IEEE Transactions on Circuits and Systems I: Regular Papers 55 (10) (2008) 3170–3177.

[109] M.A. Mabrok, A.G. Kallapur, I.R. Petersen, A. Lanzon, Generalizing negative imaginary systems theory to include free body dynamics: control of highly resonant structures with free body motion, IEEE Transactions on Automatic Control 59 (10) (2014) 2692–2707.

[110] S. Martin, A. Girard, A. Fazeli, A. Jadbabaie, Multiagent flocking under general communication rule, IEEE Transactions on Control of Network Systems 1 (2) (2014) 155–166.

[111] A. Martinoli, K. Easton, W. Agassounon, Modeling swarm robotic systems: a case study in collaborative distributed manipulation, The International Journal of Robotics Research 23 (4) (2004) 415–436.

[112] J. Mei, W. Ren, G. Ma, Containment control for multiple Euler–Lagrange systems with parametric uncertainties in directed networks, in: Proceedings of the 2011 American Control Conference, 2011.

[113] Z. Meng, W. Ren, Z. You, Distributed finite-time attitude containment control for multiple rigid bodies, Automatica 46 (2) (2010) 2092–2099.

[114] P. Menon, Short-range nonlinear feedback strategies for aircraft pursuit-evasion, Journal of Guidance, Control, and Dynamics 12 (1) (1989) 27–32.

[115] P.K. Menon, G.D. Sweriduk, B. Sridhar, Optimal strategies for free-flight air traffic conflict resolution, Journal of Guidance, Control, and Dynamics 22 (2) (1999) 202–211.

[116] M. Mesbahi, F.Y. Hadaegh, Formation flying control of multiple spacecraft via graphs, matrix inequalities, and switching, Journal of Guidance, Control, and Dynamics 24 (2) (2001) 369–377.

[117] N. Motee, A. Jadbabaie, Optimal control of spatially distributed systems, IEEE Transactions on Automatic Control 53 (7) (2008) 1616–1629.

[118] K.H. Movric, F.L. Lewis, Cooperative optimal control for multi-agent systems on directed graph topologies, IEEE Transactions on Automatic Control 59 (3) (2013) 769–774.

[119] R.R. Murphy, S. Tadokoro, D. Nardi, A. Jacoff, P. Fiorini, H. Choset, A.M. Erkmen, Search and rescue robotics, in: Springer Handbook of Robotics, Springer, 2008, pp. 1151–1173.
[120] A. Nedic, A. Ozdaglar, P.A. Parrilo, Constrained consensus and optimization in multi-agent networks, IEEE Transactions on Automatic Control 55 (4) (2010) 922–938.
[121] G.I. Nesterenko, The concept of safety of aircraft structures, Journal of Machinery Manufacture and Reliability 38 (5) (2009) 516–519.
[122] W. Ni, D.Z. Cheng, Leader-following consensus of multi-agent systems under fixed and switching topologies, Systems & Control Letters 59 (2010) 209–217.
[123] H. Nunna, S. Doolla, Multiagent-based distributed-energy-resource management for intelligent microgrids, IEEE Transactions on Industrial Electronics 60 (4) (2013) 1678–1687.
[124] K.K. Oh, M.C. Park, H.S. Ahn, A survey of multi-agent formation control, Automatica 53 (2015) 424–440.
[125] K.K. Oh, H.S. Ahn, Formation control and network localization via orientation alignment, IEEE Transactions on Automatic Control 59 (2) (2013) 540–545.
[126] A. Okubo, Dynamical aspects of animal grouping: swarms, schools, flocks, and herds, Advances in Biophysics 22 (1986) 1–94.
[127] R. Olfati-Saber, Flocking for multi-agent dynamic systems: algorithms and theory, IEEE Transactions on Automatic Control 51 (2006) 401–420.
[128] R. Olfati-Saber, R.M. Murray, Consensus problems in networks of agents with switching topology and time-delays, IEEE Transactions on Automatic Control 49 (2004) 1520–1533.
[129] R. Olfati-Saber, R.M. Murry, Flocking with obstacle avoidance: cooperation with limited communication in mobile networks, in: Proceedings of the 42nd IEEE Conference on Decision and Control, 2003.
[130] A. Ouammi, H. Dagdougui, L. Dessaint, R. Sacile, Coordinated model predictive-based power flows control in a cooperative network of smart microgrids, IEEE Transactions on Smart Grid 6 (5) (2015) 2233–2244.
[131] D.A. Paley, F. Zhang, N.E. Leonard, Cooperative control for ocean sampling: the glider coordinated control system, IEEE Transactions on Control Systems Technology 16 (4) (2008) 735–744.
[132] Z.J. Palmor, Time-delay compensation–Smith predictor and its modifications, in: The Control Handbook, vol. 1, 1996, pp. 224–229.
[133] M.N.A. Parlakçı, Improved robust stability criteria and design of robust stabilizing controller for uncertain linear time-delay systems, International Journal of Robust and Nonlinear Control 16 (13) (2006) 599–636.
[134] L.M. Pecora, T.L. Carroll, Synchronization in chaotic systems, Physical Review Letters 64 (8) (1990) 821–824.
[135] L.M. Pecora, T.L. Carroll, Master stability functions for synchronized coupled systems, Physical Review Letters 80 (10) (1998) 2109–2112.
[136] L.M. Pecora, T.L. Carroll, Chaos synchronization in complex networks, IEEE Transactions on Circuits and Systems I: Regular Papers 55 (5) (2008) 1135–1146.
[137] I.R. Petersen, A. Lanzon, Feedback control of negative-imaginary systems, IEEE Control Systems Magazine 30 (5) (2010) 54–72.
[138] H.A. Poonawala, A.C. Satici, H. Eckert, M.W. Spong, Collision-free formation control with decentralized connectivity preservation for nonholonomic-wheeled mobile robots, IEEE Transactions on Control of Network Systems 2 (2) (2015) 122–130.

[139] C. Qian, W. Lin, A continuous feedback approach to global strong stabilization of nonlinear systems, IEEE Transactions on Automatic Control 46 (7) (2001) 1061–1079.

[140] C. Qian, W. Lin, Recursive observer design, homogeneous approximation, and nonsmooth output feedback stabilization of nonlinear systems, IEEE Transactions on Automatic Control 51 (9) (2006) 1457–1471.

[141] Z. Qu, Cooperative Control of Dynamical Systems: Applications to Autonomous Vehicles, Springer Science & Business Media, 2009.

[142] Z. Qu, Cooperative control of networked nonlinear systems, in: 49th IEEE Conference on Decision and Control (CDC), IEEE, 2010, pp. 3200–3207.

[143] W. Ren, Consensus strategies for cooperative control of vehicle formations, IET Control Theory & Applications 1 (2) (2007) 505–512.

[144] W. Ren, R. Beard, Decentralized scheme for spacecraft formation flying via the virtual structure approach, Journal of Guidance, Control, and Dynamics 27 (1) (2004) 73–82.

[145] W. Ren, R.W. Beard, Consensus seeking in multiagent systems under dynamically changing interaction topologies, IEEE Transactions on Automatic Control 50 (5) (2005) 655–661.

[146] W. Ren, R.W. Beard, Consensus algorithms for double-integrator dynamics, in: Distributed Consensus in Multi-vehicle Cooperative Control: Theory and Applications, 2008, pp. 77–104.

[147] W. Ren, R.W. Beard, Distributed Consensus in Multi-vehicle Cooperative Control, Springer, 2008.

[148] W. Ren, R.W. Beard, Distributed Consensus in Multi-vehicle Cooperative Control, Communications and Control Engineering, Springer-Verlag, London, 2008.

[149] W. Ren, K.L. Moore, Y. Chen, High-order and model reference consensus algorithms in cooperative control of multivehicle systems, Journal of Dynamic Systems, Measurement, and Control 129 (5) (2007) 678–688.

[150] W. Ren, N. Sorensen, Distributed coordination architecture for multi-robot formation control, Journal of Robotic and Autonomous Systems 56 (4) (2008) 324–333.

[151] C.W. Reynolds, Flocks, herds and schools: a distributed behavioral model, in: ACM SIGGRAPH Computer Graphics, vol. 21, ACM, 1987, pp. 25–34.

[152] J.P. Richard, Time-delay systems: an overview of some recent advances and open problems, Automatica 39 (10) (2003) 1667–1694.

[153] C. Sabol, R. Burns, C.A. McLaughlin, Satellite formation flying design and evolution, Journal of Spacecraft and Rockets 38 (2) (2001) 270–278.

[154] I. Saboori, K. Khorasani, H_∞ consensus achievement of multi-agent systems with directed and switching topology networks, IEEE Transactions on Automatic Control 59 (11) (2014) 3104–3109.

[155] A. Sarlette, R. Sepulchre, Consensus optimization on manifolds, SIAM Journal on Control and Optimization 48 (1) (2009) 56–76.

[156] J.A. Sauter, R. Matthews, H. Van, D. Parunak, S.A. Brueckner, Performance of digital pheromones for swarming vehicle control, in: Proceedings of the Fourth International Joint Conference on Autonomous Agents and Multiagent Systems, ACM, 2005, pp. 903–910.

[157] L. Schenato, G. Gamba, A distributed consensus protocol for clock synchronization in wireless sensor network, in: Proceeding of 46th IEEE Conference on Decision and Control, IEEE, 2007, pp. 2289–2294.

[158] M. Schwager, D. Rus, J.J. Slotine, Decentralized, adaptive coverage control for networked robots, The International Journal of Robotics Research 28 (3) (2009) 357–375.

[159] E. Semsar-Kazerooni, K. Khorasani, An LMI approach to optimal consensus seeking in multi-agent systems, in: 2009 American Control Conference, IEEE, 2009, pp. 4519–4524.
[160] J. Shin, H.J. Kim, Nonlinear model predictive formation flight, IEEE Transactions on Systems, Man and Cybernetics. Part A. Systems and Humans 39 (5) (2009) 1116–1125.
[161] R. Simmons, D. Apfelbaum, W. Burgard, D. Fox, M. Moors, S. Thrun, H. Younes, Coordination for multi-robot exploration and mapping, in: AAAI/IAAI, 2000, pp. 852–858.
[162] O.J.M. Smith, A controller to overcome dead time, ISA Journal 6 (2) (1959) 28–33.
[163] Z. Song, A. Lanzon, S. Patra, I.R. Petersen, A negative-imaginary lemma without minimality assumptions and robust state-feedback synthesis for uncertain negative-imaginary systems, Systems & Control Letters 61 (12) (2012) 1269–1276.
[164] Z. Song, A. Lanzon, S. Patra, I.R. Petersen, Robust performance analysis for uncertain negative-imaginary systems, International Journal of Robust and Nonlinear Control 22 (3) (Feb. 2012) 262–281.
[165] D.M. Stipanović, G. Inalhan, R. Teo, C.J. Tomlin, Decentralized overlapping control of a formation of unmanned aerial vehicles, Automatica 40 (8) (2004) 1285–1296.
[166] D.M. Stipanović, P.F. Hokayem, M.W. Spong, D.D. Šiljak, Cooperative avoidance control for multiagent systems, Journal of Dynamic Systems, Measurement, and Control 129 (5) (2007) 699–707.
[167] H. Su, X. Wang, Z. Lin, Flocking of multi-agents with a virtual leader, IEEE Transactions on Automatic Control 54 (2) (2009) 293–307.
[168] Y. Su, J. Huang, Cooperative adaptive output regulation for a class of nonlinear uncertain multi-agent systems with unknown leader, Systems & Control Letters 62 (6) (2013) 461–467.
[169] C. Sun, H. Duan, Y. Shi, Optimal satellite formation reconfiguration based on closed-loop brain storm optimization, IEEE Computational Intelligence Magazine 8 (4) (2013) 39–51.
[170] J. Sun, Z. Geng, Y. Lv, Adaptive output feedback consensus tracking for heterogeneous multi-agent systems with unknown dynamics under directed graphs, Systems & Control Letters 87 (2016) 16–22.
[171] H.G. Tanner, A. Jadbabaie, G.J. Pappas, Stable flocking of mobile agents, Part I: fixed topology, in: Proceedings of the 42nd IEEE Conference on Decision and Control, 2003.
[172] H.G. Tanner, A. Jadbabaie, G.J. Pappas, Stable flocking of mobile agents, Part II: dynamic topology, in: Proceedings of the 42nd IEEE Conference on Decision and Control, 2003.
[173] H.G. Tanner, A. Jadbabaie, G.J. Pappas, Flocking in fixed and switching networks, IEEE Transactions on Automatic Control 52 (5) (2007) 863–868.
[174] H.G. Tanner, A. Boddu, Multiagent navigation functions revisited, IEEE Transactions on Robotics 28 (6) (2012) 1346–1359.
[175] Y. Tian, C. Liu, Robust consensus of multi-agent systems with diverse input delays and asymmetric interconnection perturbations, Automatica 45 (5) (2009) 1347–1353.
[176] Y.P. Tian, Y. Zhang, High-order consensus of heterogeneous multi-agent systems with unknown communication delays, Automatica 48 (6) (2012) 1205–1212.
[177] A. Tiwari, J. Fung, J.M. Carson, A framework for Lyapunov certificates for multi-vehicle rendezvous problem, in: Proceedings of the 2004 American Control Conference, 2004.
[178] H. Tnunay, Z. Li, C. Wang, Z. Ding, Distributed collision-free coverage control of mobile robots with consensus-based approach, in: Proc. 13th IEEE International Conference on Control & Automation, IEEE, 2017, pp. 678–683.
[179] T. Vicsek, A. Czirók, E. Ben-Jacob, I. Cohen, O. Shochet, Novel type of phase transition in a system of self-driven particles, Physical Review Letters 75 (6) (1995) 1226–1229.

[180] C. Wang, Z. Ding, H_∞ consensus control of multi-agent systems with input delay and directed topology, IET Control Theory & Applications 10 (6) (2016) 617–624.

[181] J. Wang, Z. Duan, Z. Li, G. Wen, Distributed H_∞ and H_2 consensus control in directed networks, IET Control Theory & Applications 8 (3) (2014) 193–201.

[182] J. Wang, Z. Duan, G. Wen, G. Chen, Distributed robust control of uncertain linear multi-agent systems, International Journal of Robust and Nonlinear Control 25 (13) (2015) 2162–2179.

[183] J.N. Wang, A. Lanzon, I.R. Petersen, A robust output feedback consensus protocol for networked negative-imaginary systems, in: 2014 IFAC World Congress, Cape Town, South Africa, Aug. 2014.

[184] X. Wang, S. Li, Nonlinear consensus algorithms for second-order multi-agent systems with mismatched disturbances, in: Proceeding of 2015 American Control Conference (ACC), July 2015, pp. 1740–1745.

[185] X. Wang, G. Zhong, K. Tang, K.F. Man, Z. Liu, Generating chaos in Chua's circuit via time-delay feedback, IEEE Transactions on Circuits and Systems. I, Fundamental Theory and Applications 48 (9) (2001) 1151–1156.

[186] X.F. Wang, G.R. Chen, Synchronization in scale-free dynamical networks: robustness and fragility, IEEE Transactions on Circuits and Systems. I, Fundamental Theory and Applications 49 (1) (2002) 54–62.

[187] G. Wen, Z. Duan, G. Chen, W. Yu, Consensus tracking of multi-agent systems with Lipschitz-type node dynamics and switching topologies, IEEE Transactions on Circuits and Systems I: Regular Papers 61 (2) (2014) 499–511.

[188] G. Wen, Z. Duan, W. Yu, G. Chen, Consensus in multi-agent systems with communication constraints, International Journal of Robust and Nonlinear Control 22 (2) (2012) 170–182.

[189] G. Wen, G. Hu, W. Yu, G. Chen, Distributed H_∞ consensus of higher order multiagent systems with switching topologies, IEEE Transactions on Circuits and Systems II: Express Briefs 61 (5) (2014) 359–363.

[190] G. Wen, G. Hu, W. Yu, J. Cao, G. Chen, Consensus tracking for higher-order multi-agent systems with switching directed topologies and occasionally missing control inputs, Systems & Control Letters 62 (12) (2013) 1151–1158.

[191] G. Wen, W. Yu, M.Z.Q. Chen, X. Yu, G. Chen, H_∞ pinning synchronization of directed networks with aperiodic sampled-data communications, IEEE Transactions on Circuits and Systems I: Regular Papers 61 (11) (2014) 3245–3255.

[192] W. Wu, W. Zhou, T. Chen, Cluster synchronization of linearly coupled complex networks under pinning control, IEEE Transactions on Circuits and Systems I: Regular Papers 56 (4) (2009) 829–839.

[193] L. Xiao, S. Boyd, Fast linear iterations for distributed averaging, Systems & Control Letters 53 (1) (2004) 65–78.

[194] M. Xin, S.N. Balakrishnan, H.J. Pernicka, Position and attitude control of deep-space spacecraft formation flying via virtual structure and θ-D technique, Journal of Dynamic Systems, Measurement, and Control 129 (5) (2007) 689–698.

[195] J.L. Xiong, I.R. Petersen, A. Lanzon, A negative imaginary lemma and the stability of interconnections of linear negative imaginary systems, IEEE Transactions on Automatic Control 55 (10) (2010) 2342–2347.

[196] S. Xu, J. Lam, Improved delay-dependent stability criteria for time-delay, IEEE Transactions on Automatic Control 50 (3) (2005) 384–387.

[197] Y. Xu, Nonlinear robust stochastic control for unmanned aerial vehicles, in: Conference on American Control Conference, 2009.

[198] A. Yamashita, M. Fukuchi, J. Ota, T. Arai, H. Asama, Motion planning for cooperative transportation of a large object by multiple mobile robots in a 3D environment, in: Proceedings of IEEE International Conference on ICRA, IEEE, 2000, pp. 3144–3151.
[199] B. Yang, W. Lin, Homogeneous observers, iterative design, and global stabilization of high-order nonlinear systems by smooth output feedback, IEEE Transactions on Automatic Control 49 (7) (2004) 1069–1080.
[200] H. Yang, Z. Zhang, S. Zhang, Consensus of second-order multi-agent systems with exogenous disturbances, International Journal of Robust and Nonlinear Control 21 (9) (2011) 945–956.
[201] X. Yang, K. Watanabe, K. Izumi, K. Kiguchi, A decentralized control system for cooperative transportation by multiple non-holonomic mobile robots, International Journal of Control 77 (10) (2004) 949–963.
[202] S.Y. Yoon, P. Anantachaisilp, Z. Lin, An LMI approach to the control of exponentially unstable systems with input time delay, in: Proceeding of IEEE 52nd Annual Conference on Decision and Control (CDC), IEEE, 2013, pp. 312–317.
[203] S.Y. Yoon, Z. Lin, Predictor based control of linear systems with state, input and output delays, Automatica 53 (2015) 385–391.
[204] W. Yu, G. Chen, M. Cao, Consensus in directed networks of agents with nonlinear dynamics, IEEE Transactions on Automatic Control 56 (6) (2011) 1436–1441.
[205] W. Yu, G. Chen, M. Cao, J. Kurths, Second-order consensus for multiagent systems with directed topologies and nonlinear dynamics, IEEE Transactions on Systems, Man and Cybernetics. Part B. Cybernetics 40 (3) (2010) 881–891.
[206] H. Zhang, T. Feng, G. Yang, H. Liang, Distributed cooperative optimal control for multiagent systems on directed graphs: an inverse optimal approach, IEEE Transactions on Cybernetics 45 (7) (2014) 1315–1326.
[207] W. Zhang, J. Hu, Optimal multi-agent coordination under tree formation constraints, IEEE Transactions on Automatic Control 53 (3) (2008) 692–705.
[208] W.-A. Zhang, L. Yu, Stability analysis for discrete-time switched time-delay systems, Automatica 45 (10) (2009) 2265–2271.
[209] X. Zhang, H. Duan, Y. Yu, Receding horizon control for multi-UAVs close formation control based on differential evolution, Science China Information Sciences 53 (2) (2010) 223–235.
[210] X.Y. Zhang, H.B. Duan, Differential evolution-based receding horizon control design for multi-UAVs formation reconfiguration, Transactions of the Institute of Measurement and Control 34 (2–3) (2012) 165–183.
[211] Z. Zhao, Z. Lin, Global leader-following consensus of a group of general linear systems using bounded controls, Automatica 68 (2016) 294–304.
[212] Y. Zheng, Y. Zhu, L. Wang, Consensus of heterogeneous multi-agent systems, IET Control Theory & Applications 5 (16) (2011) 1881–1888.
[213] Q.C. Zhong, Robust Control of Time-Delay Systems, Springer Science & Business Media, 2006.
[214] B. Zhou, Z. Lin, G.R. Duan, Truncated predictor feedback for linear systems with long time-varying input delays, Automatica 48 (10) (2012) 2387–2399.
[215] K. Zhou, C.D. John, Essentials of Robust Control, vol. 104, Prentice Hall, NJ, 1998.
[216] K. Zhou, J.C. Doyle, K. Glover, Robust and Optimal Control, Prentice Hall, Englewood Cliffs, NJ, 1996.
[217] Z. Zuo, Z. Lin, Z. Ding, Prediction output feedback control of a class of Lipschitz nonlinear systems with input delay, IEEE Transactions on Circuits and Systems II: Express Briefs (2016).

Index

Printed in the United States
By Bookmasters